Airborne Particles

The project that is the subject of this report was approved by the Governing Board of the National Research Council, whose members are drawn from the Councils of the National Academy of Sciences, the National Academy of Engineering, and the Institute of Medicine. The members of the Committee responsible for the report were chosen for their special competences and with regard for appropriate balance.

This report has been reviewed by a group other than the authors according to procedures approved by a Report Review Committee consisting of members of the National Academy of Sciences, the National Academy of Engineering, and the Institute of Medicine.

The work on which this publication is based was performed pursuant to Contract No. 68-02-1226 with the Environmental Protection Agency.

Airborne Particles

Subcommittee on Airborne Particles

Committee on Medical and Biologic Effects of Environmental Pollutants

Division of Medical Sciences
Assembly of Life Sciences
National Research Council

University Park Press
Baltimore

UNIVERSITY PARK PRESS
International Publishers in Science and Medicine
233 East Redwood Street
Baltimore, Maryland 21202

Typeset by American Graphic Arts Corporation.
Manufactured in the United States of America by The Maple Press Company.

Library of Congress Cataloging in Publication Data

National Research Council. Subcommittee on Airborne
Particles.
Airborne particles.

Bibliography: p.
Includes index.
1. Air—Pollution. 2. Particles. I. Title.
[DNLM: 1. Air pollutants. 2. Aerosols. WA754 N284a]
TD884.5.N37 1978 363.6 78-18189
ISBN 0-8391-0129-5

Contents

Committee on Medical and Biologic Effects of Environmental Pollutants

Reuel A. Stallones University of Texas, Houston, Texas, *Chairman*

Martin Alexander Cornell University, Ithaca, New York

Andrew A. Benson University of California, La Jolla, California

Ronald F. Coburn University of Pennsylvania School of Medicine, Philadelphia, Pennsylvania

Clement A. Finch University of Washington School of Medicine, Seattle, Washington

Eville Gorham University of Minnesota, Minneapolis, Minnesota

Robert I. Henkin Georgetown University Medical Center, Washington, D.C.

Ian T. T. Higgins University of Michigan, Ann Arbor, Michigan

Joe W. Hightower Rice University, Houston, Texas

Henry Kamin Duke University Medical Center, Durham, North Carolina

Orville A. Levander Agricultural Research Center, Beltsville, Maryland

Roger P. Smith Dartmouth Medical School, Hanover, New Hampshire

T. D. Boaz, Jr. Division of Medical Sciences, National Research Council, Washington, D.C., *Executive Director*

Subcommittee on Airborne Particles

Ian T. T. Higgins University of Michigan, Ann Arbor, Michigan, *Chairman*

Roy E. Albert New York University Medical Center, New York, New York

Robert J. Charlson University of Washington, Seattle, Washington

Ellis F. Darley University of California, Riverside, California

Benjamin G. Ferris Jr. Harvard University School of Public Health, Boston, Massachusetts

Robert Frank University of Washington, Seattle, Washington

Kenneth T. Whitby University of Minnesota, Minneapolis, Minnesota

John Redmond Jr. Division of Medical Sciences, National Research Council, Washington, D.C., *Staff Officer*

Preface

The air we breathe contains many particles. In the unpolluted countryside there may be mass concentrations of up to 50 $\mu g/m^3$, while at the height of pollution episodes concentrations have sometimes reached several milligrams per cubic meter, roughly 100 times as high. These particles may result from natural processes (dust, spray, natural decay, etc.) or from the activities of man (combustion or comminution). An important source is conversion of a gas to an aerosol either by cooling or by oxidation.

Particles come in all shapes and sizes. They may be of animal, vegetable, or mineral origin. They may be living or inanimate. They can affect plants, animals, or materials. Particles may affect a person's health in a variety of ways: some are inert, producing changes in the body only by their passive accumulation and inducing little or no tissue reaction; others are intensely irritant or toxic, causing changes that may result in serious illness—even death when inhaled in sufficient quantity. Some particles are known to produce cancer. Evidence indicates that the action of particles may be modified by the presence of other particles. Particles may also interact with gases that may be present in the air. These interactions may either enhance or moderate the effect of either substance when inhaled alone.

The purpose of this report is to consider those particles that are most often associated with general types of air pollution. Specific particles, such as lead, arsenic, or asbestos, which may often contaminate the air, are not discussed. These substances have been the subject of other publications of the National Academy of Sciences. Nor will living particles be considered. Bacterial and viral particles are excluded and the infections resulting therefrom are mentioned incidentally and only in very general terms. Similarly, vegetable particles, such as pollens, are considered, if at all, only in relation to their importance in pollution-induced asthmatic attacks. Possibly the most important source of airborne particles from the standpoint of health is tobacco smoking. Cigarette smoking in particular should be considered carefully in any health effects study. Full allowance must be made for any effects of smoking before conclusions can be drawn about the effects of air pollution. No attempt, however, is made here to deal specifically with aerosols arising from cigarette smoke.

The emphasis of this report, then, is on particles that result from man's activities. The types of particles and their distribution are considered. The origins, behavior, and fate of such particles, their physical and chemical characteristics, and their interactions, transport, and removal from the ambient air are discussed in chapters 1 through 4. Routine and special monitoring and trends are reviewed in chapter 5. In chapters 6 and 7, there is detailed discussion of the deposition, clearance, and retention of particles, and their effects on man and on other animals. Chapter 8 is devoted to the available epidemiologic evidence from which conclusions can be drawn about the effects of particulate pollution on man. It indicates the gaps in our knowledge about exposure/response relationships and suggests studies that should be conducted to remedy these deficiencies. Chapters 9 and 10 consider the effects of particulate matter on vegetation and materials. Summaries, conclusions, and recommendations, prepared by the authors of the individual chapters, are consolidated in Chapter 11.

Acknowledgments

This document was prepared by the Subcommittee on Airborne Particles under the chairmanship of Dr. Ian T. T. Higgins. Although the initial drafts of the various sections were prepared by individuals, the entire document was extensively reviewed by the entire Subcommittee and represents a group effort.

The Preface was written by Dr. Higgins. Chapters 1 through 5, which define particulates and describe their characteristics, measurements, sources, and effects on atmospheric processes, were written by Dr. Kenneth T. Whitby and Dr. Robert J. Charlson with assistance from Dr. M. B. Baker, Dr. S. S. Butcher, Dr. D. S. Covert, Dr. L. F. Radke, and Dr. T. V. Larson. Special credit for these five chapters should go to Dr. Larson, who coordinated the material and shaped it into its final form.

The effects of inhaled particles on both humans and animals are described in chapter 6 by Dr. Morton Lippmann and Dr. Roy E. Albert, who were assisted by Dr. D. B. Yeates.

Chapter 7, which covers the effects on lung function of sulfur dioxide and aerosols, both alone and combined, was written by Dr. Robert Frank. Epidemiologic studies on the effects of airborne particles on human health are reviewed in chapter 8 by Dr. Higgins and Dr. Benjamin G. Ferris, Jr.

Dr. Shimshon L, Lerman and Dr. Ellis F. Darley collaborated on Chapter 9, in which the effects of particulate air pollutants on vegetation are discussed.

Extensive material on the effects of atmospheric particulate matter on building materials is summarized in Chapter 10. This section was prepared by the Building Research and Advisory Board, a unit of the National Research Council.

Summaries of these chapters and the conclusions and recommendations of the Subcommittee are contained in the final chapter, to which all Subcommittee members contributed.

The preparation of the document was assisted by the comments from anonymous reviewers designated by the Assembly of Life Sciences and from members of the Committee on Medical and Biologic Effects of Environmental Pollutants. The Subcommittee is particularly indebted to Dr. Joe W. Hightower, who served as Associate Editor.

Dr. Robert J. M. Horton of the Environmental Protection Agency gave invaluable assistance by providing the Subcommittee with various documents and translations. Informational assistance was obtained from the National Research Council Advisory Center on Toxicology, The National Academy of Sciences Library, The Library of Congress, the Department of Commerce Library, and the Air Pollution Technical Information Center.

The staff officer for the Subcommittee on Airborne Particles was Mr. John Redmond, Jr. Special recognition is given to Mrs. Louise Mulligan, who verified the many references. The document was edited by Mrs. Frances M. Peter.

Airborne Particles

1

Aerosols: Characteristics, Behavior, and Measurement

ATMOSPHERIC AEROSOLS

Atmospheric aerosols, consisting of solid or liquid particles suspended in the air, range in size from a few tens of Ångstroms to several hundred micrometers. They may be described by their size range, composition, origin, effects, or in terms of what is measured by a particular technique. Special names such as fog, dust, smoke, and Aitken nuclei have arisen from various disciplines. However, because most of these names are imprecise, they are used sparingly in this report. Some of these names and their relation to the more fundamental physical and chemical terminology are tabulated in Table 1-1.

All aerosol particles are formed either by condensation of gases or vapors or by mechanical or comminutive processes. Aerosol particles may then be transformed by coagulation and condensation at the same time that they are transported by air movement and dilution. All aerosol particles eventually disappear from the atmosphere and settle on some surface that acts as a sink. Because aerosol concentrations are ordinarily very small (tens of micrograms per cubic meter) compared to the mass of the surfaces that serve as their sinks, aerosols have effects on surfaces only under conditions where a very small mass can produce a large effect because of chemical composition. In contrast, large *in situ* effects, such as visibility reduction, can result from very small aerosol mass concentrations.

The residence time of aerosol particles in the atmosphere varies tremendously, from a few minutes for either 0.005- or 100-μm diam particles near the earth's surface in the troposphere, to years for 0.5-μm diam particles in the stratosphere. The significance of these different lifetimes is that for equal masses of small and large particles emitted into a given large volume of air per unit time, the effective suspended mass of the small particles is as much as 10 to 100 times that of the large ones. For this reason, emission inventories or formation rates expressed in tonnages without regard to particle size, composition, or point of origin in the atmosphere have little utility for judging effect. A million tons of sand blowing along the desert surface is not as significant as a few

Table 1-1. Names and characteristics of aerosol particles

Name	Unique physical characteristics	Effects	Origin	Predominant size range (μm)
Coarse particle			Mechanical process	>2
Fine particle			Condensation process	<2
Dust	Solid	Nuisance, ice nuclei	Mechanical dispersion	>1
Smoke	Solid or liquid	Health and visibility	Condensation	<1
Fume	Solid	Health and visibility	Condensation	<1
Fog	Water droplets	Reduce visibility	Condensation	2–30
Mist	Water droplets	Reduce visibility; cleanse air	Condensation or atomization	5–1,000
Haze	Exists at lower RH than fog—hygroscopic	Reduce visibility		<1
Aitken or condensation nuclei (CN)		Nuclei for condensation at supersaturation >300%	Combustion, atmospheric chemistry	<0.1
Ice nuclei (IN)	Have very special crystal structure	Cause freezing of supercooling water droplets	Natural dusts	>1
Small ions	Stable particle with an electric charge	Carry atmospheric electricity	All sources	>0.0015
Large particles	Special name			0.1-1
Giant particles	Special name			>1

milligrams of plutonium smoke emitted from a fuel-reprocessing plant's stack.

It has been known for years that toxic materials such as lead and asbestos are health hazards when high concentrations are breathed by industrial workers. A relatively new realization is that low concentrations of such compounds as sulfate in the form of fine particles may also be potential health hazards. Sulfate is now known to be quite extensively distributed in the atmosphere because of the ubiquitous emission of sulfur dioxide and sulfates into the atmosphere from the combustion of sulfur-containing fuels.[770] The rising concern over the potential health hazards of sulfate aerosols has focused new attention on the mechanism of formation, distribution, transport, effects, and health hazards of secondary aerosols formed from chemical reactions in the atmosphere.

During the past five years there has been a revolution in both our knowledge and our understanding of atmospheric aerosols as a result of both monitoring and special aerosol characterization studies.[328, 330] Large monitoring efforts, such as those carried out by the Environmental Protection Agency (EPA), although they have used relatively simple total particulate sampling techniques, have enhanced our understanding of the distribution of aerosols at the surface even though the absolute accuracy can be questioned.[353]

The development of automatic physical, chemical, and gas analyzers and automatic aerosol-particle-sizing instruments has permitted the construction and extensive use of several large mobile laboratories on the earth's surface as well as the more recent application of aircraft to the three-dimensional study of the physical and chemical properties of atmospheric aerosols and gases. These new tools are accumulating a large amount of data that are only now being analyzed and understood. However, the early results have already upset many old ideas.

Sophisticated collaborative research efforts on aerosols, such as those conducted in California[328, 330] and more recently in St. Louis,[143, 808] have provided much good data on the complete physical and chemical size distributions of atmospheric aerosols at a great variety of sites.

Because atmospheric sciences, particle technology, industrial hygiene, and health effects studies of pollutants are separate fields, the latest information and understanding about aerosols has not been transferred from one field to another as rapidly as it should. It is hoped that this report, because it has been compiled by an interdisciplinary team, will aid in this diffusion of knowledge.

DEFINITIONS AND BASIC PROPERTIES

Atmospheric aerosols can best be described within the framework of the most recent concepts concerning their physical and chemical size distributions.

A *particle* is any minute piece of solid or liquid. Many particles that are important in studies of air pollution are unstable; they can change or even disappear on contact with a surface. Examples are a raindrop striking a surface and coalescing, a loose aggregate of carbon black disintegrating on contact with a surface, and an ion losing its charge after contact with a surface or an oppositely charged particle.

An *aerosol* is a mixture of gases and particles that exhibits some stability in a gravitational field.[332] In atmospheric aerosol, this gravitational stability excludes particles with a diameter greater than several hundred micrometers.

Other Names for Particles and Aerosols

Because aerosols are important in many fields of study, a great variety of identifying terms have come into use describing unique physical characteristics of the particles, effects caused by the particles, and features related to the origin or size of the particles. Particle names and their derivations are given in Table 1-1.

The definition of *coarse* and *fine particles* as larger and smaller than 1 μm radius, respectively, originated from the fact that atmospheric aerosol size distributions are bimodal: one mode occurs above 1 μm radius, the other below 1 μm radius (Figure 1-1). In atmospheric distributions, coarse particles are produced by mechanical processes, whereas fine particles result almost entirely from condensation and coagulation.

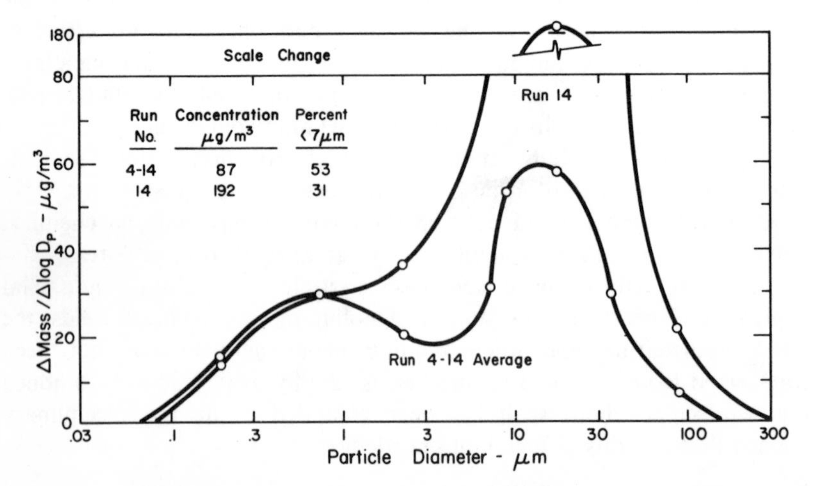

Figure 1-1. Bimodal mass distributions measured with a set of special impactors and a cascade impactor. Run 14 contains many more coarse particles than the average because of construction activity upwind. Note the negligible effect of this increased concentration of coarse particles on the fine particle mode. From Whitby, 1975, p. III–21.[798]

Certain particles serve as nuclei to initiate further processes. *Aitken* or *condensation nuclei*, *cloud condensation nuclei*, and *ice nuclei* are named for their effects.

Small and large ions carry electrical charges. Small ions are clusters of molecules around a charge. Their stability depends on their having a charge. Large ions are merely particles that carry an electric charge. *Large particles* (0.1 to 1.0 μm diam) and *giant particles* (>1.0 μm diam) are terms also applied by atmospheric physicists.

Particle Properties

Particles may be classified by a variety of their properties or formation methods. For example, *primary particles* do not change form after emission. Examples are road dust, salt spray from oceans, and cement dust. In contrast, a substantial fraction of the mass of *secondary particles* is formed by *in situ* chemical reactions involving gases. Photochemically produced sulfate and photochemical smog are examples of secondary aerosols.

Particles and aerosols may also be classified by size, area, mass, light-scattering, chemical composition, and by such geometric characteristics as sphericity or other particle shape characteristics.

Because of the almost explosive growth in data during the past few years, many of the references are to reports and papers that are new or just being published. During the past five years, methods for describing aerosol size distributions have also begun to change in keeping with the new data resulting from field studies.

THE PHYSICAL AND CHEMICAL SIZE DISTRIBUTION OF ATMOSPHERIC AEROSOLS

Until the last two decades most atmospheric aerosol data were obtained by individual investigators who collected limited data during a short time in only a few locations. Despite the "snapshot" nature of most of this work, these studies elucidated the size-dependent features of atmospheric aerosols.

In the 19th century, Aitken[5] showed that most particles in the atmosphere were smaller than 0.1 μm diam and that their concentration varied from a few hundred per cubic centimeter over the ocean to millions per cubic centimeter in cities. Rayleigh's[740] 19th century work showed that most light-scattering by atmospheric aerosols was caused by 0.1- to 1.0-μm diam particles. These early workers also discovered that the origin of particles smaller than about 1.0 μm diam is different than the origin of larger particles. The smaller particles were called smokes, fumes, or hazes; the larger ones, dusts, mists, fogs, or ashes. These early hints of the multimodal nature of atmospheric aerosols remained dormant for a long period because most size distribution data were obtained simply by counting particles with the light microscope. The overwhelming number of small particles compared to the larger particles hid the modal nature of the aerosol.

After the Second World War, Junge[399, 402, 403] began a series of studies using a variety of measurement methods. He elucidated the continuous nature of the atmospheric aerosol number size distribution and concentration in urban and nonurban locations as functions of altitudes and location. From Junge's work came the standard form for plotting size distribution data: log of dN/dD_p versus log D_p, where N = number and D_p = particle diameter. Junge[398] observed that this plot was

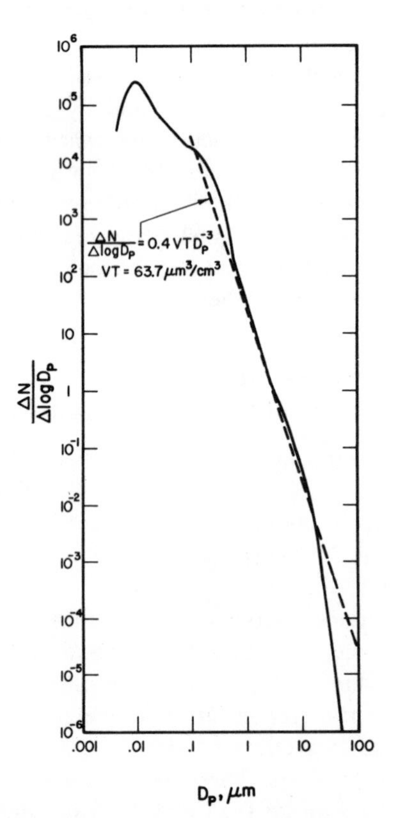

Figure 1-2. Grand average number aerosol size distribution from the 1969 Los Angeles smog experiment[805] compared with a Junge power law distribution calculated with the constants of Clark and Whitby.[153] Agreement is reasonably good from 0.1 to 10 μm diam but not above or below these sizes.

a straight line that could be described by the equation: $dN/dD_p = A\ D_p^{-k}$ where A and k were constants. He also noted that in the range from 0.1 to 10.0 μm diam, k was approximately equal to 4.0.

This distribution model, known variously as the Junge distribution or the power law distribution, has been widely used. Its acceptance was furthered by Friedlander,[259] who showed that by balancing aerosol source and removal rates a portion of the resultant theoretical number distribution steady state could be fitted reasonably well by the Junge distribution. Clark and Whitby,[153] by fitting the Junge distribution to 52 atmospheric distributions, found that the constant A was equal to 0.4 multiplied by the aerosol volume concentration in cubic micrometers per cubic centimeter. This agrees with the value predicted by Friedlander.[259]

In 1972, Whitby and his colleagues,[805] while analyzing smog size distribution data from Pasadena, found that although the Junge distribution was a reasonable fit to the number distribution, it was not a good model for the surface or volume distributions. They found that the mass size distribution of Los Angeles smog was normally bimodal, with one mode at about 0.3 μm diam and the other from 5.0 to 15.0 μm diam.

In 1947, May[517] developed the cascade impactor. Although aerosol mass concentration had been measured by filtration long before, this instrument was the first to measure directly the aerosol mass size distribution over the range of approximately 0.5 μm diam to over 10 μm diam. Unfortunately, most investigators using impactors adopted the

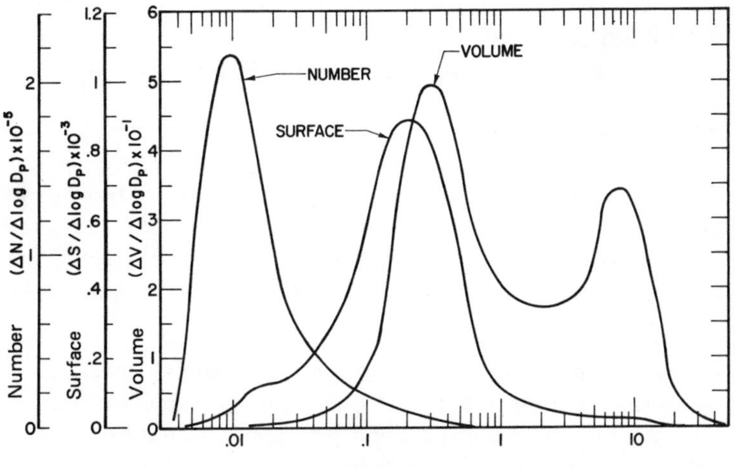

Figure 1-3. Normalized frequency plots of number, surface, and volume (particle volume times particle density) distributions for the grand average 1969 Pasadena smog aerosol. Note the bimodal distribution of mass. Each weighting shows features of the distribution not shown by the other plots. From Whitby, 1975, p. II–11.[798]

powder technology method to plot their data in cumulative form on log probability paper. This method is useful for determining the geometric mean and geometric standard deviation of *unimodal* distributions having no sampling or measurement biases. However, the impactor's sampling biases affect the value of the integral used to normalize the data.[204, 802] As a result, the true *bimodal* nature of the mass distribution is often hidden, and characteristics of the impactor, such as particle bounce and sampling bias, may be interpreted as significant characteristics of the mass size distribution.

Recent studies[506, 627, 794] have made it possible to apply the cascade impactor under many conditions to validate the measured mass size distributions. For example, Lundgren[495] has used impactors to verify directly the bimodal nature of atmospheric aerosol mass distributions (Figure 1-1).

Figure 1-2 is a conventional plot of the particle size distribution of the total average aerosol measured in Los Angeles in 1969.[805] Figure 1-3 presents the same data, but in such a manner that the apparent area under the curves is proportional to the number, surface area, or volume in a given size range. These curves were constructed assuming uniformly dense, spherical particles. This is a fairly typical urban distribution (see Chapter 3). The following important observations can be made from this figure:

Most particles are approximately 0.01 μm diam. Most of the surface area is provided by particles averaging 0.2 μm diam. The volume (or mass) distribution is bimodal; one mode is about 0.3 μm diam, the other is about 10.0 μm diam.

The mass of fine particles (<2 μm) is almost equal to the mass of coarse particles (>2 μm).

The $dN/d \log D_p$ versus D_p plot on log coordinates (Figure 1-2) shows that the number of particles decreases sharply with increasing size.

Atmospheric aerosol size distributions consist basically of three separate modes:[797, 801] nuclei (≤ 0.1 μm diam), accumulation (<2.0 μm but ≥ 0.1 μm diam), and coarse particle (>2.0 μm diam). Depending on their source there may be from one to three distinct maximums in the surface or volume distributions.

Figure 1-4 shows the surface distribution, assuming spherical particles, of an aerosol measured in Ft. Collins, Colorado under conditions in which all three modes are distinct and separate. Washout by rain greatly reduced the nuclei and coarse particle mode but had little effect on the accumulation mode.

A distinct trimodal volume distribution was measured during the General Motors Sulfate Study in Milford, Michigan during October

1975 (Figure 1-5). The nuclei mode was distinct because the background concentration of aerosol in the accumulation mode was low and because the catalyst-equipped cars on the test track emitted nuclei-mode-sized aerosols.

The origin of each mode of atmospheric aerosol size distribution can be associated with various aerosol formation mechanisms (Figure 1-6). (See Chapter 2 for further discussion of these mechanisms.)

An important transformation mechanism for fine particles is the Brownian motion of particles smaller than a few tenths of a micrometer in diameter. This Brownian motion, resulting from gas-molecule impact, causes the particles to diffuse. If the number concentration is high enough, this diffusion results in collision and coagulation to larger sizes.

Diffusion-caused coagulation has several important effects on atmospheric aerosols. It limits the maximum concentration to less than a few million particles per cubic centimeter at distances further than a few hundred feet from sources, and it causes particles smaller than approximately a few hundredths of a micrometer in diameter to coagulate rapidly with particles a few tenths of a micrometer in diameter. This tends to generate multimodal distributions near strong sources of submicrometer primary aerosols. Coagulation can also affect the shape and chemical composition of particles.

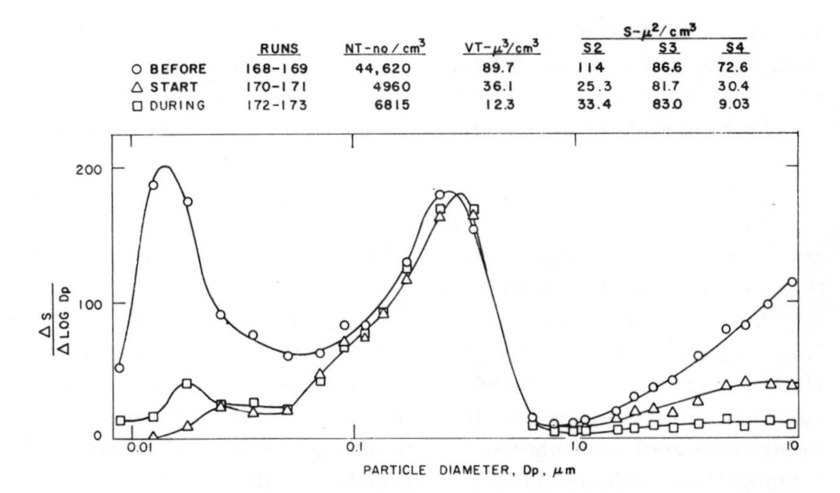

Figure 1-4. Surface area aerosol size distributions measured before, at the beginning of, and during a rain on August 14, 1970 in Ft. Collins, Colorado. The three aerosol mode peaks [nuclei (0.015 μm diam), accumulation (0.3 μm diam), and coarse particle (>10 μm diam)] are visible. As the rain continued, the nuclei and coarse particle modes were removed but the accumulation mode was unchanged. From Whitby, 1975, p. II-12.[798]

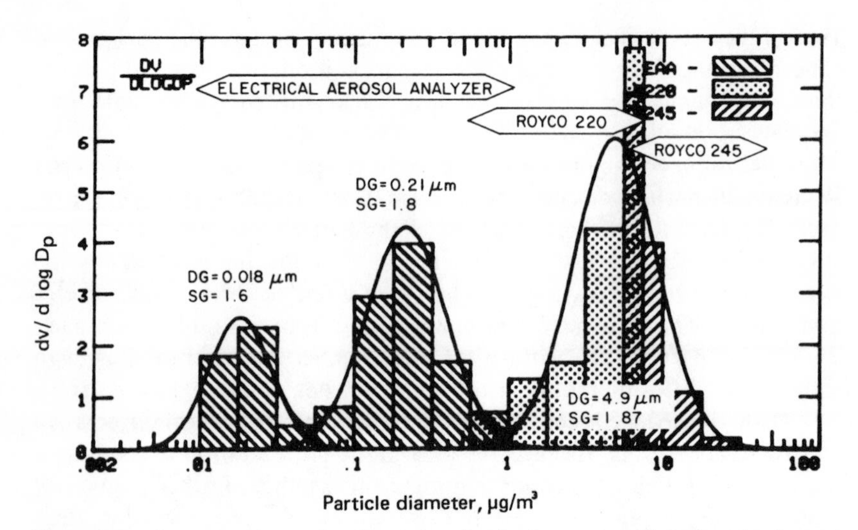

Figure 1-5. Computer-prepared volume size distribution measured during the General Motors Sulfate Study in Milford, Michigan in the fall of 1975. The range of measurement of the three instruments used to obtain the size distribution from which the distribution was calculated is shown. The distinct trimodal nature of this distribution is the result of the low aerosol concentration (<6 $\mu g/m^3$) and the strong nuclei-mode-sized emissions from catalyst-equipped cars. From Whitby *et al.*, 1976.[806]

Nuclei or Aitken Nuclei Mode

The nuclei or Aitken nuclei mode[814] accounts for most of the Aitken nuclei count and orginates primarily from the condensation and coagulation of hot, highly supersaturated vapors during combustion.

There is evidence that a prominent nuclei mode in the size distribution indicates the presence of substantial amounts of fresh combustion aerosol.[198, 803, 806] Many particles in the nuclei mode raise the Aitken nuclei count. Usually, however, they do not greatly increase the aerosol mass concentration because the nuclei mode rarely accounts for more than a few percent of the total mass.

An important exception is aerosol near heavily traveled multiple-lane roadways. Whitby *et al.*[803] found that in regions bordering the Harbor Freeway in Los Angeles the nuclei mode may contain over 25 $\mu g/m^3$ of aerosol. More recently, investigators studying emissions from catalyst-equipped cars observed distributions in which the nuclei mode contained more volume than the accumulation mode.[806]

Because particles may serve as nuclei for the condensation of water vapor, condensation is an important growth mechanism for submicrometer aerosols. Examples are fogs and hazes formed when the humidity exceeds 60%.

Accumulation Mode

The twin mechanisms of coagulation and heterogeneous nucleation (condensation of one material on another) tend to accumulate submicrometer aerosol mass in this mode.[801, 814] However, because of the sharp decrease in particles larger than 0.3 μm diam, little mass is transferred from the accumulation mode to the coarse particle size range.[332, 738] Sedimentation and impaction tend to increase the relative concentration of the smallest mechanically produced particles. Therefore, these smallest particles also accumulate in this mode.

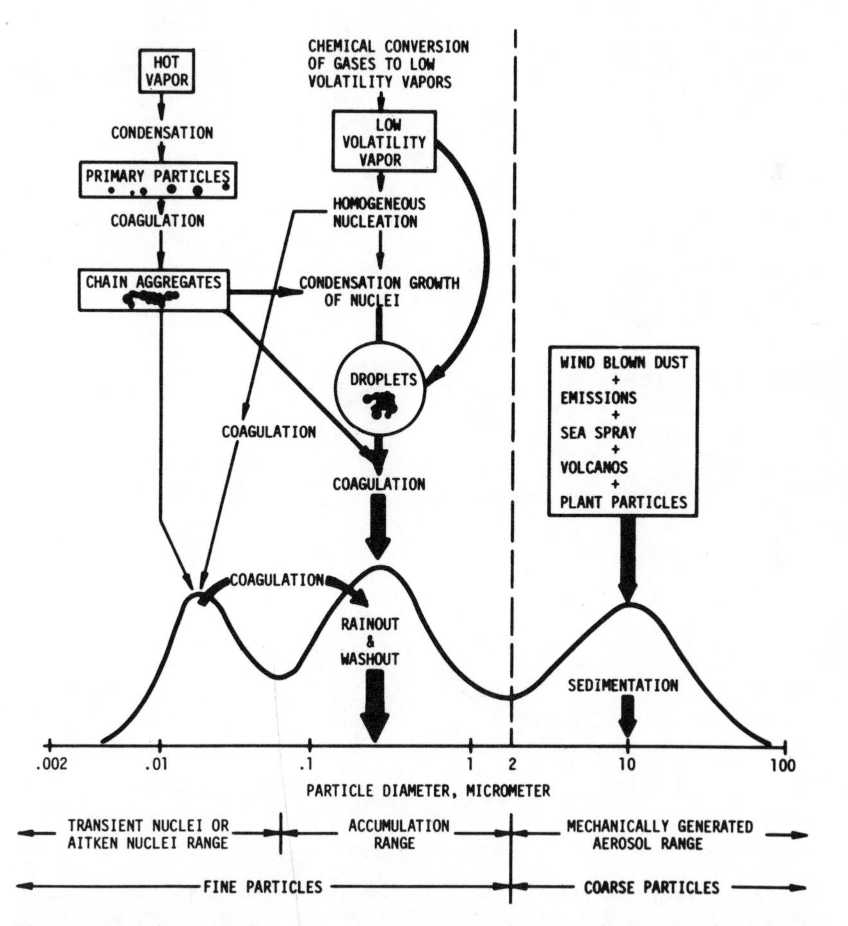

Figure 1-6. Schematic of an atmospheric aerosol surface area distribution showing principal modes, main sources of mass for each mode, and the principal processes involved in inserting mass in each mode and the principal removal mechanisms. From Whitby and Cantrell, 1976.[801]

Coarse Particle Mode

In the arbitrarily named coarse particle mode, practically all aerosol particles at relative humidities below 100% originate from mechanical processes.[328, 382, 814] Most fine particles originate from condensation

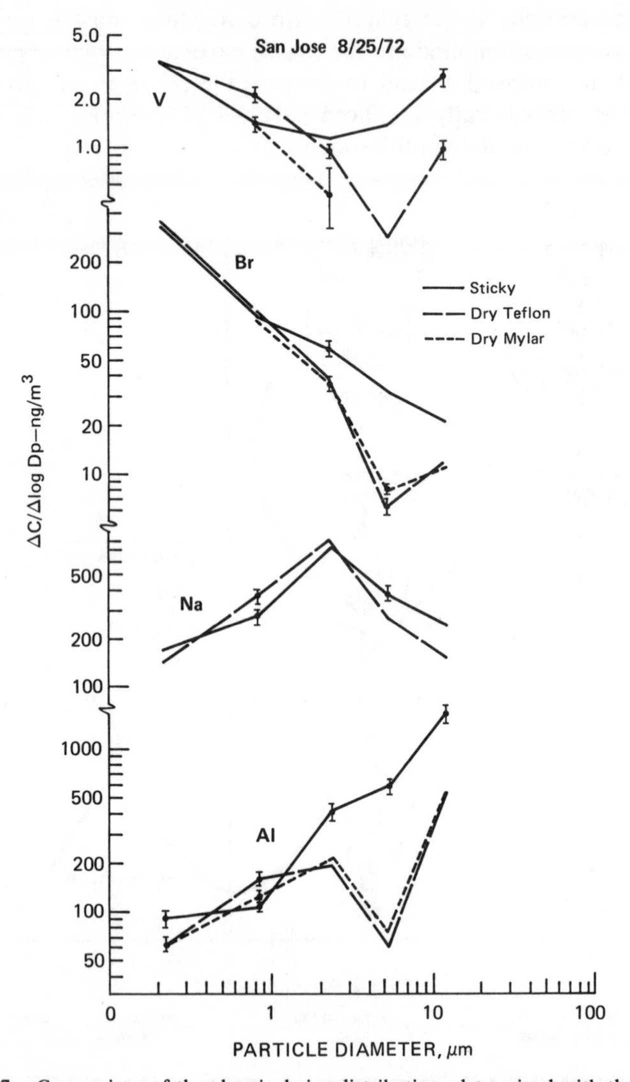

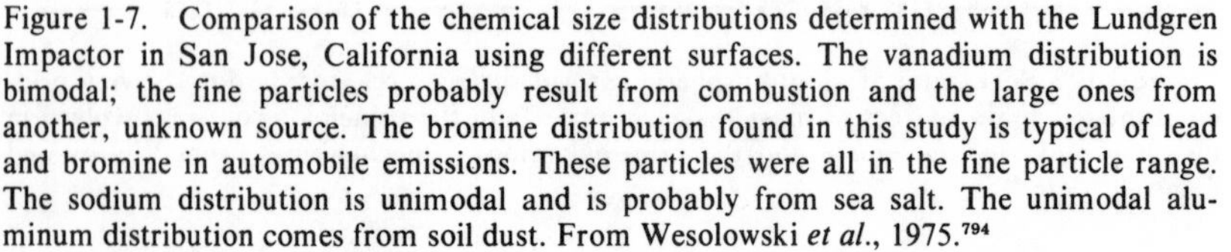

Figure 1-7. Comparison of the chemical size distributions determined with the Lundgren Impactor in San Jose, California using different surfaces. The vanadium distribution is bimodal; the fine particles probably result from combustion and the large ones from another, unknown source. The bromine distribution found in this study is typical of lead and bromine in automobile emissions. These particles were all in the fine particle range. The sodium distribution is unimodal and is probably from sea salt. The unimodal aluminum distribution comes from soil dust. From Wesolowski *et al.*, 1975.[794]

processes occurring both in the atmosphere and in source processes; coarse particles are produced from natural and anthropogenic mechanical processes. The origin, behavior, and removal processes of fine particles are almost entirely independent of those of coarse particles. Therefore, the control measures for each group are quite different and the effects on health, visibility, and meterology can be segregated.

The multimodal nature of the size-mass distribution is also supported by evidence from the size distribution of chemical elements or compounds in aerosols (see Chapter 4).

Figure 1-7 shows the chemical size distribution of several elements measured in San Jose, California. Elements that are emitted as smokes or fumes, such as lead, exist mostly in the fine particle size range.[212] These aerosols eventually coagulate and become mixed throughout the accumulation mode. Table 1-2 shows that the geometric mean size of lead and sulfate aerosol is nearly equal to the mass or volume geometric mean size of the accumulation mode. Most chemical elements are found either in the fine or coarse particle modes and are only rarely distributed between them. Elements in the soil, such as silicon, calcium, and iron, are mostly in the coarse particle range; elements such as sulfur and lead, which are produced by condensation processes from anthropogenic combustion-related sources, are principally in the fine particle range.[205,207] The size of such chemical elements as silicon from soil dust found in the mechanically generated mode is similar to the mass distribution of the coarse particle mode.[807]

INTEGRAL PROPERTIES OF SIZE DISTRIBUTIONS

Any property that is a size-dependent function of a size distribution is called an integral property of the distribution. These are classified and discussed below.

Table 1-2. Comparison of the geometric mean diameter of the fine particle mass distribution mode with chemical distribution of lead and sulfate

Distribution	Geometric mean (μm)	Geometric standard deviation	Reference
Average of 14 volume distributions of atmospheric aerosols	0.34	2.05	797
Lundgren mass distribution by impactor, average of 10 distributions	0.35	1.90	495
Lead—wind from freeway	0.27	2.46	803
Lead—wind away from freeway	0.29	2.62	803
Sulfate, average of four cities	0.39	2.03	777

System Integral Properties (SIP)

Three important integral properties of aerosol size distributions are called system integral properties because they depend only on the size distribution. These are the total aerosol number, surface, and volume or mass concentration. In a perfect sampling system, an Aitken nuclei counter would count number concentration, and a high efficiency filter sampler, evaluated by weighing, would measure the mass concentration.

Weighted Integral Properties (WIP)

When a size distribution is multiplied by a size-dependent function of the distribution, and then integrated, the resultant value is a weighted integral property. An important WIP is light-scattering, where the light-scattering cross section per volume is integrated with the volume distribution to obtain the total particle scattering coefficient, b_{sp}. Assuming uniformly dense spherical particles, the average normalized surface and volume distributions and the computed particle light-scattering coefficient for Los Angeles aerosol are shown in Figure 5-1, page 82. (See Chapter 5 for further discussion of the optical properties of aerosols.)

Pulmonary deposition is another WIP property. Fine particles are more efficiently deposited in the lung than coarse particles. (Deposition is discussed in greater detail in Chapter 6.)

Another significant WIP is the value reported by the filter stain measurement, in which a stain, left on a filter after a known quality of aerosol has been filtered, is measured by optical transmission or reflectance. Whitby *et al.*[800] and Pedace and Sansone[603] have shown that the stain test essentially measures the optically absorbing particles but does not respond to the transparent or liquid particles in the air.

Dust fall is another WIP, in which the settling velocity of particles is sufficient to allow them to reach the ground from the height at which they are launched into the atmosphere.

AEROSOL SAMPLING AND MEASUREMENT METHODS

As a result of the great diversity of application, the size range of atmospheric aerosol physical and chemical concentration variations, and the wide variety of measurement principles available, there are literally thousands of usable combinations of application and measurement methods and procedures. It is impossible to review all these different methods and studies in one report; therefore, this document concentrates on the most important methods in use and cites the most recent references on each. Because of the breadth of this field and the specific interests of its investigators, the literature usually reflects the specializations of the authors. The best general reviews of aerosol measurement techniques and technology appear in conference proceedings.

Recent Literature

Green and Lane[293] have reviewed work prior to 1955. More recently, Mercer[534] and Cadle[111] have discussed aerosol sampling and measurement techniques in considerable detail and breadth. Mercer's (1973) book emphasizes the techniques used to assess radioactive aerosols for biologic studies. Cadle's (1975) book stresses classic methods, such as the microscope, that are used to assess general atmospheric aerosols. Both books were written before the recent revolution in the understanding of atmospheric aerosols. The proceedings of the 1972 symposium, *Assessment of Airborne Particles*, edited by Mercer *et al.*,[535] contain detailed reviews of many of the methods mentioned in Mercer's earlier work.

The proceedings of several conferences contain good discussions of important aerosol measurement techniques. The Illinois Institute of Technology in Chicago sponsored several particle technology conferences[376] that were slanted toward powder technology. A National Bureau of Standards conference on aerosol measurement[129] produced some good papers on recent developments in optical measurement of aerosols. The proceedings of a recent conference on fine particles[478] at the University of Minnesota constitute a most comprehensive recent review on fine particle measurement. During an international conference on environmental assessment[378] a number of papers reviewed developments in aerosol measurement. These proceedings are especially valuable for their information on international developments. The proceedings of the Gesellschaft für Aerosol Forschung (GAF) conferences, held annually in Bad Soden, West Germany since 1973,[75] also contain useful information on aerosol measurement techniques.

Important Characteristics of Aerosol Measurement Methods

Before discussing the characteristics of the many different methods, the suitability of a method for a given purpose must be assessed.

In situ versus collection Such methods as filters and cascade impactors collect the aerosol onto a surface. The collected sample must then be evaluated for size and composition. Other methods, such as optical techniques, sense the aerosol *in situ* without collecting it. Because accumulation mode aerosols (fine particles) contain a substantial fraction of liquid at normal temperatures and humidities, these fine particles must be sized *in situ* without precipitation. In some extreme cases, such as Los Angeles smog, the liquid content may be as high as 75% or 80% of the total mass.[344] Furthermore, evaluation techniques that subject the aerosol particles to a high vacuum, such as electron microscopy, may yield

Table 1-3. Major integral sampling and measurement methods used for measuring atmospheric aerosols

Method or instrument	Effective size limit (μm)		Lower concentration limit	Measurement principle
	Lower	Upper		
Hi-vol filter and other filters—measures all of the nonvolatile mass in accumulation mode and most of that in the coarse particle mode depending on sampling efficiency.[726]	None	25	2 $\mu g/m^3$	Filter-weighing
Integrating nephelometer—essentially responsive to the *in situ* mass in the accumulation mode only.[775]	0.15	2	1% of air-scattering (about 1 $\mu g/m^3$)	Light-scattering
Condensation or Aitken nuclei counter (ANC)—for combustion-dominated aerosols, measures the nuclei mode; for aged aerosols, measures number in accumulation mode.[480,665]	0.0035	0.05–0.2	1 cm^3	Condensation—high supersaturation
Electrostatic charger—with diffusion charging upper limit is about 2 μm; senses both nuclei and accumulation modes.[479]	0.007–0.02	2–25	1 $\mu g/m^3$	Charging and precipitation
Cloud condensation nuclei counter (CNC)—lower size limit depends on supersaturation and design; measures part of nuclei mode and most of accumulation mode.[390,626]	0.01	1	100 cm^3	Condensation—low supersaturation

Ice nuclei counter (INC)—measures only those few particles capable of causing supercooled water drops to freeze; responsive to part of the accumulation mode sizes and most of the coarse particle mode sizes.[72,345,727]	None	Unknown	10^{-5}/liter	Ice nucleation
Quartz crystal—electrostatic version measures nuclei and accumulation modes and part of coarse particle mode; impaction version measures two-thirds of accumulation mode and most of the coarse particle mode.[496]	0.01 electrostatic; 0.3 impactor	10;20	1 μg/m^3	Frequency charge of crystal
β *attenuation*—performance dependent on counting period, collection surface mass.[502]	None (on filter)	20	2–5 μg/m^3	β attenuation
Contact electrification—useful only for high concentrations; sensitivity varies over 50/1 range for different materials.[391]	10	1,000	100 μg/m^3	Impact and charge measurement
Dust fall—relation to concentration is very dependent on location, wind velocity, and collector design.[261]	5	None	None	Settling and weighing
Acoustic counter—has been used primarily to count ice crystals as a sensor on ice nuclei counter.[405]	5	100	1/liter	Acoustic pulse from particle transit

Table 1-4. Major size-resolving methods used for measuring atmospheric aerosols

Method or instrument	Effective size limit (μm)		Maximum number of size intervals practical	Lower concentration limit (μm)	Measurement principle
	Lower	Upper			
Cascade impactors—measures one-half to two-thirds of accumulation mode, all of coarse particle mode; particle bounce is a serious problem in some applications.[794]	0 (filter) 0.3 (last stage)	30	8	1 $\mu g/m^3$/stage	Inertial impaction
Dichotomous virtual impactors—one filter collects fine particles and one collects coarse particles; no bounce problem.[205,484]	0 (filter) 2 (for cut on 2nd filter)	30	2	1 $\mu g/m^3$/stage	Virtual impaction
Crystal microbalance impactor—measures one-half of nuclei mode, all of accumulation mode and coarse particle mode; subject to particle bounce on dry aerosols.[731]	0 (filter) 0.06 (last stage)	30	10	0.02 μg/stage 1 $\mu g/m^3$/stage	Impaction, quartz crystal sensing
Spiral centrifuge—measures one-half of nuclei mode, all of accumulation mode, and part of coarse particle mode; sensitivity depends on method of evaluation; mass determination is difficult.[729]	0.06	10	20	10 $\mu g/m^3$	Centrifuge onto filter
Optical particle counters—single counter useful for only 1 decade of size; requires calibration; sensitive to shape and refractive index.[162,813]	0.3 (commercial) 0.1 (research)	3–100	>200 possible 20 useful	1 particle/liter	Light-scattering from single particles
Electric mobility—can measure nuclei and accumulation modes; can resolve σ_g of 1.4 without correction and 1.25 σ_g with correction.[481,799]	0.006	1	15	1 $\mu g/m^3$	Diffusion charging and electric mobility measurement

extremely misleading results when used on fine atmospheric aerosols that contain significant quantities of liquid *in situ*.

Integral versus differential measurement of particle size Many aerosol measuring methods, such as the condensation nuclei counter, integrating nephelometer, and filter collectors, integrate some response function of the instrument with the particle size distribution. The weighting of the size distribution that the integral method senses can give greatly different results. For example, although a filter collects all particles, the results obtained depend greatly on the method of evaluation. If particle counting is used, the response is essentially equal to the integral of the number distribution weighting. On the other hand, if the filter is weighed, then the integral measurement is the integral of the mass weighting of the distribution. Furthermore, integral methods are always sensitive to the modification of the size distribution by the sampling inlets and transport lines used in the technique. To illustrate, Lundgren,[495] using impactors designed with inlet cutoffs well above those usually used for atmospheric sampling, showed that particles up to several hundred micrometers in diameter were present in the atmosphere under certain conditions. However, most mass sampling methods, including the high-volume sampler,[726] truncate the distribution, thereby giving concentrations less than those actually existing.

Table 1-3 lists the most important integral sampling and measurement methods with their characteristics and a recent key reference. Table 1-4 gives the most important differential, size-resolving methods used to sample and measure atmospheric aerosols. The part of the size distribution and the modes that dominate the sensitivity of the methods are indicated. The upper and lower size limits are nominal values for the most commonly used forms of the technique.

Practical Considerations

Cost, complexity, operational requirements, calibration problems,and the demands of the particular evaluation to be used also affect the choice of methods. For example, chemical analysis usually requires that a sample be collected, then taken to the evaluation device. This normally requires a collection-type method, such as the high-volume filter collectors used in the National Air Sampling Network. Simplicity during sampling may mandate the use of a particular collection method despite its other problems.

The rapid development of modern electronics and inexpensive data acquisition systems has led to *in situ*, automatic, and complex monitoring techniques. Collection methods are being replaced with both physical and chemical analysis with *in situ*, automatic methods.

2

Aerosol Cycles

The production and disposition of aerosol particles are outlined in Figure 2-1. When studying the sources, transformations, transport, and sinks of particles, the nuclei ($\lesssim 0.1$ μm diam) and accumulation (between 0.1 and 2.0 μm diam) modes, which together comprise fine particles, should be considered separately from the coarse (> 2.0 μm diam) mode. The distinction between coarse and fine particles is shown in Figure 1-6 of the previous chapter.

PARTICLE PRODUCTION PROCESSES

Coarse Particles

Most particles with a radius larger than 1.0 μm are produced from mechanical processes. Natural, mechanically originated aerosols arise from windblown dust and sea spray. Man-made mechanical aerosols are produced by a wide variety of industrial processes, for example, the grinding of coal and cement dust and the handling of powdered materials.

Fine Particles

Most fine particles are formed in the nuclei made by various condensation processes. Coagulation and gas-particle interactions (heterogeneous nucleation) are intimately linked with fine particle formation (especially in the accumulation mode). (See the section on TRANSFORMATION PROCESSES on page 26.) Nucleation phenomena can be broadly defined as processes in which aerosol particles are formed in the same size range as large molecules from reactant gases. Nucleation implies the genesis of new particles. Two types of nucleation phenomena are important in the atmosphere. The formation of aerosols from the condensation of a hot, supersaturated vapor is exemplified by the particles resulting from condensation of metallic vapors, low vapor pressure oils, and other low vapor pressure materials formed by combustion. Most aerosols formed from the condensation of cold, supersaturated vapors result from *in situ* conversions of high vapor pressure gases into low vapor pressure reaction products. Examples are the formation of photochemical smogs from hydrocarbons and of sulfuric acid aerosol from the oxidation of sulfur dioxide.

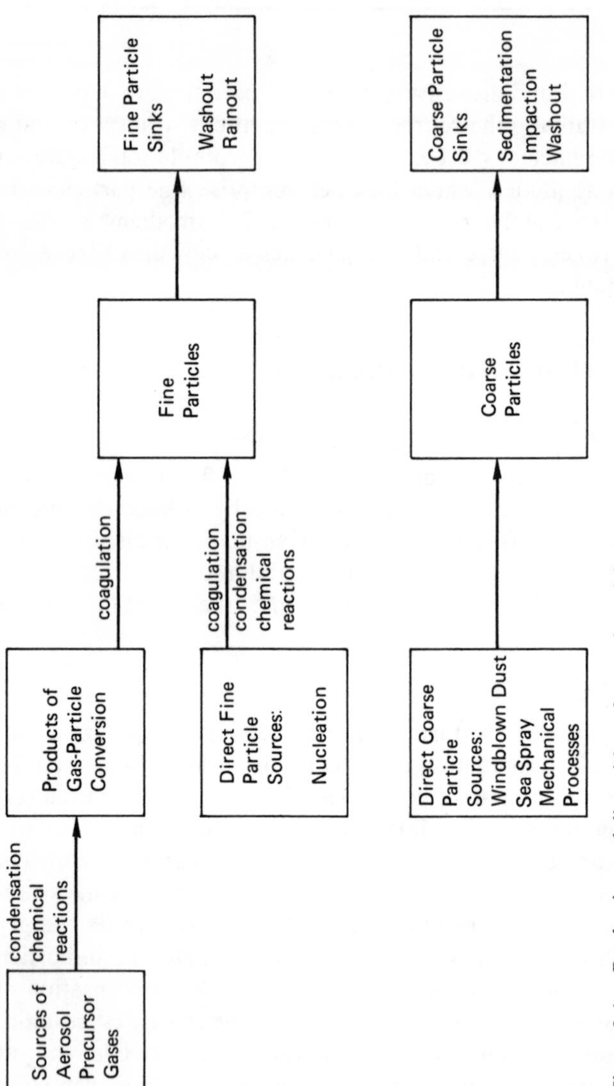

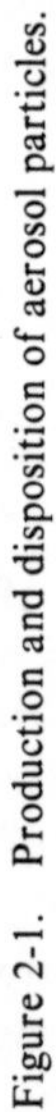
Figure 2-1. Production and disposition of aerosol particles.

Condensation of hot, supersaturated vapors The rapid cooling of many hot, supersaturated vapors produces fine particles (Figure 1-6). For example, metallic vapor leaving the vicinity of a hot source cools rapidly when mixed with air. Under these conditions homogeneous nucleation of the supersaturated vapor occurs. This leads to the formation of high concentrations of fine particles.

Condensation of cold, supersaturated vapors Oils and other low vapor pressure materials do not provide enough vapor by evaporation at ambient temperatures to form any detectable quantities of aerosol. However, *in situ* chemical processes that convert high vapor pressure gases to low pressure reaction products can result in supersaturation of the vapor. The vapor then condenses into an aerosol. The two most important examples in the atmosphere are the photochemical smog formed from the reaction of hydrocarbons with nitrogen oxides and the sulfuric acid droplets and sulfate particles resulting from the oxidation of sulfur dioxide.

Theoretical treatment of the condensation growth of fine atmospheric aerosols is difficult because condensation and coagulation usually occur simultaneously in natural conditions. Only recently has much attention been given to this problem. Brock[89] has shown that the main features of the size distribution can be predicted. Walter[783] has shown that the fluctuation in the number of particles in a chamber that is initially particle-free can be predicted qualitatively. However, quantitative prediction of production rates in laboratory experiments requires further knowledge of the complex transformation processes.

Figure 2-2 shows the development of various aerosol parameters over time during photooxidation of sulfur dioxide in a laboratory. In Period 1, the photooxidation converts sulfur dioxide to sulfuric acid vapor. This rapidly results in supersaturation because the vapor pressure of sulfuric acid is only approximately 10^{-8} mm of mercury. Homogeneous nucleation occurs, resulting in a rapid growth in the number concentration of the aerosol. As the number concentration of particles increases, more and more of the vapor condenses on existing particles. In Period 2, the number concentration reaches a maximum when the rate of new nuclei formation equals the rate of coagulation. At this point, nucleation, coagulation, and condensation all proceed simultaneously. In Period 3, nucleation decreases to a low value and the primary mechanisms of aerosol change are coagulation and condensation. Little light is scattered before the particles grow into the accumulation mode ($>0.15\ \mu m$).

The laboratory studies of photochemically produced aerosols fall into two general categories. Smog chamber studies emphasize the formation and growth of light-scattering aerosols. Most of these investigations entail detailed knowledge of the initial chemical species involved in the

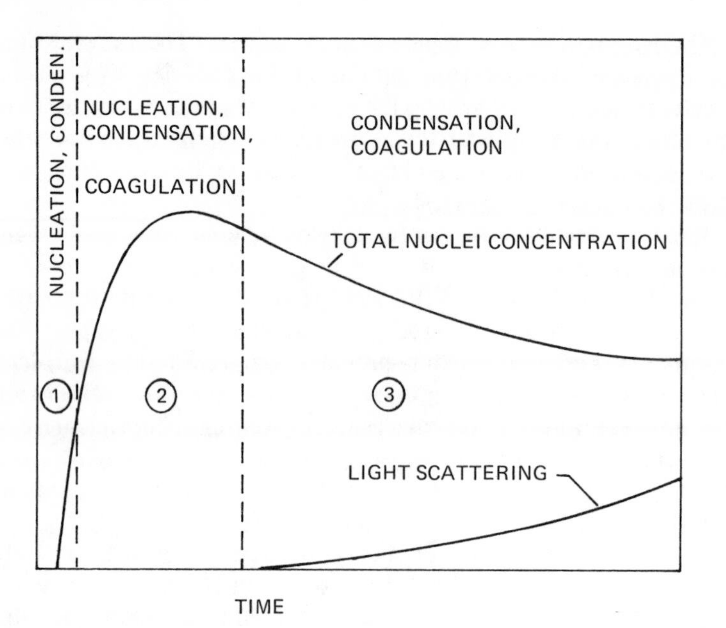

Figure 2-2. Time development of condensation nuclei concentration and light-scattering in an initially particle-free chamber containing an aerosol produced by a photochemically reacting mixture.

reaction. Early investigators concentrated on measuring the increase in light-scattering resulting from irradiating automobile exhaust with ultraviolet light for a variety of engine conditions and fuel compositions. In recent smog chamber studies, attempts have been made to determine the influence on aerosol formation of specific hydrocarbons in the urban atmosphere. Most of these studies have centered on one or two hydrocarbons with various proportions of nitric oxide, nitrogen dioxide, sulfur dioxide, and water vapor. Smog chamber studies on automobile exhaust and individual hydrocarbons have been reviewed by Wilson *et al.*[818] for the American Petroleum Institute. Bufalini[101] and Altshuller and Bufalini[21] have written extensive reviews of the literature concerning the photochemistry of the atmosphere, with discussions of photochemically produced aerosols.

In addition to smog chamber studies, investigators have studied the formation of aerosols resulting from the irradiation of particle-free ambient air. These experiments have emphasized the formation of condensation nuclei from ambient air, including natural and man-made trace gases, from which only particles had been filtered. These studies have not generally included detailed analyses of all the trace gases involved.

Renzetti and Doyle,[641] Prager *et al.*,[618] and Stevenson *et al.*[728] studied systems in which hydrocarbons, NO_x, and sulfur dioxide were irradiated and the various gases, nuclei, and light-scattering aerosols monitored. Irradiation in the reaction chamber causes the condensation nuclei count to rise rapidly to a maximum value and then to decrease (Figure 2-2). Initially, homogeneous nucleation occurs, but eventually the coagulation rate equals the aerosol formation rate and the Aitken nuclei number concentration decreases.

An extensive series of experiments was conducted by Renzetti and Doyle[641] to establish which specific hydrocarbons could produce a light-scattering aerosol in the presence of nitric oxide and in the presence and absence of sulfur dioxide. For the conditions selected, only branched-chain internal olefins, cycloolefins, and diolefins produced light-scattering aerosol in the absence of sulfur dioxide. In the presence of sulfur dioxide, light-scattering was greater for the olefins and diolefins and about the same for cyclohexene. For some hydrocarbons the addition of sulfur dioxide to the system had little or no effect on the aerosol formation.

Altshuller *et al.*[22] monitored the formation of photochemical aerosols for a number of hydrocarbon/nitric oxide reactions during a study of the products and biologic effects resulting from irradiation of nitric oxide with hydrocarbons or aldehydes. These investigators found that 4-carbon olefins in mixtures with nitric oxide produced condensation nuclei but no light-scattering aerosols, and that mixtures of olefin-aromatic hydrocarbons and nitric oxide produced very little condensation nuclei and no aerosols. On the other hand, aromatic hydrocarbons in mixtures with nitric oxide produced both condensation and light-scattering aerosols.

Wilson and Levy[817] observed that the addition of sulfur dioxide to the reaction system produced little effect on the light-scattering aerosols formed by aromatic compounds. However, for alkanes and olefins, the amount of light scattering was greater when sulfur dioxide was present but was less than that obtained from the same amount of sulfur dioxide irradiated in clean, humid air. This discrepancy in earlier studies was attributed to differences in air circulation and residence times.

In 1971, Wilson *et al.*[819] reported that stirring the reactions in a smog chamber during irradiation substantially reduced the formation of light-scattering aerosols. The effect of stirring depended on the composition of the reactants under study as well as the stirring rate. The authors concluded that the difference in sensitivity to stirring found for olefins and aromatic hydrocarbons and for systems that included or excluded sulfur dioxide accounted for many of the discrepancies in previous smog chamber studies.

The 1973 studies by Kocmond *et al.*,[419] which extend the 1971 studies by Wilson *et al.*[819] at Battelle Institute, and the 1975 unpublished work by Kittelson *et al.* (unpublished data), suggest that the reaction products of sulfur dioxide photooxidation (possibly sulfuric acid) passivate the chamber walls and thereby reduce the effects on the walls.

Aerosols resulting from the irradiation of filtered ambient air have been studied by Bricard *et al.*,[88] Vohra *et al.*,[772] and Husar *et al.*[373] Evans and Roddy[221] also added sulfur dioxide to the filtered ambient air before studying its Aitken nuclei number concentrations. These studies all show that ambient air, even in relatively pollution-free locations, contains enough gaseous precursors to form a large number of Aitken nuclei. In 1971, Husar[371] measured the aerosol volume as well as the Aitken nuclei. He showed that the volume of aerosol (and whether it is likely to grow into the light-scattering range) depends greatly on the amount and kinds of reactants present. Relatively clean urban air will produce only a few cubic micrometers per cubic centimeter per hour.

In most urban atmospheres, photochemical conversion rates should rarely exceed a few cubic micrometers per cubic centimeter per hour. Also, some photochemical conversion probably occurs throughout much of the lower troposphere and perhaps even in the stratosphere. However, the converted mass is usually so small compared to the existing mass of the aerosol that the amount of growth is undetectable.

TRANSFORMATION PROCESSES

Aerosol transformation can be generally classified into two categories: gas-particle interactions and particle-particle interactions (coagulation). These processes are discussed below for both coarse and fine particles.

Coarse Particles

These particles are not greatly transformed in the atmosphere. The surface distribution reaches a maximum in the accumulation mode (at about 0.3 μm). Gas-particle interactions are not great in the coarse mode, in which only a small fraction of the particle surface is available (see Figure 1-3, Chapter 1). The surface distribution also limits the mass transfer by coagulation from the accumulation mode to the coarse mode. The small coagulation rates within the coarse mode are attributed to the decreasing number and mobility of these larger particles.

Fine Particles

In contrast, fine particles undergo major gas-particle and particle-particle interactions.

Gas-particle interactions In this process, particulate mass is produced by accretion of reactive gases. The total number of atmospheric particles remains constant, unlike nucleation or coagulation. The mass concentrations of typical aerosol-producing gases usually far exceed the mass concentration of the particles or other substances formed from the gas. For example, 0.1 ppm of sulfur dioxide (a frequently encountered concentration) is approximately 285 $\mu g/m^3$; if oxidized to ammonium sulfate, it would make about 590 $\mu g/m^3$ of aerosol. Thus, the particles produced by only 5% to 10% of these gases are a significant addition to the mass concentration.

In liquid phase reactions, reactant gases are dissolved in aqueous or organic aerosol droplets, including cloud droplets. Most submicrometer atmospheric aerosols, whether formed by direct emission or by secondary conversion, are deliquescent or hygroscopic; thus, their particles contain substantial amounts of water. Charlson *et al.*,[144] Winkler,[825] and Ho *et al.*[344] have shown that urban submicrometer aerosols usually contain large quantities of water. Furthermore, there is always abundant water vapor to provide sources for this condensation; at 80% relative humidity the aerosol particles contain as much as 70% water. Therefore, submicrometer atmospheric aerosols must be modeled as liquid droplets rather than as dry solid particles. The submicrometer aerosols that comprise most of the atmospheric aerosol surface have sufficient water under most atmospheric conditions to present a liquid surface to the gas phase. Consequently, it is probable that the dry surface chemistry is often unimportant.

The relationship between the pressure of a gas in equilibrium with a liquid solution containing the gas and the mole fraction of the gas in solution is approximately linear for many gases. This relationship, known as Henry's Law, applies to gases present at low concentrations in the solution and surrounding medium. This constant of proportionality is discussed extensively in the literature. For water vapor pressure over dilute solutions, the constant of proportionality is the saturation water vapor pressure, which is also a function of temperature. However, for such gases as carbon dioxide, sulfur dioxide, and ammonia, the amount of gas dissolved in the liquid phase also depends on the degree of dissociation of the dissolved species. For these particular gases, the dissociation is regulated by the pH of the liquid, which can be governed by the gases themselves, by the chemical composition of the original particle, or by reaction products formed in the liquid. Therefore, the extent of gas-liquid interaction is regulated not only by the amount of liquid available and the temperature, but also by the chemical composition and reactivity of the various aerosol components.

Clark[152] and Clark and Whitby[154] obtained considerable insight into some of the mechanisms of secondary aerosol transformation through their studies of aerosols formed in a smog chamber. In the classic smog chamber experiment, an initially particle-free gas mixture is irradiated, resulting in the formation of reaction products and aerosol. Although the number of particles rises and then decreases, the surface area first increases and then becomes relatively constant. However, when sulfur dioxide is used, the aerosol volume increases at a nearly linear rate, indicating that the conversion rate is practically constant.

Clark and Whitby[154] concluded that in most photochemical systems studied, the chemical conversion process and the physical behavior of the particles formed are essentially separate. For example, the authors found a unique relationship between the equilibrium surface of the aerosol and the volume conversion rate that is independent of the chemical system. This relationship suggests that if the surface area of an aerosol for a given volumetric conversion rate is greater than the value shown in Figure 2-3, then the low vapor pressure products from chemical conversion in the atmosphere will condense on existing particles rather than form many new particles by homogeneous nucleation. Even relatively "clean" sources of combustion, such as gas-fired burners, are prolific sources of condensation nuclei. In all urban areas, therefore, the nuclei and surface area provided by combustion aerosols are probably sufficient to preclude much aerosol mass formation by homogeneous nucleation of photochemically or chemically converted vapors. Husar *et al.*[373]

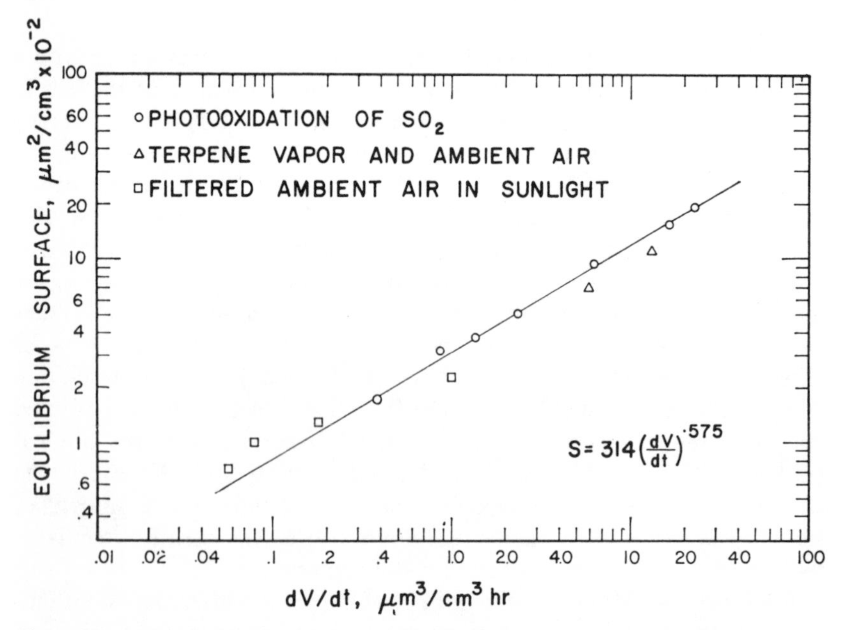

Figure 2-3. Variation of the apparent equilibrium surface area with volumetric conversion rate. From Clark and Whitby, 1975.[154]

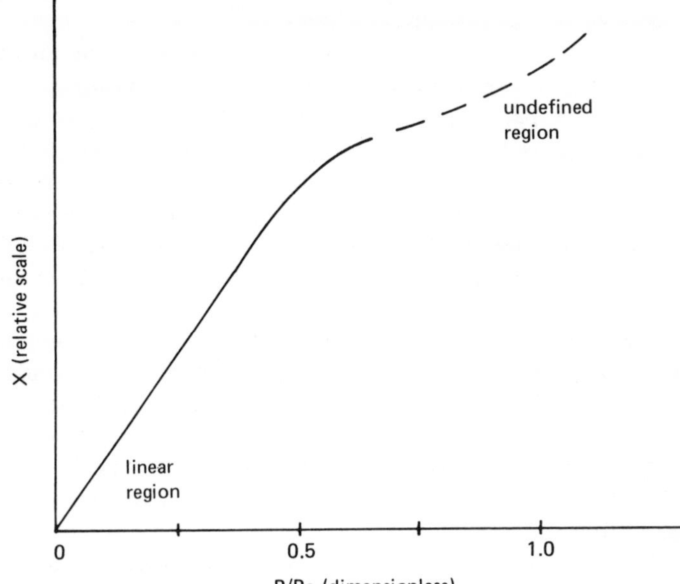

Figure 2-4. Ratio of grams of adsorbate to grams of adsorbent (X) versus relative vapor pressure (P/P_o).

concluded that if the surface area of aerosols exceeds approximately 2,000 $\mu m^2/cm^3$, homogeneous nucleation is unlikely.

Rasmussen and Went[628] have observed a blue haze over vegetation where terpene vapors are released to the atmosphere. Under these circumstances, where there are few combustion nuclei present, photochemical or chemical conversion of terpene vapors could result in homogeneous nucleation of the products. However, this condition could be transient, disappearing as soon as the surface of the aerosol increases sufficiently to accommodate all of the condensible vapors.

Solid phase reactions are defined as the reactions and sorption of gas molecules on surfaces of dry solid particles. The ratio of grams of adsorbate (gas) to grams of adsorbent (X) depends on gas pressure (P) for a given temperature. This is a complicated thermodynamic function of both solid and liquid phases and, in such cases as hygroscopic growth of nitrate aerosols, their history. For many substances, the relationship is illustrated by the Langmuir isotherm in Figure 2-4 where a and b are temperature-dependent constants. Although adsorption of typical atmospheric gas mixtures have not been studied extensively, Figure 2-4 plots the expected adsorption of all atmospheric gases except water vapor, whose values of P/P_o may be as high as 0.5 to 1.0. Thus, adsorption on aerosols may be extremely complex.

For comparison, consider the liquid and solid phase takeup of atmospheric sulfur dioxide. The concentration of sulfur dioxide dissolved in water at a given partial pressure is a function of the partial pressures of certain other gases, mainly ammonia, and the water content of the air. For 0.1 ppm (285 $\mu g/m^3$) sulfur dioxide, 0.1 ppb (76 $\mu g/m^3$) ammonia, and a water concentration of 0.5 g/m^3 (fog), the concentration of dissolved sulfur dioxide is 5.5 $\mu g/m^3$. Solid takeup of sulfur dioxide, at the same partial pressure and at 10 $\mu g/m^3$ of airborne particles, is estimated[611] to be 0.005 $\mu g/m^3$.

Particle-particle interations (coagulation) Aerosol coagulation is an important mechanism by which the large number concentration of Aitken nuclei produced by homogeneous nucleation is transferred to larger particles. This transfer is governed by Brownian diffusion. Aerosol coagulation has been discussed by many authors, including Fuchs[261] and Hidy and Brock.[333] There are two types of coagulation: homogeneous and heterogeneous.

Homogeneous coagulation, which occurs between like-sized particles, is of special importance in or near fine particle sources. Number concentrations of primary particles may approach $10^{10}/cm^3$. Because the monodisperse coagulation coefficient is on the order of $10^{-9} cm^3/sec$, the coagulation rate is very rapid at concentrations exceeding about $10^8/cm^3$. [See Equation 1, where dn/dt = coagulation rate (number/time/volume); K = coagulation coefficient (homogeneous or heterogeneous, volume/time); and n = particle number concentration (number/volume).]

$$dn/dt = -Kn^2 \tag{1}$$

In integral form (from n_o to n), this becomes

$$n = \frac{n_o}{Kn_o t + 1} \tag{2}$$

The result is that, within milliseconds, aggregates of these small primary particles are formed. The very small aggregates or primary particles contribute to a surface area mode that usually occurs at around 0.02 μm. The larger aggregates, resulting from high concentrations of primary emissions, can contribute significantly to the accumulation mode of surface area or of mass.

Figure 2-5 shows the normalized surface area size distribution of a number of combustion aerosols. Note that clean flames, such as those from methanol, give particles predominantly in the transient nuclei range, approximately a few hundredths of a micrometer. In contrast, dirty flames, such as those from a candle or from acetone, which provide more mass, result in a distinct mode in the accumulation range at a few tenths of a micrometer.

Heterogeneous coagulation refers to the coagulation of two or more sizes of aerosol particles. The rate of heterogeneous coagulation may be considerably greater than that of homogeneous coagulation.[261]

If the initial concentration × the time during which coagulation has occurred is much greater than the coagulation coefficient, then the concentration becomes a function only of time and is independent of the initial concentration (see Equation 2). That is why the maximum Aitken nuclei count in the atmosphere rarely exceeds a few million even near sources. Also, the observed Aitken counts in the atmosphere are poor indicators of the initial Aitken counts in the sources, because the concentration is usually coagulation-limited before measurement.

The aerosols from real sources and in the real atmosphere are rarely sufficiently monodispersed so that monodisperse coagulation theory applies. Naturally occurring size distributions are quite varied so that heterogeneous coagulation occurs.

The heterogeneous coagulation coefficient can be one or more orders of magnitude greater than the monodisperse coagulation rate. For example, the rate at which 0.01 μm nuclei coagulate with particles in the accumulation mode (0.2 μm) has a coefficient of about 3×10^{-8} com-

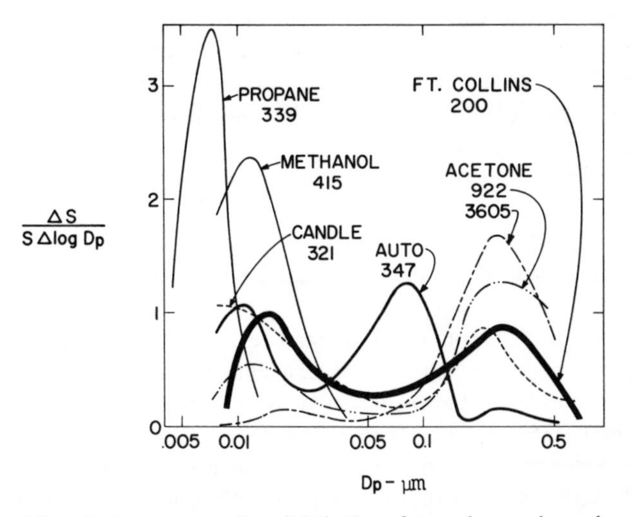

Figure 2-5. The submicrometer surface distribution of several aerosols produced by combustion compared with a distribution observed in Colorado.[804] Combustion produces an average primary particle mode of 0.01 μm and an agglomerated accumulation mode between 0.2 and 0.8 μm. The figures are the surface area in $\mu m^2/cm^3$ for each of the aerosols.

pared to 10^{-9} for 0.1 μm particles coagulating with themselves. Because coagulation of small particles with larger ones can be very rapid, substantial quantities of mass can be transferred by coagulation directly from the smallest sizes to the accumulation mode. The volume transferred by coagulation is relatively constant up to approximately 0.3 μm, after which it decreases sharply, thus producing the cutoff between the two upper modes.

The form of an aerosol size distribution resulting from pure coagulation of a relatively monodisperse aerosol has been studied extensively. Friedlander[259] has calculated that a unimodal size distribution with a geometric standard deviation of approximately 1.3 results from the coagulation of a monodisperse aerosol. This so-called self-preserving size distribution preserves its shape and spread as it grows in size by coagulation. Husar[371] obtained similar results by measuring size distribution of coagulating aerosols in the 0.01 to 0.1 μm range. No good experimental measurements of coagulating heterodisperse aerosols have been made.

PARTICLE TRANSPORT

Coarse Particles

The dispersion of large and/or dense particles, whose settling velocities equal or exceed the vertical component of the wind velocity, is affected by sedimentation. However, such theoretical treatments of their diffusion as reported by Yudine[840] and Smith[707] require extensive knowledge of particle size, shape, and density distributions, because settling velocity is a complex function of these variables.

Most anthropogenic coarse particles are emitted close to the surface, where they will be removed before they travel very far. Steady, high concentrations of coarse particles larger than about 25 μm can be observed only near their source or during unusual conditions like dust storms.[309]

Convective transport of these particles is also important. Increasing evidence indicates that significant concentrations (10 to 30 $\mu g/m^3$) of coarse particles can be transported over long distances after being injected high into the troposphere over dry deserts. Jaenicke *et al.*[383] observed concentrations of Sahara dust exceeding 100 $\mu g/m^3$ on the island of Barbados. Prospero and Bonatti[622] found dust in the eastern Pacific that apparently traveled from the Asian mainland. Bryson[94] observed concentrations of that same dust exceeding several hundred micrograms per cubic meter at 4.5 km over northwest India. Whitby *et al.*[802] observed the steady concentrations of coarse particles near Ft. Collins, Colorado that were apparently transported long distances and mixed well enough so that the characteristic rapid fluctuations of locally generated dust were smoothed out (see Chapter 3).

Fine Particles

Because sedimentation velocities of particles <1 μm radius are generally $<10^{-2}$cm/sec, sedimentation does not affect their atmospheric dispersion. The dispersion of fine particles, which most efficiently scatter and absorb light, is governed mainly by eddy diffusion and advection. It may therefore be assumed that this class of particles behaves as a gas.

Atmospheric particles are largely confined to a 2- to 3-km layer near the earth's surface. Since particle residence time is short (a few days at most) and their vertical transport is slow, particle concentrations drop quickly above 2 km and are very small at 5 km (see Figures 5-13 and 5-14, Chapter 5). The low altitude concentrations are controlled by the stability of lower atmospheric layers and the consequent disparity of mixing times and lifetimes. Clouds affect fine particles that form cloud condensation nuclei by causing them to have finite fall speeds. Thus, the volume of atmosphere actually available for dumping waste particles is limited.

SINKS FOR ATMOSPHERIC PARTICLES

Coarse Particles

Sedimentation and washout are the most important sinks for coarse particles. These sinks limit the maximum particle size in the atmosphere under a particular set of atmospheric conditions as well as the maximum mass concentrations of large particles as a function of atmospheric condition.

Fine Particles

Coagulation has already been discussed as a sink for the nuclei mode; washout is another (see Figure 1-4 in Chapter 1). Accumulation mode particles have fewer efficient atmospheric sinks; their motion is governed almost totally by air movement. Minor removal mechanisms include fallout, impaction with objects at the surface, and washout by impaction processes during precipitation. Of greater importance are incorporation into cloud droplets (nucleation processes) and subsequent rainout. In Europe, Rodhe and Grandell[647] estimated the lifetimes of accumulation-mode sulfate aerosol to be 100 to 300 hr in summer and 35 to 80 hr in winter based on the actions of the removal mechanisms in the European climate.

The importance of natural as compared to man-made aerosols must be understood in order to develop realistic control strategies. Control measures cannot bring concentrations below natural levels. In many locations natural aerosols may even dominate those produced by humans. The presence of sea salt aerosol in maritime regions, terpene hazes in forested regions, natural windblown dust in the plains areas, etc., must be quantified if practicable control measures are to be designed.

AEROSOL BUDGETS

There are several estimates of global aerosol budgets, but they cannot be applied directly to urban regions. Also, they do not distinguish between coarse and fine particles. They do, however, illustrate the role of human activity in the aerosol cycles. Table 2-1[377] provides examples of global aerosol budgets. Estimates by Robinson and Robbins[646] and Hidy and Brock[333] are similar. These budgets show that, on a global scale, anthropogenic aerosol sources constitute from perhaps 5% to 50% of the total, regardless of particle size and lifetime. Since human activity is confined to a small portion of the Northern Hemisphere, and since most human-

Table 2-1. Estimates of particles smaller than 20 μm radius emitted into or formed in the atmosphere[a]

Type of particle	Number in atmosphere (10^9 kg/yr)
Natural	
Soil and rock debris[b]	100–500
Forest fires and slash-burning debris[b]	3–150
Sea salt	(300)
Volcanic debris	25–150
Particles formed from gaseous emissions	
Sulfate from H_2S	130–200
Ammonium salts from NH_3	80–270
Nitrate from NO_x	60–430
Hydrocarbons from plant exudations	75–200
Subtotal:	773–2,200
Man-made	
Particles (direct emissions)	10–90
Particles formed from gaseous emissions	
Sulfate from SO_2	130–200
Nitrate from NO_x	30–35
Hydrocarbons	15–90
Subtotal:	185–415
TOTAL:	958–2,615

[a] From *Report of the Study of Man's Impact on the Climate*, 1971, p. 189.[377]

[b] Includes unknown amounts of indirect man-made contributions.

produced aerosol is limited to approximately 1,000 km from the source (residence time × average tropospheric wind velocity), most aerosol in urban regions is clearly dominated by human activity, particularly by the production of sulfur dioxide from the combustion of oil and coal.

Few assessments have been made of the aerosol budget on an urban or a regional basis. Miller *et al.*[543] developed an approach to this subject for the Los Angeles basin showing the relative importance of anthropogenic and naturally produced particulate matter. They included both direct and gas-particle production in their model.

No carefully prepared budgets exist showing particle size and vertical distribution, although these are important factors. As a result, it is not possible to derive exposures (concentration × time in the atmosphere) at the surface from these budgets. However, it is safe to conclude that the fractional concentrations of fine particles in the lower troposphere are at least as great as their fractional concentrations at their sources. They may even be up to 100 or more times greater due to their relative longevity.

Because aerosol residence time depends strongly on particle size, the budget in Table 2-1 predicts that the future aerosols may be increasingly dominated by the lower mode, especially in urban areas. Hidy and Brock[333] projected that the anthropogenic aerosol contribution would be dominated by sulfate and direct production.

SULFUR: AN EXAMPLE OF AEROSOL CYCLE PROCESSES

In general, aerosol particles are an important part of the life cycles of many chemical substances. The study of one such substance, sulfur, enables one to see this explicitly (Figure 2-6). Although there are thousands of detectable compounds in the atmosphere, our knowledge is sufficiently limited to preclude assessment of their contribution to observed atmospheric phenomena. There is, however, growing awareness of the role of sulfur compounds in determining the characteristics of aerosol[19] and sufficient data do exist on sulfur for it to serve as an example. As in total aerosol budgets, most sulfur data are given as global averages.[260] Since approximately 30% to 50% of the atmospheric sulfur is anthropogenic and since humans populate only a small percentage of the earth, the role of sulfur compounds in the atmosphere can only be understood by studying local or regional sulfur cycles. Although quantitative data are so far lacking, man may dominate the regional cycles in the eastern United States, in Europe, and in a few other places.

In the troposphere the total atmospheric burden is approximately 2×10^{12} g. Atmospheric sulfur exists largely in such reduced forms as hydrogen sulfide, methyl sulfide, etc., and in the oxidized forms of sulfur

Table 2-2. Sulfur dioxide reactions in the troposphere

Category	Reaction	Characteristic time	References	Remarks
Homogeneous reactions	$SO_2 + \frac{1}{2}O_2 + h\nu$	≥125 days	113, 169, 268, 310, 406, 407, 641, 698, 766, 767	Unimportant in troposphere
	$SO_2 + O + M$	10^6 days 2×10^3 days	602 113, 563	Assuming $[O] = 5 \times 10^3$ cm^{-3}
	$SO_2 + OH_2$	20–30 days	462, 602	Assuming $[OH_2] = 10^9$ cm^{-3}
	$SO_2 + OH + M$	15 hr	188	Assuming $[OH] = 10^7$ cm^{-3}
	$SO_2 + \left.\begin{matrix} CH_3O \\ CH_3O_2 \end{matrix}\right\}$ organic oxidants	20–50 hr	113	Smoggy conditions
	$SO_2 + NO_2 +$ olefins $+ h\nu$	2–30 hr	407, 641, 708, 766, 819	
	$SO_2 + O_3 +$ olefins	20 hr	170	Assuming 105 $\mu g/m^3 < [O_3] <$ 210 $\mu g/m^3$ 35 $\mu g/m^3 <$ [HC][a] $<$ 70 $\mu g/m^3$ $[SO_2] =$ 285 $\mu g/m^3$ Relative humidity = 40%

Heterogeneous reactions				
Relative humidity < 100%	SO_2 + metal oxides	—	766	
	SO_2 + fresh graphitic carbon	—	582	Slower for deactivated carbon
	$SO_2 + O_2 + H_2O$ + metal ion catalysts	—	82, 146, 392, 393, 404, 514	Important near sources of SO_2 and metal ions
	$SO_2 + O_3 + H_2O$ + metal ion catalysts	—	51	Possibly more important than comparable O_2 reaction
Relative humidity ≥ 100%	$SO_2 + O_2 + H_2O$	1 day	264, 679	Droplet pH = 5.6
	$SO_2 + O_3 + H_2O$	1 day	607, 608	Droplet pH = 5.6
	$SO_2 + O_2$ + metal ion catalysts	0.5–5 days	82, 146, 392, 393, 404, 514	Strongly dependent on catalyst concentration
	$SO_2 + O_3$ + metal ion catalysts	—	51	Possibly important

[a] as CH_4.

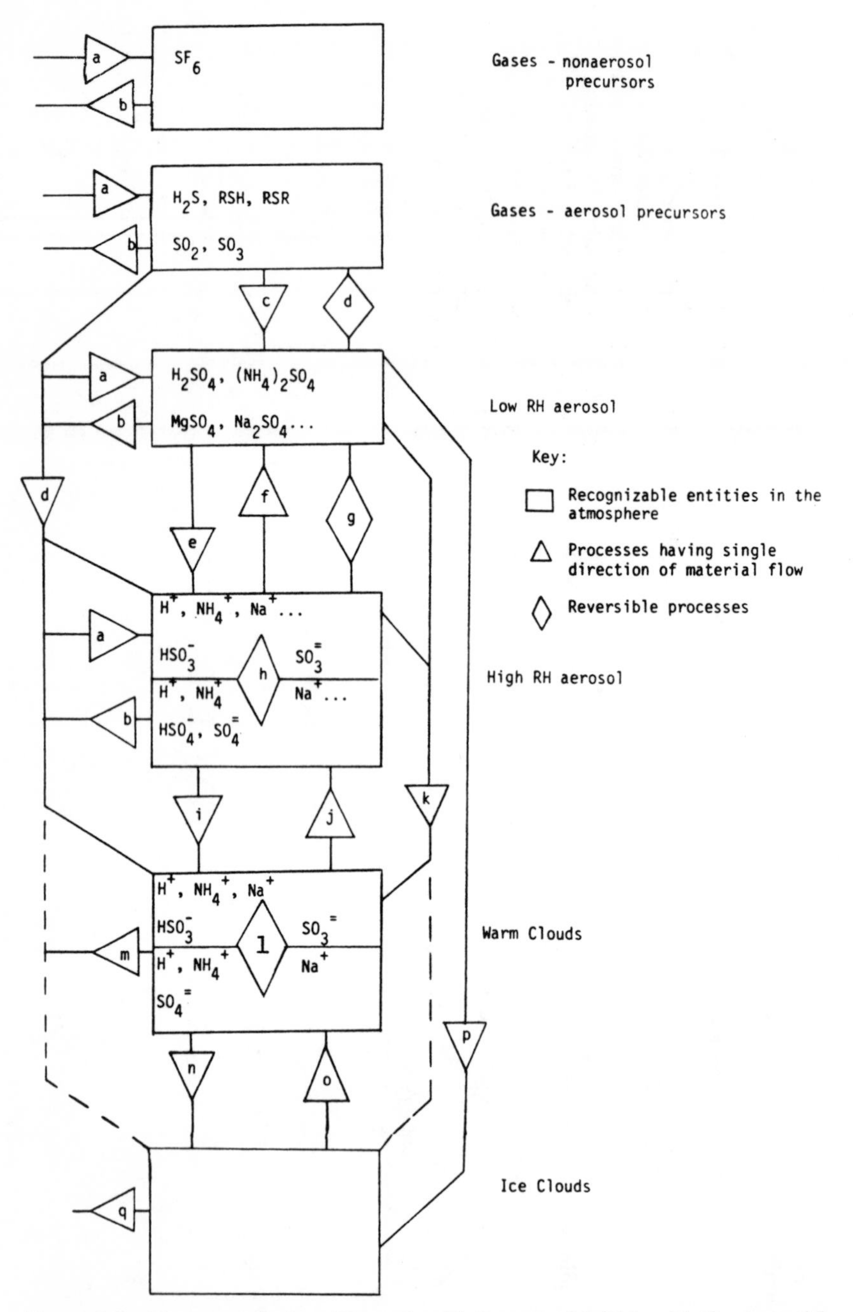

Figure 2-6. The tropospheric sulfur cycle: (a) Sources; (b) Sinks; (c) Gas-to-particle conversions; (d) Sorption; (e) Deliquescence; (f) Efflorescence; (g) Raoult's equilibrium; (h) Reaction in concentrated solution droplet; (i) Nucleation and condensation of water; (j) Evaporation; (k) Capture of aerosol by cloud drops; (l) Reaction in dilute solution; (m) Rain; (n) Freezing of supercooled drop by ice nucleus; (o) Melting; (p) Direct sublimation of ice on ice nucleus; (q) Precipitation. R = organic radical.

dioxide and sulfate ions by a variety of gas phase and liquid surface reactions. The most important reactions are listed in Table 2-2. Sulfate ions combine with available cations (e.g., hydrogen, ammonium, etc.) in both liquid and solid particles. Sulfur is transported in the atmosphere largely through its gaseous compounds and sulfate-containing particles. Thus, sulfur transport has a spatial scale of $\sim 10^3$ km. The sinks are dry deposition of sulfur dioxide gas and sulfate particles and removal by precipitation. The average residence time is roughly estimated to be a few days.

3

Measurements of Size Distribution and Concentration

Before comparing results from the many studies of tropospheric aerosols or particulates, a scheme for classification should be devised. Consideration should be given to:

Type of study
Location of study
Method and purpose of measurement
Average sampling time

There are three general types of studies. The first includes data from federal and state sampling networks that monitor aerosol concentrations in primarily urban areas. Most of these data have been obtained with the high-volume filter sampler, with the results expressed as total suspended particulate (TSP) mass. The second category of data, such as that obtained in the California Tri-City Project[662] and the Environmental Protection Agency Cascade Impactor Network,[455] contains some information on particle size distributions and chemistry. In the third category of data, intensive research studies use the latest and most advanced equipment to obtain detailed data. However, because of the complexity of these intensive studies, the number of sites and time over which measurements can be made are limited. Examples are the 1969 Pasadena smog experiment[805] and the most recent aerosol characterization experiment (ACHEX) sponsored by the California Air Resources Board,[330, 331] the Regional Air Pollution Study (RAPS) program in St. Louis,[808] and the Midwest Interstate Sulfur Transport and Transformation Study (MISTTS) (Cantrell and Whitby, unpublished data).

The second method of categorizing tropospheric aerosol data is by the location of the study. Studies may be categorized as urban, non-urban, or having a goal of determining the characteristics of the natural global background. Most air pollution studies have been conducted in areas having pollution problems, such as urban or downwind-of-urban areas. In studies of the natural background, however, investigators consciously seek the cleanest possible air, such as that over oceans, icecaps, or other remote areas.

In studies of the long distance transport of tropospheric aerosols, the vertical distribution of the aerosols and the variations in their size distribution with altitude are important. These data can be further categorized by altitude above the earth's surface.

Particulate data can also be classified by the kind of measurement used, e.g., concentration, size distribution, chemical distribution, or other special characteristics. Because of the great variety of measurable characteristics, as well as the inevitable compromises that must be made in field work, few investigators have measured a complete list of physical and chemical parameters in any given study. The most complete attempt to measure all relevent chemical and physical parameters was made during the ACHEX study in California.[330, 331]

Another method for classifying aerosol data is by average sampling time. Monitoring networks often make limited measurements of concentration over long periods. These monitoring methods usually provide only limited data on size distribution and chemical composition. Because of the difficulties of mobilizing a large array of measurement methods to measure many chemical and physical parameters, comprehensive measurements have usually been limited to short intensive studies. However, as the duration of the study is decreased, the validity of the time averages also decreases. This inevitable trade-off between depth and validity of the time averages is one of the facets of field and monitoring research that deserves very careful attention during the planning phase of such efforts.

In the following discussions of tropospheric aerosol data, location is used as the primary method of classification, with measurement and time-average characteristics adopted as the subordinate methods of classification. The order in which locations are discussed progresses from the cleanest, most natural background locations to the most polluted ones.

BACKGROUND MEASUREMENTS

Global Background Over the Oceans

Concentration measurements Because oceans occupy a large portion of the earth's surface and play an important role in the earth's radiation balance, the global background over the oceans has been studied intensively for a long time by atmospheric scientists from many countries.[402, 403] The altitude above the ocean's surface can be divided into four regions:

1. Close to the surface, that is, within a few meters above the surface, is a region where salt spray aerosols occur in relatively high concentra-

tions, depending on sea conditions. Pueschel *et al.*[624] measured coarse particle concentrations from 10 to 50 $\mu g/m^3$ several meters above the surface in the Pacific near the coast of Washington.

2. The second region extends from the upper edge of the surface layer to the top of the mixing layer. Concentrations averaging 5 $\mu g/m^3$ of salt aerosol are found in this second region. Condensation in clouds at the upper edge of the mixing layer is an effective barrier to the transport of salt aerosols to the region between the upper edge of the mixing layer and the lower edge of the stratosphere, the tropopause.
3. This region between the mixing layer and the tropopause has extremely clean air. Data obtained from the Mauna Loa background monitoring station indicate that the average concentration of aerosol is very low, with Aitken nuclei counts averaging a few hundred per cubic centimeter. Prospero[623] measured sea salt concentrations of 2 to 3 $\mu g/m^3$ in this layer over the mid-Atlantic.
4. Above the tropopause in the stratosphere, the aerosol concentration is essentially the same over both sea and land, because the removal rate is slow compared to the transport rate by stratospheric winds. Measurements by Junge,[401] Podzimek,[612] and other investigators have indicated that typical particle concentrations in the stratosphere range from 1 to 20 Aitken nuclei/cm^3, corresponding to mass concentrations from 0.001 to 0.01 $\mu g/m^3$. These concentrations are subject to large changes due to occasional injection of high aerosol concentrations into the stratosphere from volcanoes or from atomic tests.

There is also evidence of a sulfate aerosol layer just above the tropopause but it is not known whether this originates from sulfur diffusing upward through the tropopause or from volcanic injections into the stratosphere. Figure 3-1 shows a typical variation in Aitken nuclei count and calculated mass concentration with altitude based on a summary of available data.

Size distribution measurements Global background aerosols over the oceans tend to be dominated by aerosols in the surface layer that have been mechanically produced by the sea surface, and by the coarse particle and fine particle aerosols above the mixing layer that have been transported from land. Aged particles are those that have had enough time to coagulate and settle for at least a few hours. They tend to be unimodal, the nuclei mode having been removed by coagulation and the coarse particles by settling, unless large amounts of dust are present. These unimodal distributions have a geometric mean size by volume or mass ranging from 0.4 to 1 μm, with an average geometric standard deviation of 2.[383, 402] Some typical number distributions are shown in Figure 3-2.

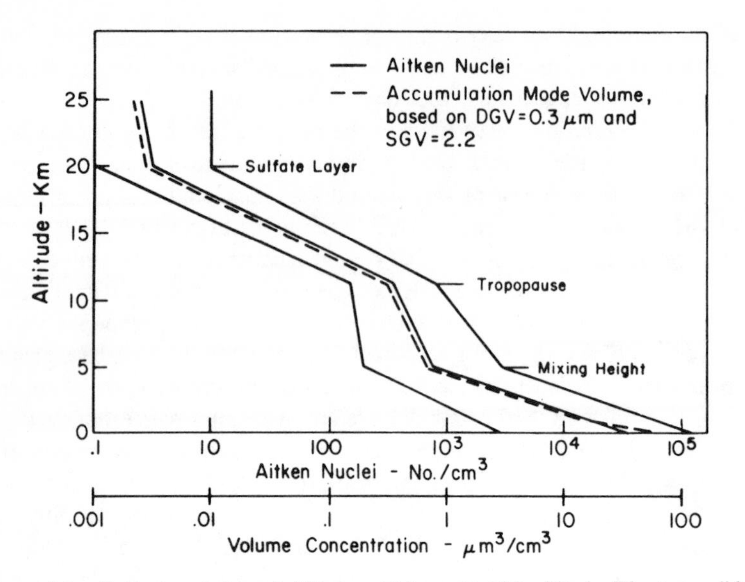

Figure 3-1. Typical variation of Aitken nuclei counts with altitude. The two solid lines above and below the central solid line represent the upper and lower boundaries as deduced from the available literature. The accumulation mode volume has been calculated assuming a nuclei count to volume ratio of 1,000 nuclei/cm³:1 μm³/cm³. From Whitby, 1975.[798]

In 1976, Flyger *et al.*[250] reported finding very small nuclei-mode-sized aerosols (<0.01 μm diam) above Greenland. They interpreted this to indicate that chemical or photochemical production of new nuclei is occurring. On occasion, they found Aitken nuclei counts of 10,000/cm³ at altitudes of several kilometers. Surface and background counts are normally between 200 and 500/cm³.

In a recent comprehensive report, Schaefer[666] has summarized many years of measurement of fine particles, especially of condensation nuclei. The results agree well with the modal concepts discussed in Chapter 1. He has categorized his condensation nuclei count data into seven categories ranging from 50 to 10 million particles/cm³.

Long-distance transport of coarse particles As discussed in Chapter 2, recent evidence[96, 97, 124, 383, 623] indicates that dust advected above the mixing layer into the midtropospheric altitudes over hot, dry desert areas is then transported long distances over the ocean. For example, studies over the Atlantic have shown that dust from the Sahara Desert is transported in measurable quantities across the entire South Atlantic. Concentrations exceeding 100 μg/m³ above the mixing layer at several kilometers in altitude were measured during the Global Atmospheric Research Program's Atlantic Tropical Experiment (GATE).

The average concentration of dust during the dry season over the island of Barbados is 10 μg/m³. During the rainy season, it is less than 1

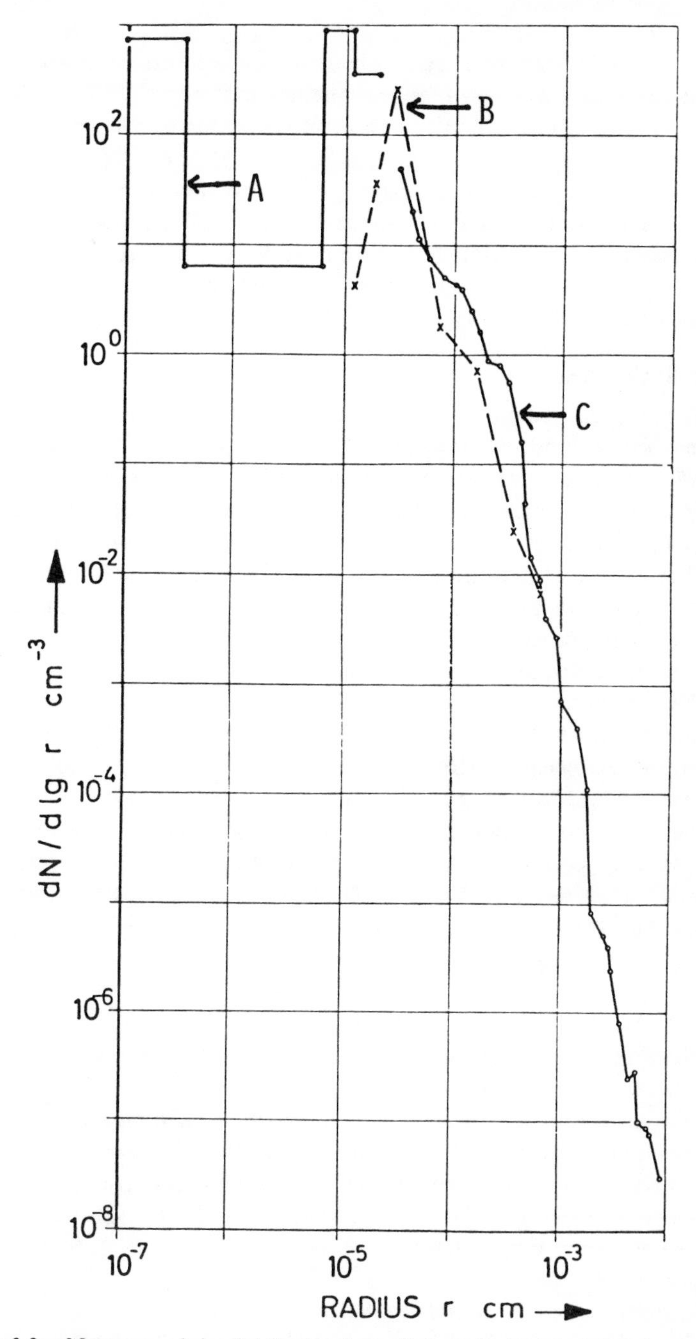

Figure 3-2. Mean aerosol size distribution from all data obtained during Spring, 1969 on the *R. V. Meteor*. From Jaenicke *et al.*, 1971.[383] A: Below 2×10^{-5} cm radius. Data obtained with diffusion battery and electrostatic precipitators. B: Data obtained with double state impactors. C: Composite data obtained with the impactor, plus an optical counter, and free air impactor plate.

$\mu g/m^3$. This Saharan dust layer is bounded on the bottom by the oceanic mixing layer at 1.2 to 1.8 km, and at the upper edge by an inversion at about 3.7 km.[623] The sea salt concentration in this dust layer averages 2 to 3 $\mu g/m^3$.

Similar dust plumes have been observed downwind of the desert areas of north India and Pakistan, and over the China Sea downwind of some deserts on the mainland. These dust plumes may play a substantial role in the earth's meteorology and cloud physics, but conclusive results are not yet available.

Clean Background Over Land

Clean background conditions over land are more complex than over the oceans. Over land, variations in surface albedo, topography, and precipitation can cause very large variations in the rates of generation by the surface and of removal. Thus, there is no single set of parameters for the background aerosol concentration and size distribution over land.

The following locations and conditions have been selected to illustrate the possible range of concentrations and the general effect of different surface and location characteristics on observed aerosol concentrations and size distributions.

Remote areas with snow and ice surfaces Studies over Antarctica[354] and the Arctic show the aerosol concentration to be as low as or lower than over the oceans. Aitken nuclei counts of $100/cm^3$ are common. Most of the aerosol in these locations has been transported over long distances, such as the aerosol observed by Flyger *et al.*[249] over the Greenland ice cap. On occasion, nuclei counts as high as $10{,}000/cm^3$ were observed at altitudes of 1 km, with accompanying accumulation mode mass concentrations of 10 $\mu g/m^3$. Although these high nuclei counts imply *in situ* production of new particles, the majority of the aerosol mass was found in the accumulation mode. It was probably transported from the North American continent.

Remote dry or desert areas As indicated earlier, these areas may be the source of large quantities of dust advected to high altitudes. Such aerosols are produced by mechanical processes and modified by sedimentation removal of the larger particles. In their measurement of dusts over the Sahara, Schütz and Jaenicke[682] found geometric mean sizes by volume to be 3 to 6 μm diam and geometric standard deviations from 2.0 to 2.5 μm diam. Concentrations can range to several hundred micrograms per cubic meter.

Gillette *et al.*[275] found that the distribution of coarse particle aerosols generated by wind erosion over land in the Midwest was similar to that found by Schütz and Jaenicke over the Sahara. Figure 3-3 shows a typical size distribution observed in the Midwest studies.

Sehmel[688] and Heinsohn *et al.*[323] have studied the dust produced by cars and trucks on roadways. This source of coarse particles is important since few areas of the United States are free from vehicles traveling on roadways, and although most roads are paved, many in desert areas are not.

Whitby[796] found evidence of such particles settling as a steady rain in the western United States. These particles were possibly carried to high altitudes by thunderstorms.

In dry areas, these particles provide a substantial coarse particle mode background. Because these are measured by such filter techniques as the standard high-volume filter method for total suspended particulates, they affect the significance of such integral mass measurements. In the western United States, these fugitive dusts often account for most of the total suspended particulates.[323]

Sverdrup *et al.*[747] reported measurements made at the Goldstone tracking station in the Mojave Desert as part of the California ACHEX study. One day after rain, when the winds were from the northwest, the total aerosol volume was 1.85, the submicrometer volume was 1.03 $\mu m^3/cm^3$, and the Aitken nuclei count was less than $1{,}000/cm^3$. Although Schaefer[666] reported somewhat lower Aitken nuclei counts in very clean

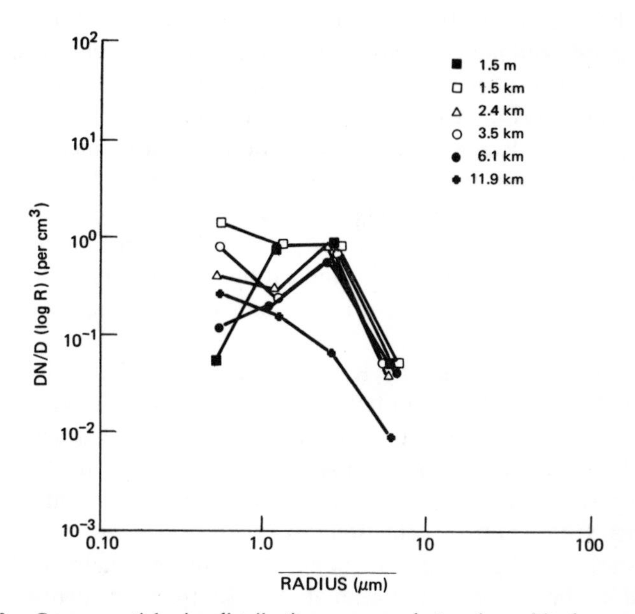

Figure 3-3. Coarse particle size distributions measured at various altitudes over the Midwest on April 30, 1971. If these number distributions are transformed to volume distributions, the coarse particle mode sizes are in general agreement with those given in Table 3-3. From Gillette *et al.*[275]

locations, such as Yellowstone National Park during winter, the Goldstone measurement on October 31, 1972 is probably the lowest that could occur in most dry areas.

Areas of average precipitation and vegetative cover The precipitation and vegetative cover like that found over the eastern two-thirds of the United States provide a quite different background aerosol situation than in the dry areas of the western United States.

First, because the area over which dust can be suspended is limited, the source of coarse particles, and, therefore, the coarse particle mode, is usually much smaller. In many cases, measurements as small as 5 or 10 $\mu g/m^3$ have been recorded. However, during periods of high wind or near dry, bare fields, the amount of coarse particles in a particular location can be increased temporarily to several hundred micrograms per cubic meter.

The amount of natural accumulation-mode-sized aerosol that exists during true background conditions is not clear. Evidence suggests that most areas in the United States are affected at some time by anthropogenic aerosol sources.

Natural accumulation mode aerosol concentrations were studied by Whitby *et al.*[806] during the General Motors Sulfate Study. On several days when the wind was out of the northwest over essentially uninhabited regions of Canada and the United States, they measured aerosol accumulation mode volume concentrations from 1 to 2 $\mu m^3/cm^3$, corresponding to 1 to 3 $\mu g/m^3$. In early morning, coarse particle volume was only 2 to 5 $\mu m^3/cm^3$. This increased to 5 to 15 $\mu m^3/cm^3$ later in the day, as the relative humidity decreased and the traffic in the area increased. These observations suggest that most of these coarse particles had a local, anthropogenic origin, even though they were not emitted from a stack. This air transversed vegetated, but relatively uninhabited, areas of Canada and the United States, indicating the possible accumulation mode background in that area prior to man's arrival and his subsequent pollution.

Measurements during the RAPS study in the St. Louis area and during the General Motors Sulfate Study in central Michigan, when the wind was from the southwest, south, or east, indicate that accumulation mode volume concentrations ordinarily ranged from 5 to 20 $\mu m^3/cm^3$. This suggests that most of this aerosol was aged and had been transported from anthropogenic sources.

Altshuller,[20] by using the concentrations of various chemicals measured in impactor and National Air Sampling Network (NASN) Hi-Vol studies, has recently calculated the concentration of fine particles that should exist in different parts of the country. He calculated fine particle mass concentrations (FPF) for various urban and nonurban areas in the United States (Table 3-1). The lowest value of FPF is 2.7 $\mu g/m^3$.

The differences in FPF and coarse particle mass concentrations (CPF) between the eastern and western United States can be clearly seen by comparing the ratios of the average eastern concentrations to the average western concentrations. ["East" includes East Coast, Southeast, and Midwest (E)]. The ratio for urban FPF is 1.7:1; nonurban FPF is 2.2:1; urban CPF is 1.2:1; and nonurban CPF is 1.3:1. Although the average concentrations of coarse particles in the East and West are similar, the East has nearly twice as much fine particle mass in both urban and nonurban areas as the West. This is evidence that the fine particle concentrations, both in urban and nonurban areas, are much more directly related to human activity than are coarse particle concentrations.

MEASUREMENTS OF ANTHROPOGENIC INFLUENCE

Evidence from the many measurements of aerosols throughout the world indicates that most places on earth are influenced at some time by anthropogenic activity. In any meaningful discussion of anthropogenic influence, one must first specify which aerosol characteristics are to be discussed; then the significance of specific concentrations should be defined.

To present the data, the amounts by particle size will be broken down into fine, coarse, or, where available, the amounts in the nuclei, accumulation, or coarse particle modes. The data will be further categorized in a number of different ways, because there is no general agreement as to what constitutes an urban aerosol or a background aerosol.

Urban Aerosols

Aerosols in urban areas differ greatly in size, source, meteorology, and other aspects; therefore, only the most general characteristics will be discussed here. Because urban areas have many aerosol sources, various sections or individual sampling stations can be strongly influenced by local sources. Only data thought to be relatively uninfluenced by local sources have been included below.

In Table 3-2, the fine and coarse particle concentrations in several urban areas are compared to very clean values. The results of Table 3-2 compared with Table 3-1 indicate that the concentrations determined by the two different methods are in reasonable agreement. These tables also indicate that fine particle concentration in urban areas is considerably higher than in clean areas. During polluted conditions, the fine particle

Table 3-1. Concentrations of fine particulate species at urban and nonurban sites in various regions in the United States during 1968–1970 (or 1967–1969, where noted)[a]

Region	Location	Concentration ($\mu g/m^3$)								
		$[SO_4^{-2}]$ + NH_4	BSO[b]	$[NO_3^{-2}]$ + NH_4	Pb	Fe (1967–1969)	Cu, Mn, Ni, V, Zn (1967–1969)	FPF[c]	FPF/TSP[c]	CPF[c]
East Coast	Urban	16.1	6.0	1.4	1.0	0.4	0.75	25.5	0.25	76.5
	Nonurban	8.4	1.6	0.7	0.08	0.05	0.15	10.8	0.31	24.0
Southeast	Urban	9.1	5.7	1.8	1.0	0.4	0.3	18.4	0.18	83.8
	Nonurban	7.7	1.3	0.8	0.07	0.08	0.15	10.2	0.22	36.2
Midwest (E)	Urban	14.6	5.6	2.1	1.1	0.84	0.6	25.5	0.19	108.7
	Nonurban	9.0	1.7	0.7	0.08	0.13	0.25	11.4	0.28	29.3
Midwest (W)	Urban	6.7	4.4	1.4	0.7	0.4	0.25	13.9	0.13	93.0
	Nonurban	3.2	1.2	0.4	0.03	0.04	0.1	5.0	0.24	15.8
Southwest	Urban	4.9	4.5	1.6	1.0	0.25	0.15	12.4	0.15	70.3
	Nonurban	4.8	1.5	0.7	0.04	0.04	0.1	7.2	0.23	24.1
Mountain	Urban	3.6	4.3	1.2	0.8	0.25	0.25	10.4	0.12	76.3
	Nonurban	1.6	0.8	0.15	0.00	0.01	0.1	2.7	0.23	9.0
West Coast	Urban	7.6	6.4	3.0	1.7	0.25	0.25	19.2	0.23	64.3
	Nonurban	3.4	1.15	0.35	0.00	0.03	0.1	5.0	0.10	45.0

[a] From Altshuller, 1976.[20]

[b] BSO = Benzene-soluble organics.

[c] FPF = Fine particle mass concentration; TSP = total suspended particulate mass; CPF = coarse particle mass concentration (TSP − FPF).

Table 3-2. Fine and coarse aerosol concentrations from some urban measurements compared to clean areas

Location	Condition	Concentration ($\mu g/m^3$) ($\rho_p = 1$)		Reference
		Fine particles	Coarse particles	
St. Louis	Very polluted	296.0	94.0	[a]
Los Angeles	Grand average	37.0	30.0	805
Los Angeles freeway	Wind from freeway	77.0	59.0	803
Denver	Grand average	16.6	232.0	815
Goldstone	Clean	1.5	3.0	747
Milford, Mich.	Very clean	1.03	0.82	806
Pt. Arguello (seaside)	Marine air	1.1	53.0	331

[a] Cantrell and Whitby, unpublished data from 1975 EPA MISTTS Program, St. Louis.

concentration in St. Louis reached 296 $\mu g/m^3$ on one day. In 1973, a concentration approaching 300 $\mu g/m^3$ was observed in Los Angeles. Under normal conditions, fine particle concentrations in urban areas seem to average 20 to 30 $\mu g/m^3$.

At a rural site 40 km from St. Louis and at an urban site in St. Louis, Dzubay *et al.*[207] observed that sulfur and lead dominated the fine particles and were an insignificant fraction of the coarse particles. The composition of the rural and urban aerosols were similar, indicating that the rural sample was transported from the city. Chemical composition as a function of aerosol size will be discussed in greater detail in Chapter 4.

Urban versus Nonurban Aerosols

Aerosol size distribution data from a variety of locations are listed by distribution modes in Table 3-3. These data show that the volume of aerosol in the nuclei mode is usually quite small ($<1\ \mu m^3/cm^3$) except very near intense sources of condensation nuclei, such as a freeway or roadway. In most urban areas, the accumulation mode contains upwards of 30 $\mu m^3/cm^3$. The amount in the coarse particle mode is quite variable. The highest values are obtained in dry areas where windblown dust may dominate the coarse particles.

Table 3-3 summarizes the modal parameters of a large amount of data collected in a variety of locations in the United States. Those sites classified as urban were all within cities. At the urban sites, the nuclei mode average was 0.72 $\mu m^3/cm^3$, the accumulation mode average was 41.4 $\mu m^3/cm^3$, and the coarse particle mode average was 36.9 $\mu m^3/cm^3$.

Table 3-3. Summary of modal parameters for different locations and conditions

Location and type of aerosol	Number of sites	Nuclei mode		Accumulation mode		Coarse mode	
		DGV[a] (μm)	V[b] (μm³/cm³)	DGV (μm)	V (μm³/cm³)	DGV (μm)	V (μm³/cm³)
Urban—mostly grand averages	9	0.042(0.01)[c]	0.72(0.19)	0.35(0.06)	41.4(12.8)	5.7(1.2)	36.9(7.1)
Urban—influenced by close source	4	0.033(0.008)	13.2(12.5)	0.26(0.041)	26.9(24.2)	6.7(3.8)	41.6(7.7)
Background—little urban or source influence	7	0.035(0.006)	0.035(0.044)	0.33(0.082)	4.4(2.9)	7.8(5.0)	56.8(85.4)
Background—near sources	2	0.032(0.017)	0.77(0.62)	0.22(0.014)	8.3(7.9)	4.6(1.1)	7.8(4.8)
Background—influenced by urban areas	3	—	0.047(0.081)	0.37(0.11)	42.2(32.2)	4.23(0.81)	14.1(5.4)

[a] DGV = geometric mean diameter of mode.

[b] V = volume of aerosol in mode.

[c] Figures in parentheses are standard deviations.

An example of "urban—influenced by close source" would be an urban site within 50 m of a road or freeway. In this case there is a relatively large volume in the nuclei mode, as compared to the general urban aerosol. The volume in the accumulation mode is less than the urban average. Because these data apply only to a particular site they have no great significance.

The background sites were divided into three groups. The first group, obtained from averaging data from seven sites, showed little urban or source influence. In this group, the nuclei mode concentration was very low. At four of the seven sites, it was essentially equal to 0. The average for the accumulation mode was 4.4 $\mu m^3/cm^3$, roughly one-tenth of that for the urban areas. Some of the background sites had accumulation mode volumes as low as 1 $\mu m^3/cm^3$. However, the volume in the coarse particle mode was essentially the same as for the urban sites.

The second group consisted of two background sites that had nearby sources of clean combustion aerosol. In this case the nuclei mode was about the same as the urban. However, the accumulation mode volume was much less. There was a very low volume in the coarse particle mode (7.8 $\mu m^3/cm^3$) because these measurements were made on days when there was very little windblown dust.

A third group consisted of background sites at which there was ample evidence of incursions of urban aerosol from as far as 150 km.

Figure 3-4 shows an incursion of accumulation mode air in substantial volume as measured by Sverdrup *et al.*[747] at the Goldstone tracking station in 1972. There is a large increase in volume in the accumulation mode with little increase in the coarse particle mode.

Figure 3-5 shows two size distributions measured at Hunter-Liggett Military Reservation during the 1972 ACHEX experiment, in which local anthropogenic activity caused significant brief fluctuations in the coarse particle mode without affecting the accumulation mode.

In general, the coarse particle mode is influenced mostly by local sources of coarse particles. The accumulation mode is affected by secondary aerosols, which are formed over large areas and which can be transported over distances of hundreds of kilometers.

Long-Distance Transport of Accumulation Mode Aerosols

As discussed in Chapter 2, fine particles can be transported over long distances. Evidence outlined above shows that accumulation mode aerosols have been transported from 100 to 150 km. Recent evidence indicates that accumulation mode aerosols can be transmitted at even greater distances—perhaps up to several thousand kilometers.

Brosset *et al.*[90] and others in Europe have shown that sulfate and soot-containing accumulation-mode-sized aerosols are frequently transported from northern Europe to Norway and Sweden. Waggoner

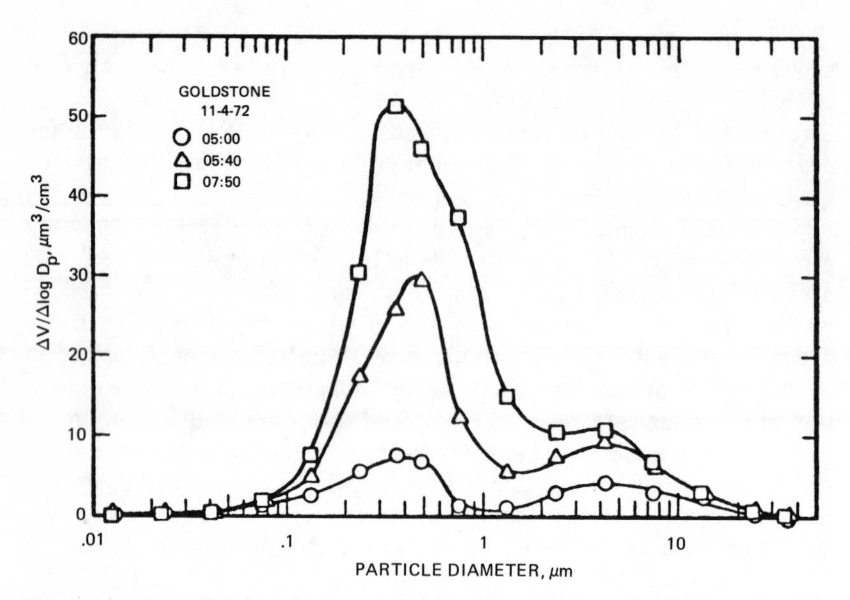

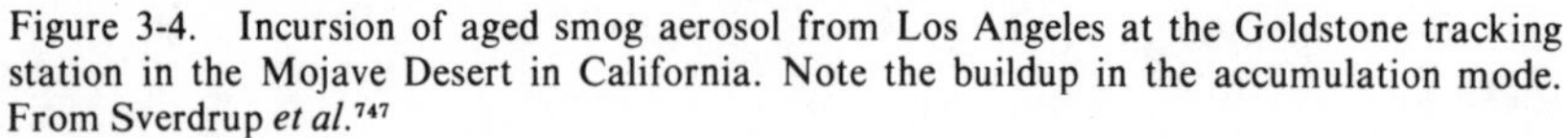
Figure 3-4. Incursion of aged smog aerosol from Los Angeles at the Goldstone tracking station in the Mojave Desert in California. Note the buildup in the accumulation mode. From Sverdrup *et al.*[747]

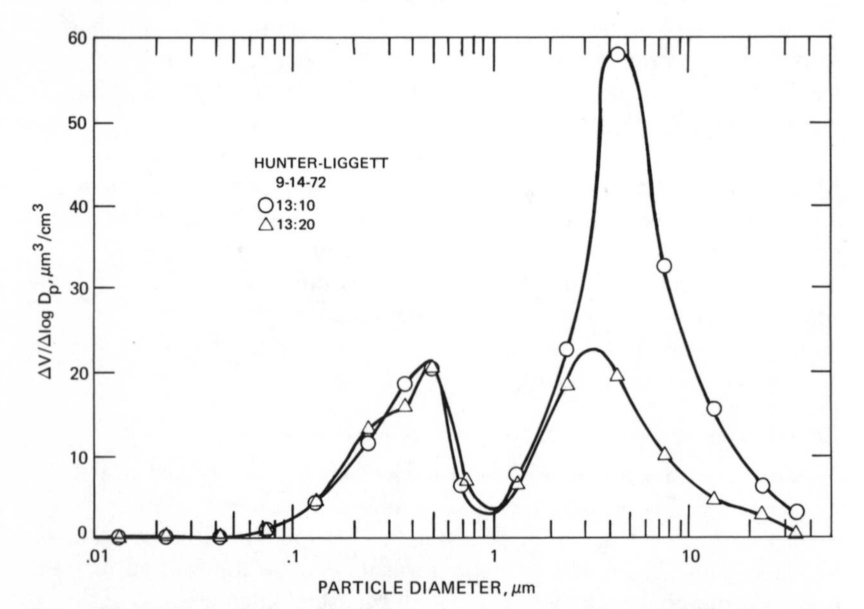

Figure 3-5. Sudden growth of the coarse particle mode due to local dust sources measured at the Hunter-Liggett Military Reservation in California. This shows the independence of the accumulation and coarse particle mode. Whitby and Sverdrup, unpublished data.

et al.[776] have reported a significant correlation between sulfate mass concentrations and particle scattering coefficient measured at 1 to 2 km over Norway.

In June 1975, Husar *et al.*[372] observed a large, high pollution episode that wandered around many states in the eastern half of the United States. They measured the light-scattering and sulfate in St. Louis during June and July 1975 (Figure 3-6). The light-scattering and sulfate values are highly correlated. The peak aerosol concentrations probably correspond to 60 to 70 $\mu g/m^3$ of total aerosol and 35 to 40 $\mu g/m^3$ of sulfate.

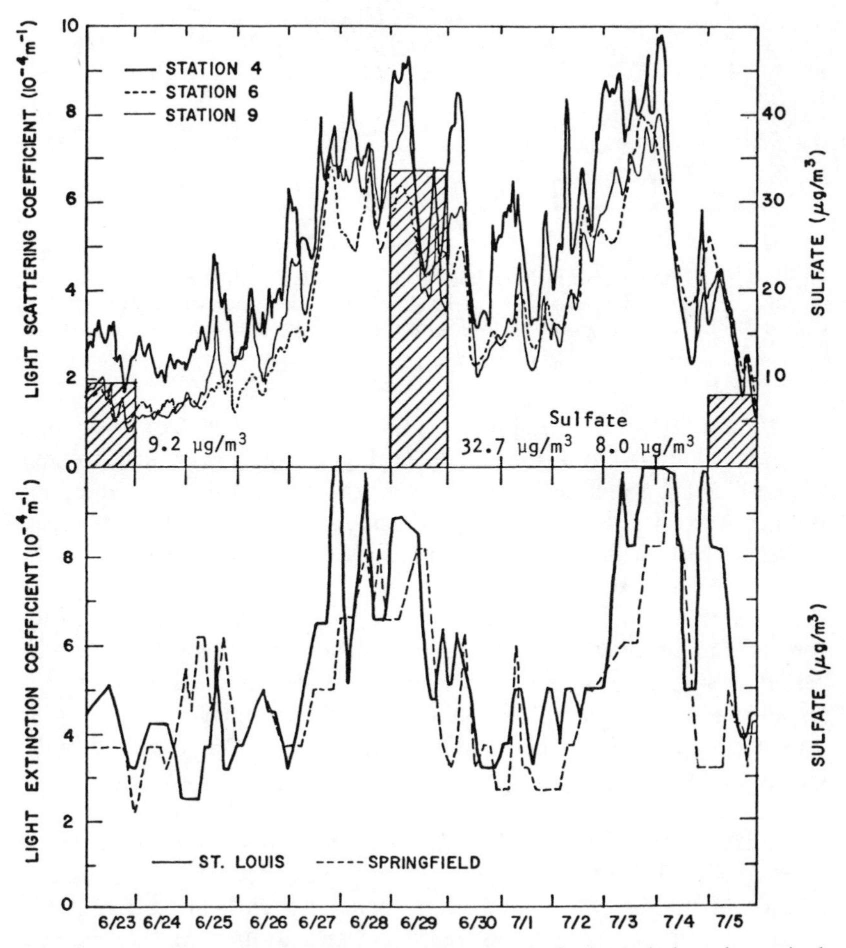

Figure 3-6. Light-scattering and sulfate measurements in St. Louis during a long episode of haziness. The upper graph shows the b_{scat} and sulfate data and the lower one the extinctions calculated from visibility data. From Husar *et al.*[372]

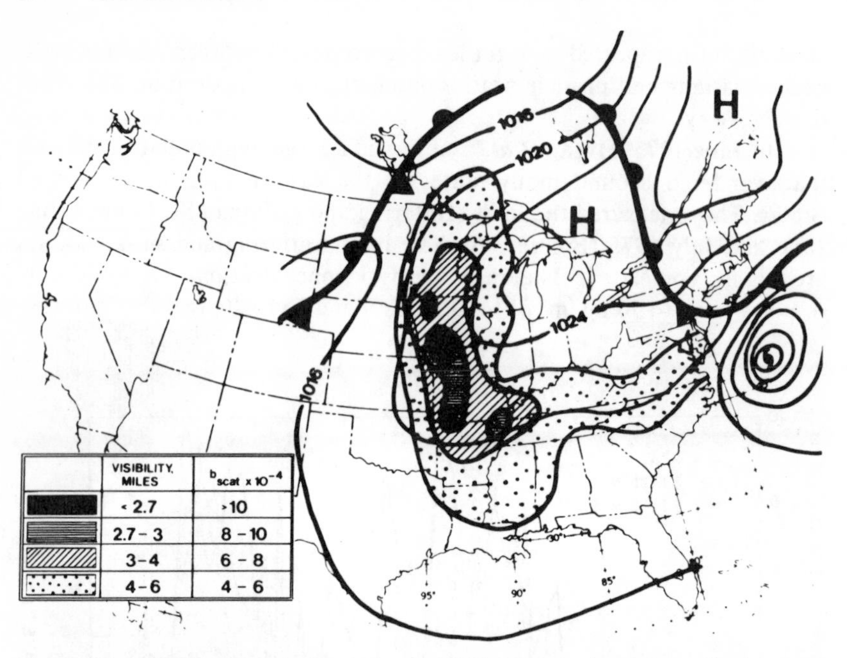

Figure 3-7. Visibility isopleths from June 30, 1975 showing a large area of haziness over the Midwest of the United States. The peak values of b_{scat}, corresponding to the lowest visibilities, correspond to the peak values of sulfate in Figure 3-6 and the most hazy areas in Figure 3-8. From Lyons and Husar, 1976.[499]

Figure 3-7 shows visibility isopleths calculated from Weather Service visibility data on June 30, 1975. Figure 3-8 shows a Synchronous Meteorological Satellite (SMS) photograph obtained on that same day. Both figures show that the region of haziness covers multiple-state areas, consistent with earlier satellite observations.[500] From Figure 3.7, it can be estimated that these hazy areas contain sulfate aerosol concentrations of approximately 35 $\mu g/m^3$. Figure 3-9 shows the correlation between sulfur content and fine particle mass measured near St. Louis for a 20-day period starting August 18, 1975. From these various measurements it is apparent that an accumulation mode aerosol containing a large proportion of sulfate frequently covers a large portion of the eastern third of the United States.

Simultaneous Measurements of Light-Scattering and Fine Particle Mass Concentration

Because of its importance, the relationship between light-scattering and either total suspended particulate mass concentration or fine particle mass concentration has been examined by a number of investigators. Chapter 5 contains a more detailed discussion of aerosol optical properties in general, and the particle scattering coefficient (b_{scat}) in particular. Charlson,[135] in a review paper, gave a relationship between total aerosol mass concentration (c) and b_{scat} as c ($\mu g/m^3$) = 4.5×10 b_{scat} (m^{-1}). If the fine particle mass is 50% of the total average mass concentration, the ratio of fine particle mass concentration to b_{scat} would be 22.5:1.

Recently, Patterson and Wagman[598] obtained samples at four different visibility levels in separate Anderson impactors (see Table 3-4). By

Figure 3-8. Photograph from the SMS satellite on June 30, 1975 showing the large area of air pollution haziness shown in Figure 3-7. This hazy air mass transported from the east and south. From Lyons and Husar, 1976.[499]

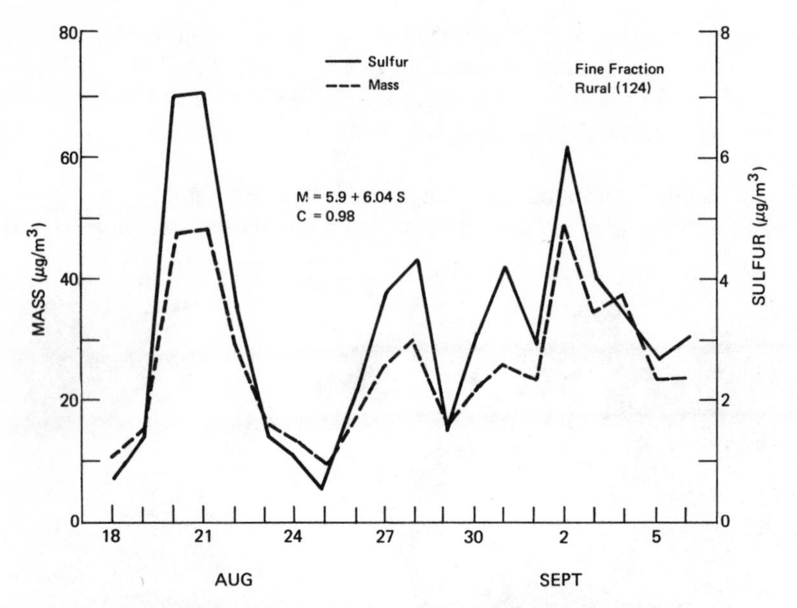

Figure 3-9. Correlation of fine particle fraction and the sulfur content of aerosol measured on a rural site in St. Louis in the summer of 1975. Data of Dzubay and Stevens.[206]

log-normal fitting, they determined the mass median diameter (MMD) in both the fine particle and coarse particle fractions, corresponding approximately to the accumulation and coarse particle modes. The MMD's in both fractions agree with those in Table 3-3, which were obtained by entirely different methods. As the visibility decreased or, in other words, as b_{scat} increased, the percent of the mass in the fine particle fraction, or accumulation mode, increased (Table 3-4). At the lowest visibility, 61% of the mass was in the fine particle fraction.

In Table 3-4, the ratio of fine particle mass to b_{scat} has been calculated from the data of Patterson and Wagman. The ratio varies from 12.2:1 to 16.8:1, but is relatively constant considering the range of visibility conditions sampled.

Recently, Sverdrup and Whitby[746] examined the relationship of condensation nuclei count to the ratio of fine particle volume to b_{scat} for the data collected during the ACHEX Program in California (see Figure 3-10). The ratio varies from around 5:1 at very low concentrations to about 35:1 at the highest concentrations near strong sources of aerosol.

Samuels *et al.*[662] made a detailed study of the relationship between particle mass concentration smaller than 3 μm diam, and prevailing visibility and b_{scat}. They obtained ratios of particle mass concentration <3 μm to b_{scat} ranging from 22.7:1 to 31:1.

Table 3-4. Summary of particle size and light-scattering parameters at various levels of visibility[a]

Visibility level	Measured b_{scat} range 10^{-4} m^{-1}	Mass concentration ($\mu g/m^3$)	Fine particle fraction[b]		Coarse particle fraction[b]		Calculated b_{scat}	Fine particle mass, b_{scat} $\mu g/m^{-3}$ 10^{-4} m^{-1}
			%	MMD[c]	%	MMD		
Background	0-1.5	44.8	30	0.48	70	7.8	0.8	16.8
A[d]	1.5-3.0	78.5	42	0.46	58	8.2	2.7	12.2
B[d]	3.0-4.5	114.0	63	0.45	37	8.3	4.9	14.7
C[d]	>4.5	212.0	61	0.42	39	8.1	8.9	14.5

[a] From Patterson and Wagman, 1977.[598]

[b] Fine particle and coarse particle fractions correspond approximately to the masses in the accumulation and coarse particle modes, respectively.

[c] MMD = Mass median diameter.

[d] Visibility categories based on arbitrary division of light-scattering coefficient data.

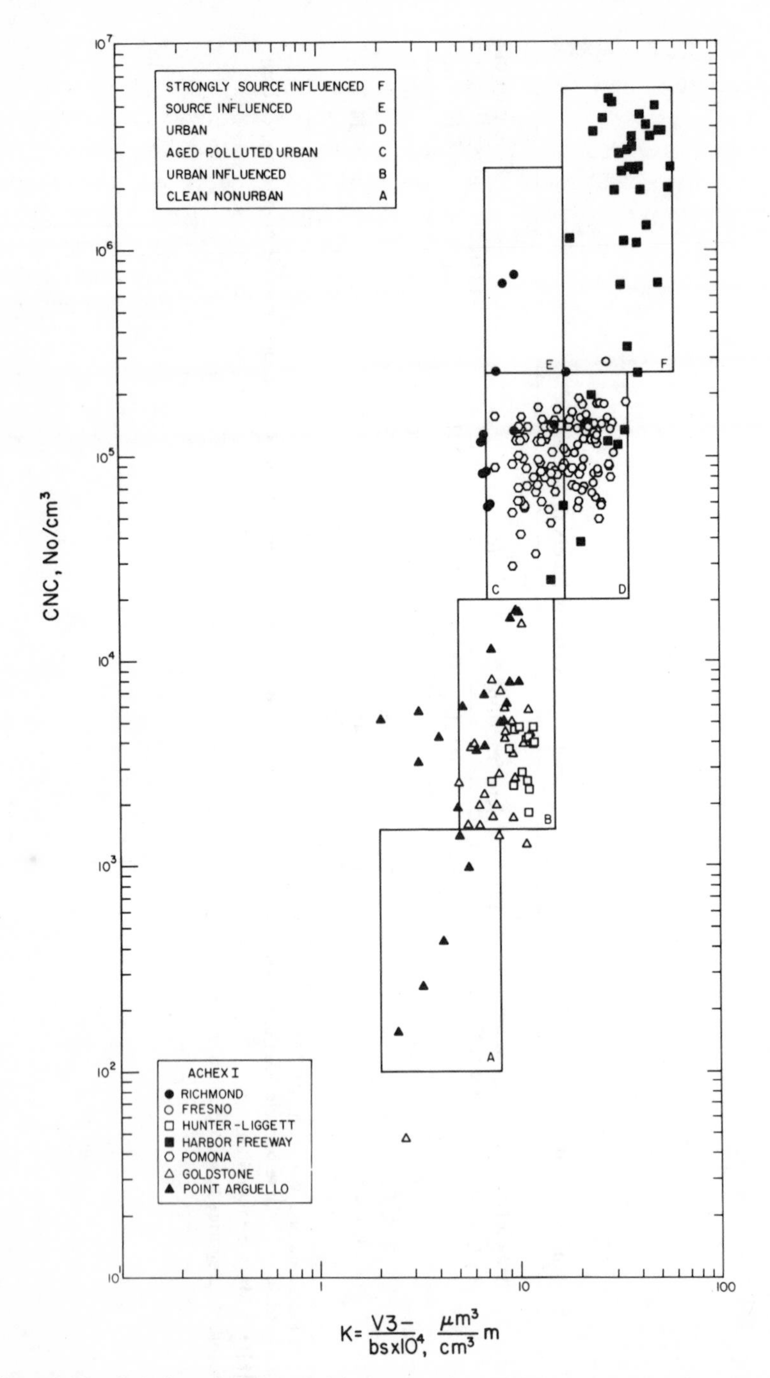

Figure 3-10. Relation of Aitken nuclei concentration (CNC) to the ratio of the aerosol (<1 μm diam) volume concentration ($V3-$) to b_{scat} (bs) for the sites in California that were measured during ACHEX I. From Sverdrup and Whitby, 1976.[746]

The above studies all show a good correlation between fine particle concentration and light-scattering, and, hence, visibility. Charlson[135] obtained correlation coefficients of 0.9 between light scattering and submicrometer volume. They also indicate a relationship between the absolute concentration and the ratio of b_{scat} to aerosol mass. The ratio of fine particle mass concentration to b_{scat} ranges from 10:1 to over 30:1, depending on concentration and site. An average ratio, for levels at which visibility degradation is important, would be approximately 20:1. Figure 3-10 shows that much of the urban data from California clusters near the ratio of 20:1.

4

Chemical and Trend Data

There are both routine monitoring programs and specific research efforts designed to collect information on the composition of atmospheric particles. By far the largest volume of data over the longest period has been accumulated by the routine monitoring operation of the National Air Sampling Network (NASN) and other federal programs. However, the analyses performed on these samples are limited to the determination of a few ionic and atomic species and to extraction with solvents such as benzene to indicate the presence of organic material. There are no routine analyses for silica, silicates, oxygen in various substances, or high molecular weight organic materials that are benzene insoluble, nor is elemental carbon measured. Because of the incompleteness of these routine chemical analyses, it is not possible to obtain an anion-cation balance[19] or specific molecular information. An average of 25 to 30% of the weighable substances on the NASN filters are routinely analyzed. Furthermore, filterable mass is usually less than suspended particulate mass due to volatility of a fraction of the aerosol. In certain situations, however, where gas-to-particle reactions occur on the filter (e.g., where sulfur dioxide is converted to sulfate), the filterable mass may be larger than the suspended mass.

Many, perhaps most, of the important chemical properties of particulate matter are due to the molecular rather than the ionic or atomic character. Toxicity, water solubility, hygroscopicity or the lack of it, deliquescence and efflorescence, refractive index, particle shape and state, acidity, basicity, and several other properties depend on the molecular nature rather than on the presence of a single atomic or ionic species. However, the ionic composition of a highly hydrated droplet aerosol that is composed of a large number of anions and cations clearly dominates the system behavior. At low humidity this same aerosol may have some definable molecular characteristics, such as deliquescence, if adequate amounts of the necessary ions are present in the particles. For example, sulfuric acid, ammonium sulfate, calcium sulfate (or gypsum), and magnesium sulfate (or epsom salts) are all sulfates but grow differently at high relative humidity (RH), appear in different size classes, and have different sources and, presumably, different effects.

More complete and molecular analyses by the large-scale routine networks in the United States are not feasible. Attempts made in various intensive research programs have been limited in time and scope (see

Chapter 3). Consequently, definitive molecular composition as a function of particle size, time, and location have yet to be determined. Because existing data are limited, it is seldom possible to relate particle properties to effects, nor is it possible to identify the predominant sources of the various particle classes.

The composition of gases can be measured specifically. These measurements provide a scientifically defensible basis for cause-effect studies. In contrast, most particulate matter is measured by its total suspended particulate content, which includes all filterable particles. Molecular and ionic species are often lumped together, precluding cause-effect studies where chemical composition is important. To understand the behavior and effects of particulate matter, analyses must be as specific as those currently used for gases.

OVERVIEW OF DATA

National Air Sampling Network (NASN)

The typical composition data from the NASN summarized in Table 4-1 show only one of the many ways in which these data can be presented. For example, the distribution of specific ions in rural areas may be mapped. From the maps in Figure 4-1 the differences in concentration and distribution of sulfate and nitrate ions are immediately evident. Given the number of sampling sites, their spatial distribution in both urban and rural areas, and the longevity of the NASN, it is evident that a large variety of data presentations can be made. Available information from the Environmental Protection Agency (EPA) consists mainly of statistical tables containing raw data. Some publications on trends have been developed from these data.[494] Further similar work is in progress both in the EPA and in various research groups. Unfortunately, the decentralization of the EPA has resulted in great delays in processing NASN samples and data, so that the newest facts in this report are several years old. Chemical analyses are at a near standstill.

Composition Data from Intensive Studies

Some of the first intensive studies of atmospheric particle composition were conducted by Junge[397] at Frankfurt, Germany, Round Hill, Massachusetts, Mauna Kea, Hawaii, and various sites in Florida in the early 1950s. His analyses showed that sulfate and ammonium ions were frequently predominant, although only a few ions were analyzed and no mass or ionic balance was attained.

Data collected at Round Hill show the daily variation of these ions for fine particles (0.08 to 0.8 μm radius) and for coarse particles (0.8 to

Table 4-1. Arithmetic mean and maximum urban particulate concentrations in the United States, biweekly samplings, 1960–1965[a]

Pollutant	Number of stations	Concentration ($\mu g/m^3$)	
		Arith. average[b]	Maximum
Suspended particulates	291	105	1,254
Fractions:			
Benzene-soluble organics	218	6.8	[c]
NO_3	96	2.6	39.7
SO_4	96	10.6	101.2
NH_4	56	1.3	75.5
Sb	35	0.001	0.160
As	133	0.02	[c]
Be	100	<0.0005	0.010
Bi	35	<0.0005	0.064
Cd	35	0.002	0.420
Cr	103	0.015	0.330
Co	35	<0.0005	0.060
Cu	103	0.09	10.00
Fe	104	1.58	22.00
Pb	104	0.79	8.60
Mn	103	0.10	9.98
Mo	35	<0.005	0.78
Ni	103	0.034	0.400
Sn	85	0.02	0.50
Ti	104	0.04	1.10
V	99	0.050	2.200
Zn	99	0.67	58.00
Gross beta radioactivity	323	(0.8 pCi/m^3)	(12.4 pCi/m^3)

[a] From U.S. Department of Health, Education and Welfare, 1969, p. 16.[768]

[b] Arithmetic averages are presented to permit comparable expression of averages derived from quarterly composite samples. As such they are not directly comparable to geometric means calculated for previous years' data. The geomtric mean for all urban stations during 1964-1965 was 90 $\mu g/m^3$; for the nonurban stations, 28 $\mu g/m^3$.

[c] No individual sample analyses performed.

8.0 μm radius) (see Figures 4-2 and 4-3). The words "large" and "giant" were used extensively in these studies. These figures indicate that there are few sodium or chlorine ions in the smaller particles and many more in the larger ones. The opposite is true of the ammonium and sulfate ions, which are usually <1 μm diam. These and similar data have formed the basis for the concept of a background of sulfate aerosol in some regions. The origin of this aerosol, containing sulfate, ammonium, and hydrogen ions, is not known. It may be dominated by man's activities, such as the oxidation of sulfur dioxide from fossil fuel combustion. Furthermore, as discussed in Chapter 3 and as indicated in Figure 3-2, this aerosol is widespread in the eastern half of the United States. At

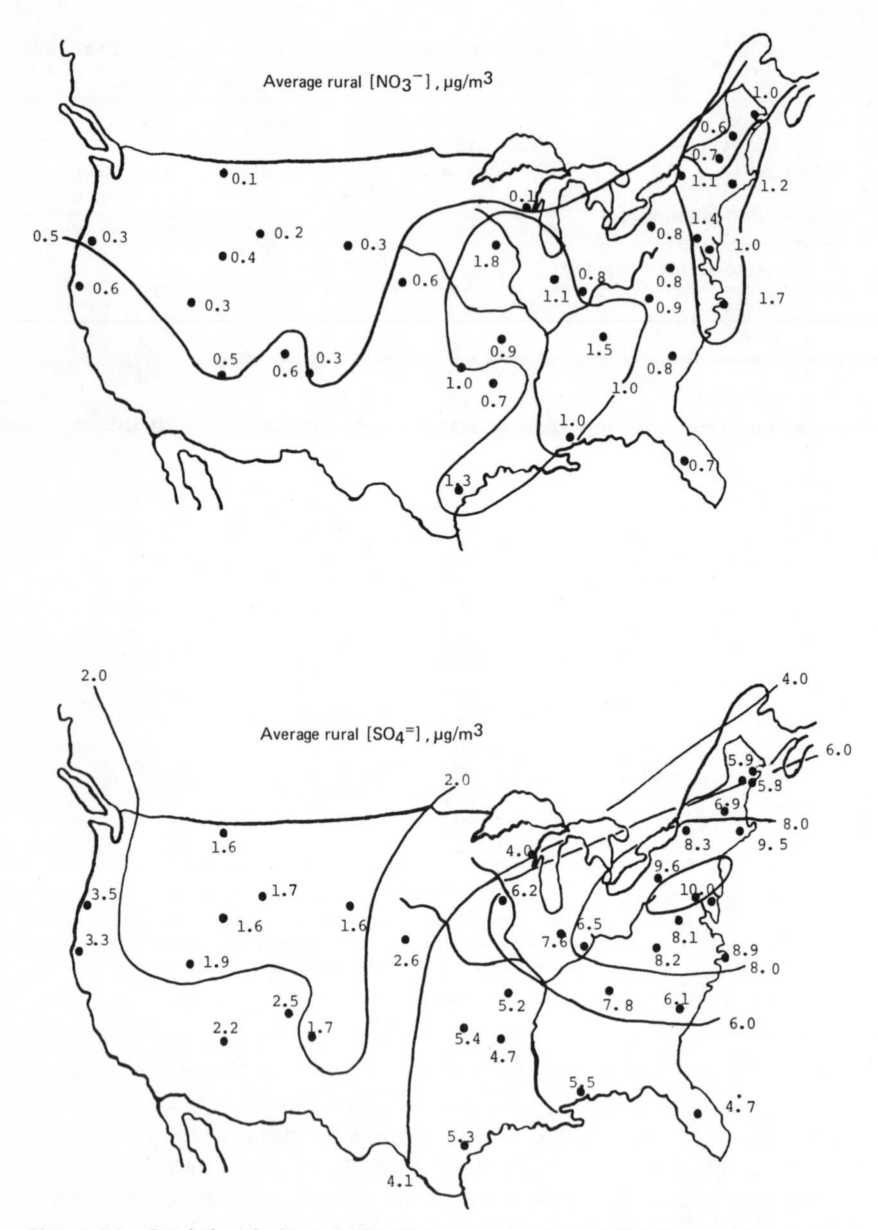

Figure 4-1. Isopleths of nitrate and sulfate concentrations drawn from measurements taken at nonurban sites by the Environmental Protection Agency (unpublished data).

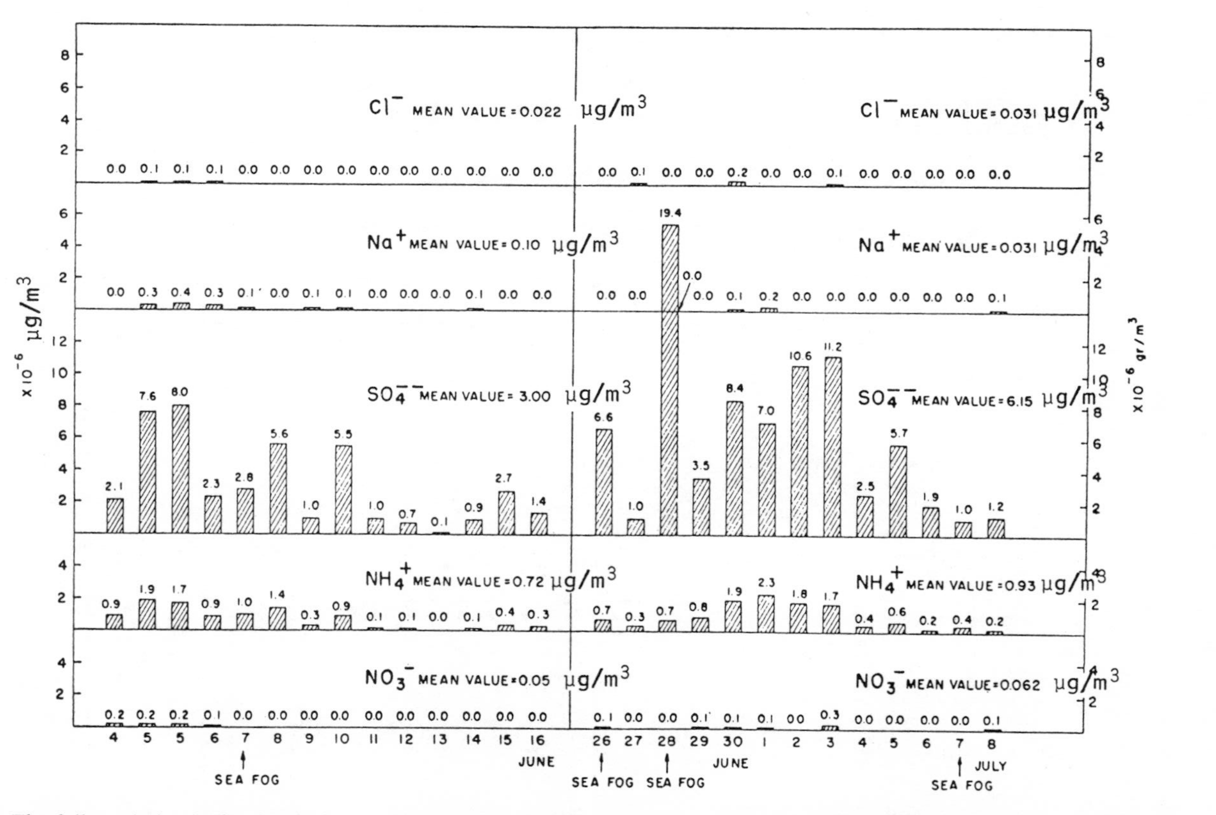

Figure 4-2. The daily variation in the chemical composition of fine particles (0.08 to 0.8 μm radius) at Round Hill, Massachusetts. From Junge, 1954.[400]

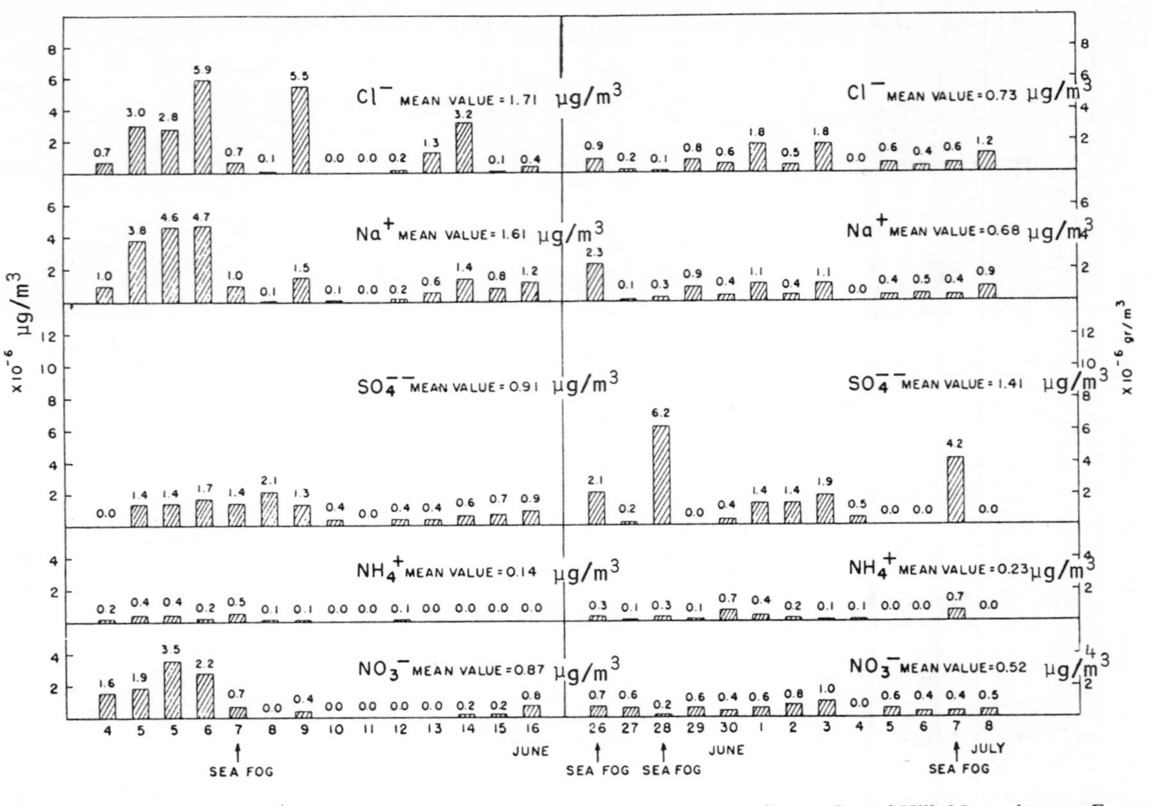

Figure 4-3. The daily variation in the chemical composition of coarse particles (0.8 to 8.0 μm radius) at Round Hill, Massachusetts. From Junge, 1954.[400]

the very least, this indicates that care must be taken in interpreting rural data on particle composition in terms of a natural background versus large spatial scale pollution.

One of the earliest, and still most complete, urban analyses was reported by Cadle[110] in Los Angeles. He noted the presence of sulfate and significantly higher amounts of nitrate ions. The lack of size segregation limited the conclusions to be drawn from these analyses. Perhaps the most useful results came from the separation of mineral and other inorganic substances (~60%), of which 45% are water-soluble, and organic compounds, which comprised approximately 25% of the mass. Although these are analyses of just one sample from one perhaps unique city, the insight provided by this sort of a separation aids the understanding of the sources and physical properties of the particulate. It also crudely indicates the molecular nature.

From their analyses of intensive studies in Pasadena, California, Miller *et al.*[543] reported compositional features similar to those found in Los Angeles (Table 4-2). The dominant substance analyzed was noncarbonate carbon (by combustion analysis). Sulfate, not given in this table, is typically in the range of 1 to 10 $\mu g/m^3$ when the total mass concentration is approximately 100 $\mu g/m^3$.

After measuring the effects of humidity on light-scattering in St. Louis, Charlson *et al.*[144] concluded that the light-scattering aerosol both inside and outside the city was dominated by one or more sulfates, such as sulfuric acid, ammonium disulfate, and ammonium sulfate. The lack of correlation with local wind direction, which might be expected from such maps as Figure 4-1, appears to be due to the long-distance (10^3 km) transport of those sulfates. Sulfates in Southern California, on the other hand, might have local sources.

Table 4-2. Percentages of chemical elements in Pasadena particulate matter[a] (total mass loading = 101.5 $\mu g/m^3$, measured at 50% relative humidity)

Element	%	Element	%
Al	0.8	I	0.006
Ba	0.04	K	0.32
Br	0.6	Mg	1.1
C (noncarbonate)	19.0	Mn	0.03
Ca	0.99	Na	1.0
Cl (dry analysis)	0.07	Pb	3.3
Cu	0.03	V	0.01
Fe	3.2	Zn	0.18

[a] From Miller *et al.*, 1972.[543]

Data from Independent Investigators

Large amounts of useful data have been acquired over the years by independent investigators studying individual substances in aerosols. Notable among these is the work of Winchester and his colleagues[820, 821] on substances detectable by various nuclear techniques. Schuetzle *et al.*[681] studied the molecular nature of both organic and inorganic materials in size-fractionated samples by high resolution mass spectroscopy. Lodge[482] deduced from electron microscopy that fine particles often contained free sulfuric acid.

Because space limitations preclude the inclusion of an exhaustive list of composition measurements, only a few examples from recent studies are mentioned.

SIZE DEPENDENCE OF COMPOSITION

Before the recent intensive studies of composition versus particle size and prior to the understanding of the bimodal volume distribution, it was common to assume a chemically well-mixed system. Current control strategies for particles are based implicitly on this as well as on the frequent assumption that all particles come directly from sources and not from production or condensation in the air. Optical models of aerosol frequently assume a uniform composition/size dependence.[422] Health-effects models, such as pulmonary deposition, often use a similar approach.

Recent data suggest that a chemically well-mixed aerosol that has a similar composition at all sizes is probably rare or nonexistent. Rather, certain classes of substances appear to dominate above the 1- to 2-μm diam range, and others below. Dzubay and Stevens[205] analyzed urban aerosol samples using a two-stage dichotomous filter, with a separation at a particle diameter (D_p) of $\cong 2$ μm (see Figures 4-4 and 4-5). They found that at least 75% of the sulfur, zinc, bromine, arsenic, selenium, and lead occur in the fine particles and at least 75% of the silicon, calcium, titanium, and iron in the coarse ones (Table 4-3).

The fine particles between 0.1 and 1.0 μm diam consist mainly of sulfates from sulfur dioxide oxidation, nitrates from NO_x reactions, ammonium from ammonia reactions, condensed organic matter (partially oxygenated), and primary emitted particles, such as lead and carbon. The coarse particles, from 2.0 to 50.0 μm diam, consist largely of mechanically produced substances such as soil or rock dust, road and tire dust, fly ash, and sea salt (see also Figure 1-7).

Table 4-4 shows similar results that are further classified by urban and rural location.[207] The percentage of fine sulfur particles is almost the same in both locations, consistent with the theory of long-distance

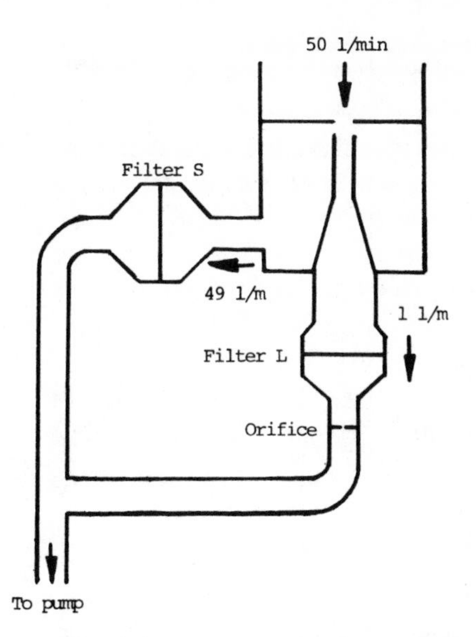

Figure 4-4. Schematic view of a dichotomous sampler that contains a virtual impactor. The flow rate at the inlet is 50 liters/min, and the flow rates at the outlets are 49 and 1 liters/min. From Dzubay and Stevens, 1975.[205] With permission from *Environmental Science and Technology*. Copyright by the American Chemical Society.

transport of sulfates discussed in Chapter 3. The optical studies of Charlson *et al.*[144] showed that the acid sulfate species sulfuric acid and/or ammonium bisulfate frequently occur in the fine particle mode, confirming the classic results of Junge.[397] Natusch *et al.*[575] showed that arsenic, antimony, cadmium, lead, selenium, and thorium emitted from coal-fired power plants occurred primarily in the fine particle mode. A size-

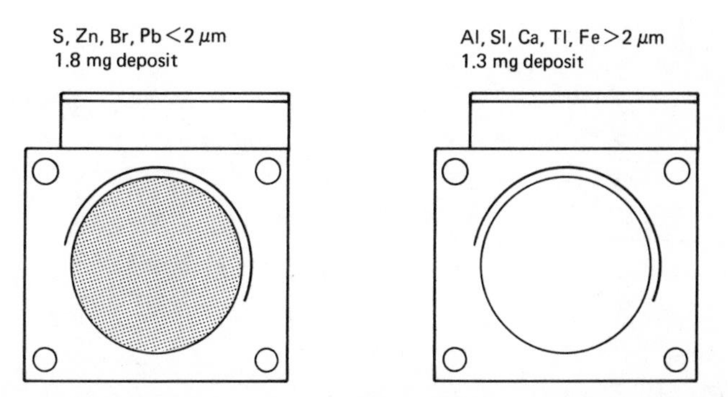

Figure 4-5. Filters used in dichotomous sampler for the 23-hr period beginning at 1015 hr, August 30, 1973 in a St. Louis residential neighborhood. The air volume sampled was 68 m^3. From Dzubay and Stevens, 1975.[205] With permission from *Environmental Science and Technology*. Copyright by the American Chemical Society.

Table 4-3. Analysis of St. Louis aerosol samples collected for a 23-hr period beginning 1015 hr August 30, 1973[a]

Element	Fine[b] (ng/m³)	Coarse[b] (ng/m³)	Element	Fine[b] (ng/m³)	Coarse[b] (ng/m³)
As	20±10	<4	Pb	460±50	110±12
Ba	<30	<30	Rb	<1	<1
Br	114±12	30±4	S	3,700±600	600±200
Ca	110±40	1,700±300	Se	7±2	<2
Cd	<13	<13	Si	600±160	2,000±700
Co	<2	<2	Sn	<14	<14
Cr	<3	<3	Sr	<2	4±2
Cu	<3	<3	Ti	64±7	210±30
Fe	130±14	400±40	V	4±2	<3
K	160±60	240±80	Zn	45±6	15±2
Mn	5±2	6±2	Gravimetric	26,600±800	19,200±800
Ni	<2	<2			

[a] From Dzubay and Stevens, 1975.[205] With permission from *Environmental Science and Technology*. Copyright by the American Chemical Society.

[b] The collected size ranges are 0 to 2 μm for fine particles and 2.5 to 10.0 μm for coarse particles.

segregating elemental analysis network is currently being operated by the State of California Air Resources Board Monitoring Program. Preliminary results confirm the 1976 studies of Flocchini *et al.*[245]

Some of the elements mentioned above can affect human health. Size-segregating chemical analysis coupled with information regarding the molecular state of these elements should aid in the understanding of these effects (see Chapter 6).

Two types of particle mixtures can be defined:

External mixtures, in which the individual particles are pure or nearly pure compounds and composition varies from particle to particle.

Internal mixtures, in which each particle contains all the substances in the same proportions.

The small Brownian displacement and collision frequencies of coarse particles and their shorter atmospheric lifetimes suggest that the coarse particles of the mechanical mode probably exist as discrete particles of identifiable substances. The larger Brownian displacement and high collision frequencies of fine particles suggest that this mass mode is internally mixed. Because the interaction between these two modes is relatively weak, they form an external mixture.

HYGROSCOPICITY AND DELIQUESCENCE

Several effects of atmospheric particulates depend on the growth of the particles at high RH. Growth and impaction in the respiratory tract, visibility degradation, and cloud condensation are notable examples. In turn, the properties controlling such growth are controlled by the molecular composition. Most of the hygroscopic and deliquescent substances present in atmospheric particles appear in particles $<1\ \mu m$ diam, with the notable exception of sea salt particles, which may be larger.

Deliquescent salts undergo a sudden phase transition from a dry crystal to a solution droplet when the RH exceeds that of the saturated solution of the highest hydrate of the salt. The deliquescence points of several atmospheric substances are: ammonium sulfate, 80%; sodium sulfate, 86%; sodium chloride, 75%; and ammonium nitrate, 62%.

Because of their polar molecular nature, hygroscopic compounds absorb water until they become solutions at equilibrium with the ambient humidity. They exhibit monotonic growth in size as RH is increased, in contrast to the deliquescent salts. Examples of hygroscopic atmospheric substances are sulfuric acid, glycols, sugars, organic acids, and alcohols.

Internal mixing affects the hygroscopic and deliquescent growth characteristics of fine particles. Clear-cut deliquescent growth has been observed optically for two atmospheric substances: sea salt and ammo-

Table 4-4. Comparison of the 18-day-average percentage composition of fine and coarse particles at an urban and a rural site in St. Louis, measured between August 18 and September 7, 1975[a]

Particle	Urban (%)			Rural (%)		
	Fine (29 $\mu g/m^3$)	Coarse (22 $\mu g/m^3$)	Fine/Coarse	Fine (26 $\mu g/m^3$)	Coarse (15 $\mu g/m^3$)	Fine/Coarse
Ca	0.7	8.2	0.09	0.5	4.2	0.12
Fe	1.4	4.8	0.29	0.3	1.3	0.23
K	0.4	1.2	0.33	0.3	0.9	0.33
Pb	2.2	0.6	3.67	0.51	<0.11	>5.0
S	12.5	1.4	8.90	12.6	0.9	14.0
Si	1.0	8.0	0.13	0.5	4.0	0.13
Ti	1.1	2.0	0.55	<0.1	0.2	<0.5

[a] From Dzubay *et al.*, 1977.[207]

nium sulfate.[167] Frequently, interval mixing masks the deliquescence of minor substances.

The quantitative aspects of this growth with RH can be studied optically, by weighing techniques, and with microwave absorption. Atmospheric fine particles are almost always sufficiently hygroscopic or deliquescent to exhibit growth below 100% RH. The amount of growth that occurs as RH is raised from a low of 20% to 30% to 95+% depends on the fraction of the material that is hygroscopic or deliquescent and on its composition. For specific pure substances, such as sodium chloride or sulfuric acid, it is possible to calculate the size increase with RH, assuming the solution behavior is known. The ratio of particle radius (r) at the given RH to that of the dry particle, $r\,(RH)/r(O)$, assuming ideal solution behavior, is typically 2 at ~90% RH for a hygroscopic salt. The

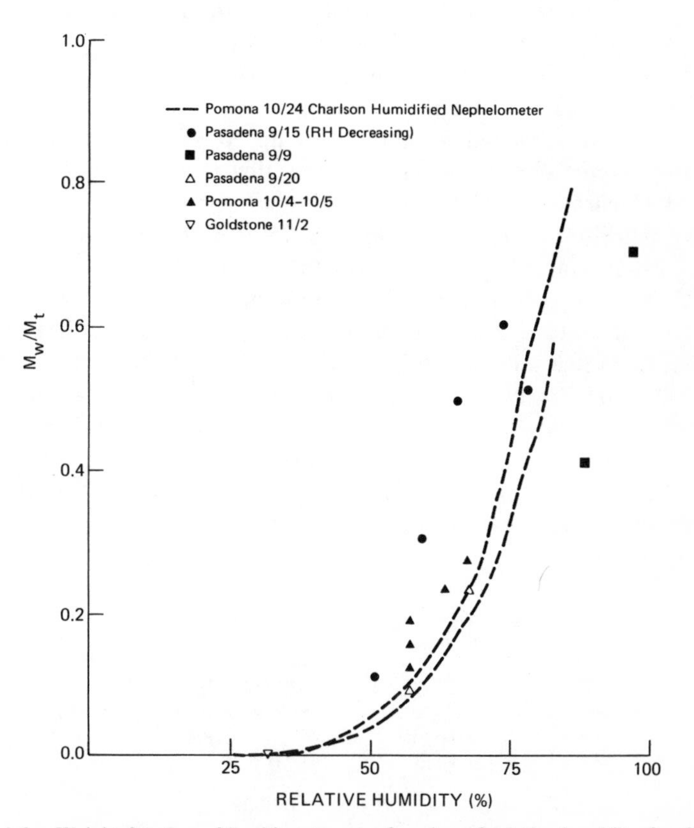

Figure 4-6. Weight fraction of liquid water as a function of relative humidity for the Los Angeles Basin area. The data points indicate results obtained with a microwave technique and the dotted curves indicate the results obtained optically with nephelometer. From Ho *et al.*, 1974.[344]

nonideality of solutions actually alters the real behavior slightly; however, the magnitude of the growth is adequately demonstrated.

Measurement of impure atmospheric particles by optical, mass, or microwave techniques indicates that in most urban locations, the fine particles are often hygroscopic, occasionally deliquescent, and only rarely hydrophobic. Covert's[167] data from eight sites in California, three sites in and near St. Louis, and two sites in Denver demonstrate these three behavior classifications. In California the aerosol was rarely deliquescent. A comparative study of optical and microwave techniques produced the results shown in Figure 4-6 for the mass of water in submicrometer aerosol versus RH. At 80% RH, approximately 50% of the mass of the fine particles is water. Furthermore, the particles clearly contain water at 50% RH and below. The fact that such particles are usually wet suggests that it is necessary to consider the presence of the water for studies of aerosol chemisorption, such as sulfur dioxide-particle synergism.[526] It must also be concluded that the fine particles are often composed of hygroscopic or deliquescent substances.

A notable exception to this observation that fine particles are usually wet occurred in Denver where, from 5% to 10% of the time, limited growth occurred at humidities up to 93%. That aerosol was probably composed of water-insoluble, hydrophobic substances. An aerosol of carbon, certain metal oxides, or partially burned oil would behave in this manner. From 90% to 95% of the time, the dominant behavior was simple hygroscopic growth.

If the fine particles are uniform in composition with size, it is possible to calculate the change in mass or volume distribution as they equilibrate at some increased humidity. Figure 5-7 shows the calculated distribution of aerosol volume as a function of particle size, as the particle grows hygroscopically to 1.5 and 2.0 times its initial size. The effect of hygroscopic/deliquescent growth of particles $<1\ \mu m$ diam dramatically increases the mass median diameter of these particles and the total amount of material in the condensed phase.

Table 4-5. Growth rates of sea salt aerosol

Dry aerosol radius (μm)	Time (sec) to reach 97% of equilibrium radius at 90% RH
0.1	0.0022
1.0	0.22
5.0	5.4

The final aspect of hygroscopic/deliquescent growth is the response time of the particles. Keith and Arons[410] and Azarniouch *et al.*[44] have calculated the time required for such particles to attain various fractions of their equilibrium size when a step increase of humidity is imposed. Table 4-5 illustrates this response time, suggesting that growth is sufficiently rapid to occur in the upper airways of humans. However, the response time for particles >1 μm diam and for humidities approaching 100% becomes sufficiently long that complete equilibrium may not be attained in the breathing cycle.

TRENDS IN DATA ON PARTICULATE MATTER

Although the measurement methodologies of the past decades have many shortcomings, they do indicate important trends. Notable among these long-term data are the measurements of "settleable dust" or "dust fall" made in some cases since 1925 and in many cities since 1955, and of the total filterable particulate matter obtained by high-volume samplers in 58 urban and 20 nonurban sites since 1957.[494]

Calculations of dust fall [mass/(area $\times$ time)] show that Pittsburgh, Cincinnati, Chicago, and New York in particular have enjoyed decreases in settleable dust nearing a factor of 2 between 1925 and 1964 (Figure 4-7). The data of Ludwig *et al.*[494] on the composition of the coarse particle mode indicate that settleable dust consists mainly of mechanically produced substances, some of which are amenable to simple control measures. However, decreased dust fall should not be attributed to control alone since oiling or paving of roads, changes in agricultural practices, and location could produce the same result. Much of the decrease in dust fall is no doubt due to the control of industrial

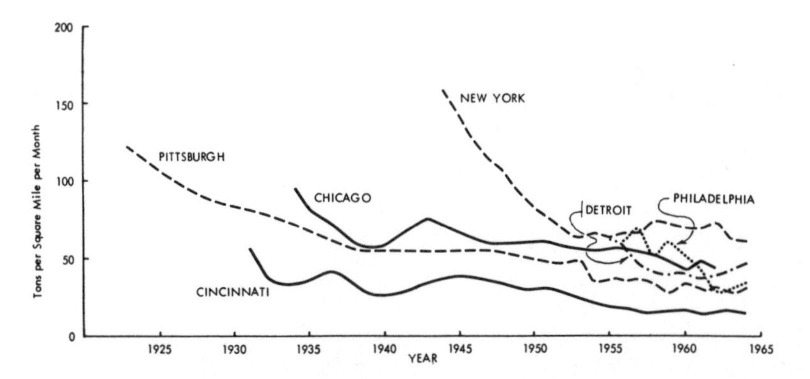

Figure 4-7. Trends in settleable dust in six cities. From Ludwig *et al.*, 1974.[494]

emissions of very large particles ($>1\ \mu m$ diam), but no quantitative relationship can be deduced from the data and from the lack of analysis of long-term composition records.

Statistical analysis of measurements of the total filterable particulate matter (TFP) suggests that urban values have decreased approximately 20% between 1960 and 1971. The most recent data available from the EPA were collected in 1973. From their analyses of these same data, minus that for the year 1971, Ludwig *et al.*[494] suggested that the filterable mass concentration in nonurban areas increased slightly over the period of record in spite of the improvement in cities.

In view of the bimodal mass or volume distribution and of the usual location of the NASN high-volume sampler on dusty roofs, the decrease in urban TFP cannot be uniquely interpreted in terms of decreases in emissions from man's activities. The lack of size resolution and chemical composition precludes a detailed understanding of the trends.

5

Effects on Atmospheric Processes

Particulate matter influences atmospheric physical processes in two basic ways. First, it scatters and absorbs light. This results occasionally in dramatic reduction of visual range, colored hazes, decreased sunlight at the ground, and other related effects. These optical effects are probably the only effects of air pollution that are directly perceived by most people. These may therefore be the most socially or politically important.

Second, the interaction of particulates with water in condensation and cloud processes implies that particulates may cause changes in the amount and composition of precipitation. This chapter will summarize the state of knowledge concerning both these influences.

RELEVANT OPTICAL PROPERTIES

Basic Principles

The angular distribution (β) of light scattered by a unit volume of aerosol can be described by the function $\beta_{sp,\lambda}(\phi)$. This is called the volume phase function. The fraction of light of wavelength λ scattered into all directions per unit pathlength is called the scattering component of the particulate extinction coefficient, and is denoted $b_{sp,\lambda}$.

The total extinction coefficient of a real atmosphere is the sum of two terms:

$$\begin{aligned} b_{ext,\lambda} &= b_{scat,\lambda} + b_{abs,\lambda} \\ &= b_{Rg,\lambda} + b_{sp,\lambda} + b_{ap,\lambda} + b_{ag,\lambda} \end{aligned}$$

where

$b_{ext,\lambda}$ = extinction coefficient at the wavelength
b_{scat} = scattering component of extinction
b_{abs} = absorption component of extinction
b_{Rg} = Rayleigh scattering component of extinction due to gas molecules
b_{sp} = scattering component of extinction due to particles
b_{ap} = absorption component of extinction due to particles
b_{ag} = absorption component of extinction due to gases

For brevity, the subscript λ is omitted in the following discussion, but wavelength dependence is an important measurement consideration. Both b_{sp} and b_{ap} define the role of atmospheric particulate matter in light-extinction processes. The scattering of light by atmospheric particles is maximum at angles (ϕ) near 0°. In other words, most of the scattered light energy is redirected only slightly and only from approximately 10% to 20% of the scattered light is reflected in the backward hemisphere ($90° \leq \phi \leq 180°$). We can define a hemispheric backscatter coefficient b_{bsp}, where

$$\Delta b_{bsp} = 2\pi \int_{\pi}^{\pi} /2 \ \beta_{sp} (\phi) \sin \phi d\phi$$

The amount of light scattered out of the atmosphere depends on the solar zenith angle; but, where the earth is flat and the sun is directly overhead, the intensity of light reflected by particles contained in a layer of air of thickness Δx is proportional to $b_{bsp} \Delta x$.

Other and more complex optical variables can also be identified, such as those related to angular scattering dependence, polarization, etc.; however, much more data exist on quantities b_{sp}, b_{ap}, and b_{bsp}, which are most directly related to atmospheric effects.

b_{sp}—The Light-Scattering Coefficient Due to Particles

b_{sp} is a relevant integral property of the atmosphere that usually controls both the transmission of sunlight and the visual range. It is also often highly correlated with the volume concentration of particles between 0.1 and 1.0 μm diam. Since the submicrometer volume mode contains a large portion of the respirable particles and contains substances that have the potential to affect health, the measurement of b_{sp} should provide a useful index for health-related studies. b_{sp} is easily measured with an integrating nephelometer.

b_{sp} depends on the following aerosol properties:

Fine particle mass concentration
Total mass concentration
Particle size distribution
Refractive index (determined by the chemical composition)
Particle shape
Relative humidity
Relative humidity history (regarding hysteresis)
Wavelength of light

While it is difficult to assign an absolute order of dominance of the aerosol properties in determining b_{sp}, the order in which they are listed reflects their dominance under such conditions as low relative humidity (RH) in urban atmospheres.[137] The effects of nonspherical particle shape and, to some extent, the effects of refractive index on the value of $b_{sp,\lambda}$ are relatively small. The main factor controlling $b_{sp,\lambda}$ for atmospheric size distributions is the mass or volume, or number of particles between 0.1 and 1.0 μm diam for $\lambda \sim 0.5$ μm. Values of b_{sp} at 550 nm as low as $1 \times 10^{-7} m^{-1}$ have been measured with the integrating nephelometer in clean upper tropospheric air at the Mauna Loa Observatory[74] and as high as $2 \times 10^{-3} m^{-1}$ in Los Angeles smog.[218] Still higher values occur when the RH exceeds about 80%,[167] demonstrating that b_{sp} is an extremely variable atmospheric property.

b_{sp} as a function of particle mass concentration There is a significant relationship between light-scattering and both the size and volume or mass concentration of particles. This is apparent in studies of the dependence of light scattering on particle size. Figure 5-1 shows the calculated size dependence of scattering for an average measured size distribution,[805] assuming uniformly dense spherical particles, a real part of the particulate refractive index equal to 1.5, and an imaginary part equal to 0.001. Also shown are the volume and surface area distribution functions $\Delta V/\Delta \log D_p$ and $\Delta S/\Delta \log D_p$ versus $\log D_p$.

Figure 5-1 also shows that both light-scattering and the volume concentration are dominated by particles between 0.1 and 1.0 μm diam. This suggests a correlation between light-scattering and fine particle volume or mass concentrations. This figure also shows that the upper size limit for light-scattering is governed by the surface area. Therefore, a correlation of light-scattering with the concentration of particles >1.0 μm diam or with total mass concentration is not expected unless the fine mode correlates with the coarse volume mode. It is somewhat surprising, in view of this, that the measured correlation coefficient between b_{sp} and total aerosol mass concentration is as high as the observed range between 0.5 and 0.9. While the former value is not impressive nor particularly useful, the latter is sufficiently high to allow inference of mass concentration from b_{sp}. Aerosol Characterization Experiment (ACHEX) II data[330] (Figure 5-2) yield a correlation coefficient of 0.865 between b_{scat} and total particle volume ($b_{scat} \approx b_{sp}$ under the conditions of these measurements). Table 5-1 summarizes the various published correlations of b_{sp} and mass. The ratio of b to b_{sp} is relatively insensitive to variations in modal particle diameter, D_m, in the range from 0.3 to 0.8 μm (Figure 5-3),[219] which largely accounts for the observed high correlations between v and b_{sp}.

There are few extensive measurements relating the mass concentration of fine particles to light-scattering. Measurement of b_{scat} and submicrometer volume concentration as part of the ACHEX II program (Figures 5-4 and 5-5) (Clark, 1974, personal communication) yielded correlation coefficients of 0.94 for b_{scat} and total submicrometer volume, and of 0.95 for b_{scat} and volume in the size range of 0.1 to 1.0 μm diam.

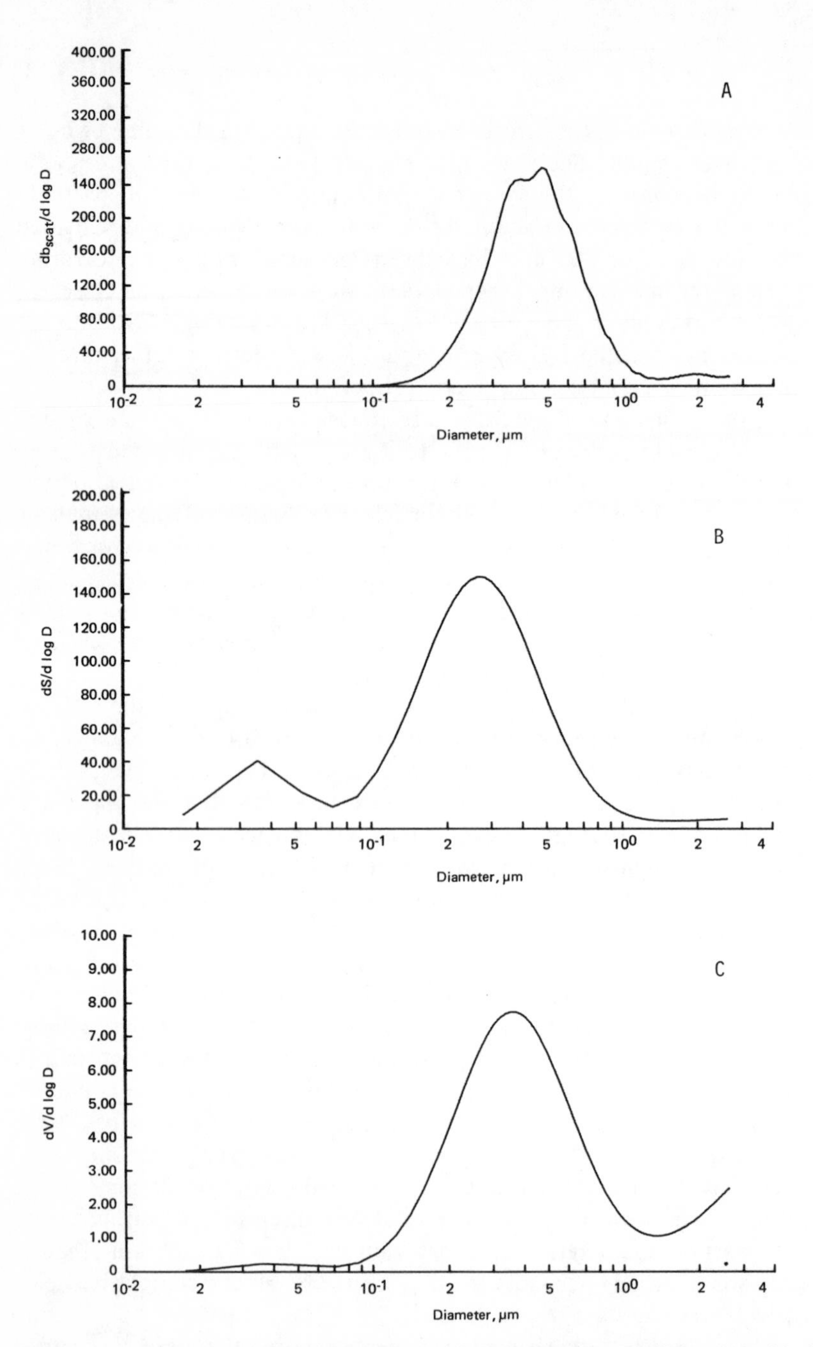

Figure 5-1. Comparison of the computed particle light-scattering coefficient (A) with the surface (B) and volume (C) distributions of the 1969 Los Angeles grand average aerosol size distribution.[805] The light-scattering coefficient was computed assuming a real part of the refractive index (1.5) and an imaginary part (0.001) at a wavelength of 552 nanometers. The particle light-scattering is limited at the lower end by transition to Rayleigh scattering and at the upper end by a sharply decreasing surface area of the distribution.

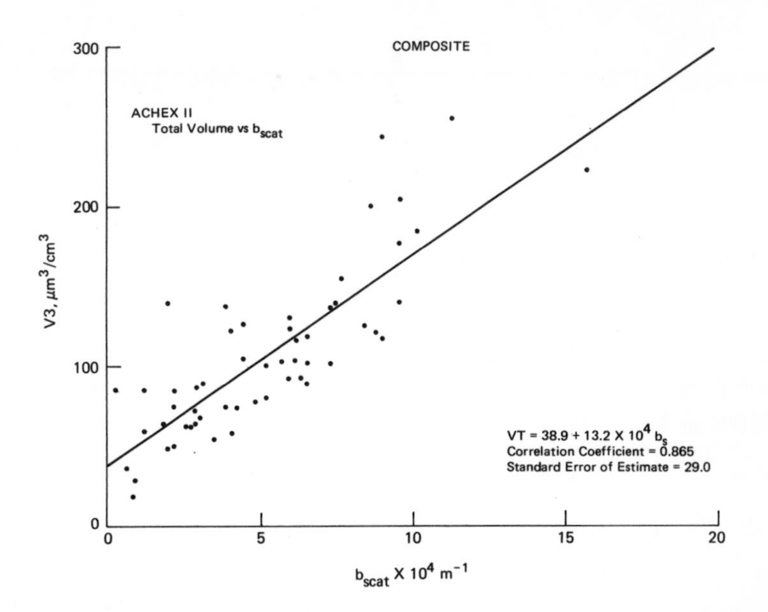

Figure 5-2. Two-hour averages from West Covina, Pomona, Rubidoux, and Dominguez Hills. Correlations between b_{scat} and total aerosol volume. From Hidy *et al.*, 1974, p. 3–29.[330]

Results in other types of locations produced somewhat different values of the ratio v to b_{sp} and of the correlation coefficient due mainly to variations in the nuclei mode. Locations near Aitken nuclei concentration (CNC) sources such as freeways may show poorer and/or different correlations of fine particle mass and light-scattering (Figure 5-6). These measurements are discussed in more detail in Chapter 3.

Compounds that dominate b_{sp} Classes of compounds that dominate the accumulation mode (0.1 and 2.0 μm diam) are far more important in determining b_{sp} than compounds that occur in the coarse particle mode (>2 μm diam). Sulfates (sulfuric acid, ammonium bisulfate, ammonium sulfate, and perhaps others), nitrates (ammonium nitrate), and organic compounds (condensed hydrocarbons and oxidized organic matter) appear to be important. In contrast, soil dust, tire dust, road debris, airborne rock-crushing products, fly ash, etc., contribute little to b_{sp} except in the relatively rare situation of dust storms.

Substances of greatest importance to atmospheric light-scattering occur predominantly in the fine particles that dominate optical effects. Many of these same compounds (sulfuric acid, ammonium sulfate, etc.) are hygroscopic or deliquescent. They cause more light-scattering at elevated RH than they would if they were hydrophobic, like silica and some soil dust.

Table 5-1. Summary of nephelometric light-scattering: mass concentration studies by location and mass sampling method[a]

Location	Mass sampling method: 2.5 cm diameter open face, glass fiber filter			2.5 cm diameter nucleopore filter			High-volume air sample			Glass fiber filter behind Lippman Harris Cyclone			Reference
	r	A	B	r	A	B	r	A	B	r	A	B	
Los Angeles, CA	0.83	−0.57	2.4	—	—	—	0.53	−0.09	2.2	0.83	0.33	3.7	662
Oakland, CA	0.69	0.40	1.3	—	—	—	0.86	−0.61	2.4	0.79	0.34	3.2	662
Sacramento, CA	0.95	0.0	2.2	—	—	—	0.93	−0.56	2.8	0.98	0.13	4.4	662
New York, NY	—	—	—	0.92	−0.33	3.0	—	—	—	—	—	—	136
San Jose, CA	—	—	—	0.56	1.5	1.7	—	—	—	—	—	—	136
Seattle, WA	0.83	−0.08	3.5	[b]	—	—	0.73	−0.26	3.6	—	—	—	136
Boston, MA	—	—	—	—	—	—	0.86	0.15	2.0	—	—	—	702

[a] r = linear correlation coefficient; A, B = regression coefficients in equation; $b_{sp,500nm}$ (10^{-4} m^{-1}) = $A + 10^{-2} B$ (mass concentration, μg/m^{-3}).
[b] Both types of filters were used.

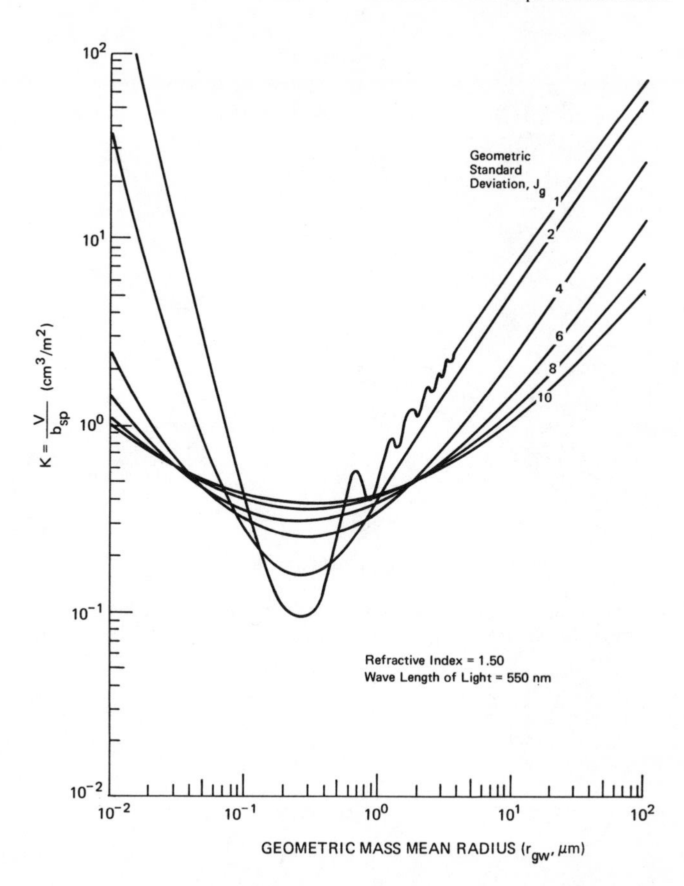

Figure 5-3. The ratio V/b_{sp} as a function of the log-normal size distribution parameters for a white aerosol. From Ensor and Pilat, 1971, p. 500.[219]

One approach to estimating the importance of various substances is to calculate their effect via a combination of measurements and theory. Gartrell and Friedlander[267] estimate that in Pasadena and Pomona smog visibility deterioration can be ascribed largely to nitrates, sulfates, organics, ammonium ions, and water, converted from the gas phase to a condensed phase. Patterson and Wagman[598] measured aerosol composition, mass, and size distributions for four different ranges of b_{scat}. Their results suggested an important role for carbon compounds.

Shape effects Light-scattering may be influenced by particle shape. Although several theoretical approaches have been attempted, very few experimental efforts have been made to isolate shape effects in atmospheric scattering. Such atmospheric particles as, for example, sea salt at high RH, are nearly homogeneous and spherical; in other cases, a particle sample resembles a bowl of popcorn, and may be impossible to

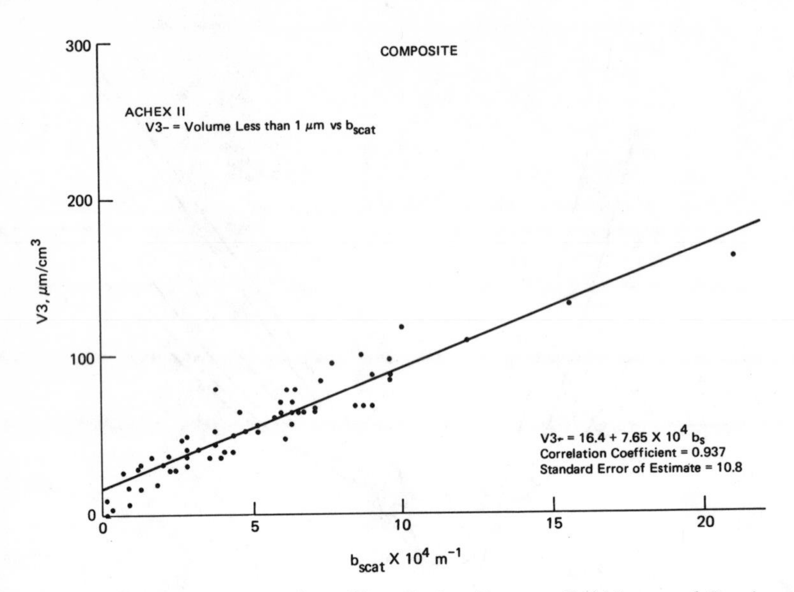

Figure 5-4. Two-hour averages from West Covina, Pomona, Rubidoux, and Dominguez Hills. Correlations between b_{scat} and aerosol volume <1 μm. From Hidy *et al.*, 1974, p. 3–29.[330]

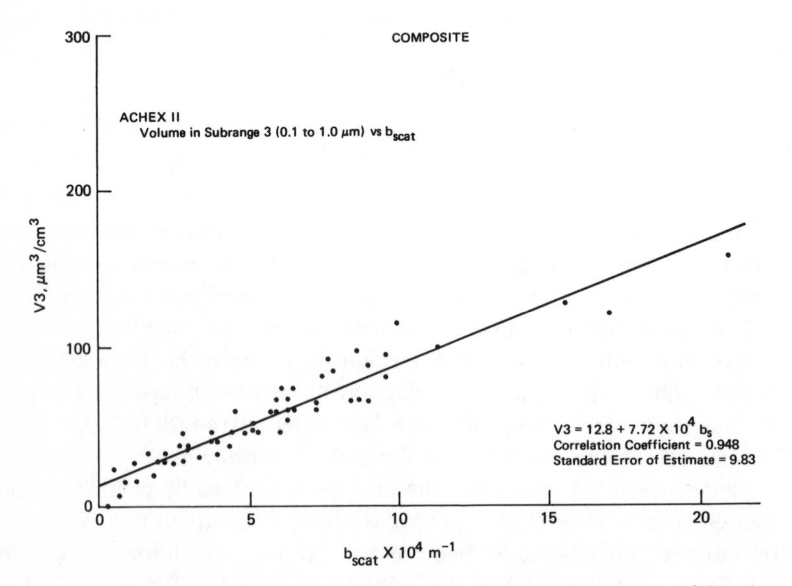

Figure 5-5. Two-hour averages from West Covina, Pomona, Rubidoux, and Dominguez Hills. Correlations between b_{scat} and aerosol volume in subrange from 0.1 to 1.0 μm. From Hidy *et al.*, 1974, p. 3–29.[330]

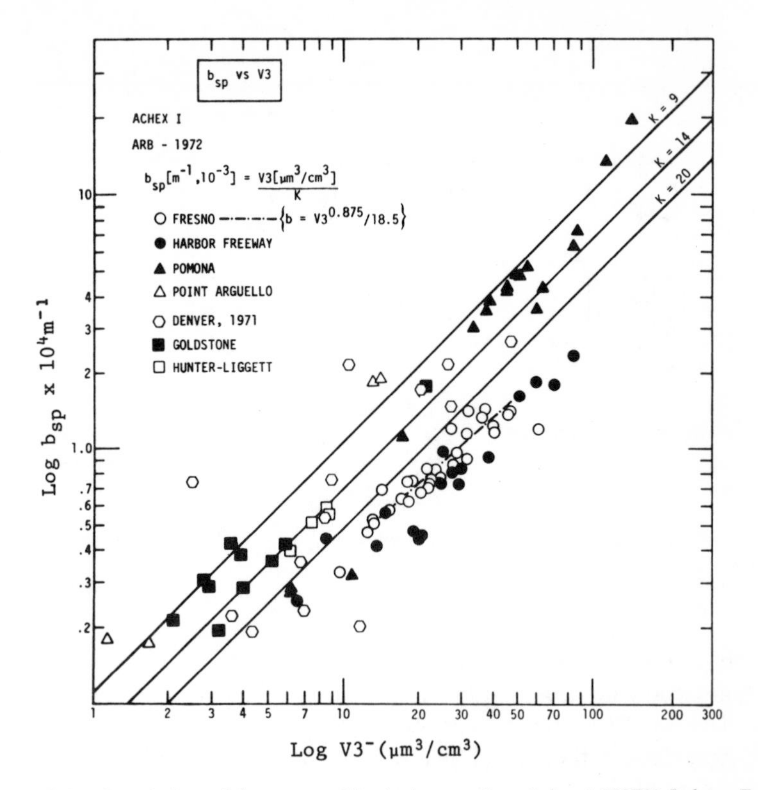

Figure 5-6. Correlation of b_{sp} versus $V3^-$ (volume <1 μm) for ACHEX I data. From Hidy *et al.*, 1974, p. 3–32.[330]

analyze theoretically. One effect of nonspherical particle shape and lack of homogeneity is deformation of the phase function, $\beta_{sp,\lambda}(\phi)$. This is particularly important for scattering angles over $\pi/2$. Holland and Gagne[356] show such results for silica dust. A more definitive statement regarding particle shape effects must await the results of current research.

b_{sp} as a function of relative humidity Also of importance are the changes that occur in the light-scattering to mass or volume relationships with hygroscopic aerosol growth (see Figures 3-4 and 3-5). Figures 5-7 and 5-8[167] illustrate in another way the calculated changes in the distributions due to radii increasing by 1.5 and 2.0 times. The normalized distributions of $\Delta V/\Delta\log D_p$ and $\Delta b_{sp}/\Delta\log D_p$ were calculated (using refractive index m = 1.33) and plotted versus the original (i.e., low humidity, diameter) on a semilogarithmic graph. This illustrates clearly the particle size intervals that are responsible for these changes. Changes in light-scattering are due primarily to the growth of the particles that were initially in the optical subrange (0.1 to 1.0 μm diam) and not due to

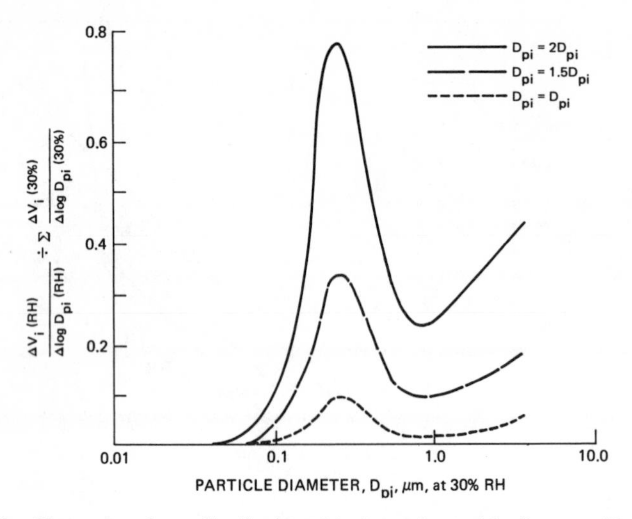

Figure 5-7. Change in volume distribution with change in particle diameter. From Covert, 1974.[167]

growth of particles whose initial diameters were <0.1 μm. Measurement of hygroscopic aerosol growth by optical techniques senses primarily the size increase of particles between 0.1 and 1.0 μm diam.

An extensive theoretical investigation of the relationships among the parameters of aerosol light-scattering, volume, and mass as a function of RH has been made by Hänel.[315-317] His calculations of the ratio b_{sp} (RH)/b_{sp} (RH = 30%) agree closely with results of direct measurements.

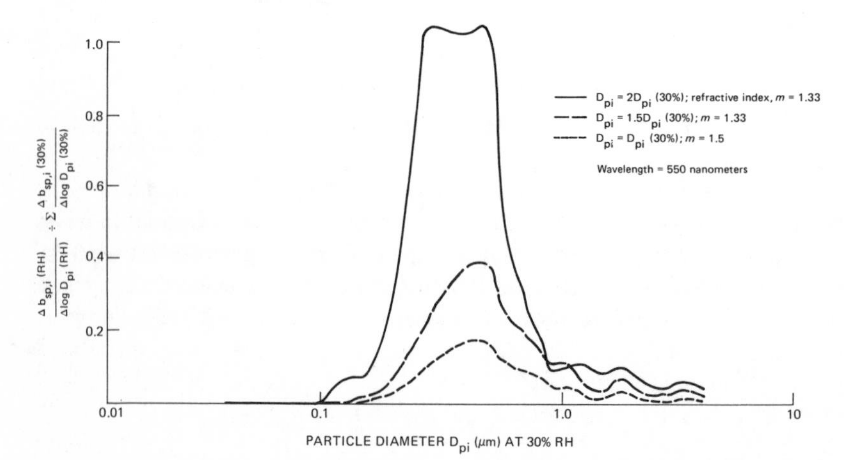

Figure 5-8. Change in scattering distribution with change in particle diameter. From Covert, 1974.[167]

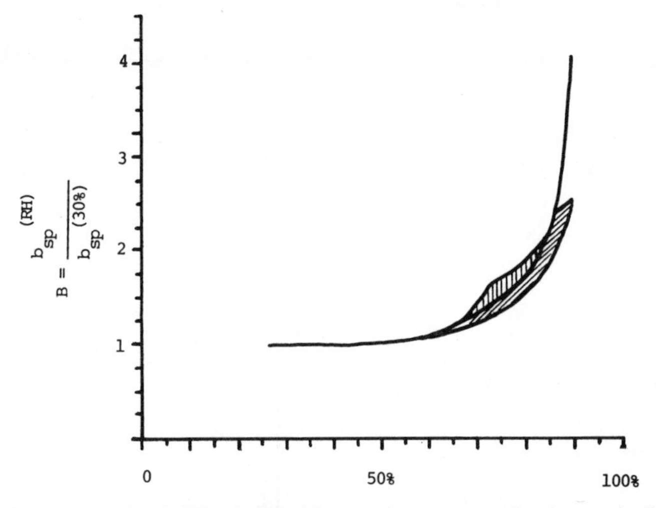

Figure 5-9. Range of variability in Humidogram data averaged by site; vertically hatched area includes strongly deliquescent aerosol at Pt. Reyes, California and Tyson, Missouri. From Covert, 1974.[167]

By preconditioning the aerosol sample to regulate the RH of the aerosol before entry into an integrating nephelometer, and then measuring the temperature and dew point inside the nephelometer, this instrument was modified to monitor b_{sp} as a function of RH.[167] A summary of measured curves is shown in Figure 5-9.

b_{sp} as a function of wavelength The scattering coefficient of a medium depends on the wavelength of the incident light. If the light is polychromatic (like sunlight), this may result in vivid coloration of the scattered light, depending on scattering angle and size distributions. For single scatterers and a Junge size distribution ($dn/dr = Cr^{-(v+1)}$, where dn/dr is the number of particles per unit volume with radii between r and $r + dr$), it can be shown that $b_{sp} \sim \lambda^{-\alpha}$ where $\alpha \equiv 2 - v$. The results of atmospheric measurements regarding the wavelength dependence of b_{sp} fall into two categories:

Normal wavelength dependence where $0.5 \leq \alpha \leq 2$, with a mean value of approximately 1.2. Figure 5-10 shows the frequency of occurrence of α measured in Pasadena, California, by Charlson *et al.*[138]

Anomalous wavelength dependence where $\alpha < 0$.

Normal wavelength dependence results in the attenuation of blue light from a direct beam and its scattering into 4π steradians around the scattering volume. Of course, Rayleigh scattering always occurs simultaneously and has a wavelength dependence that is similar:

$$b_{Rg} \propto \lambda^{-4}$$

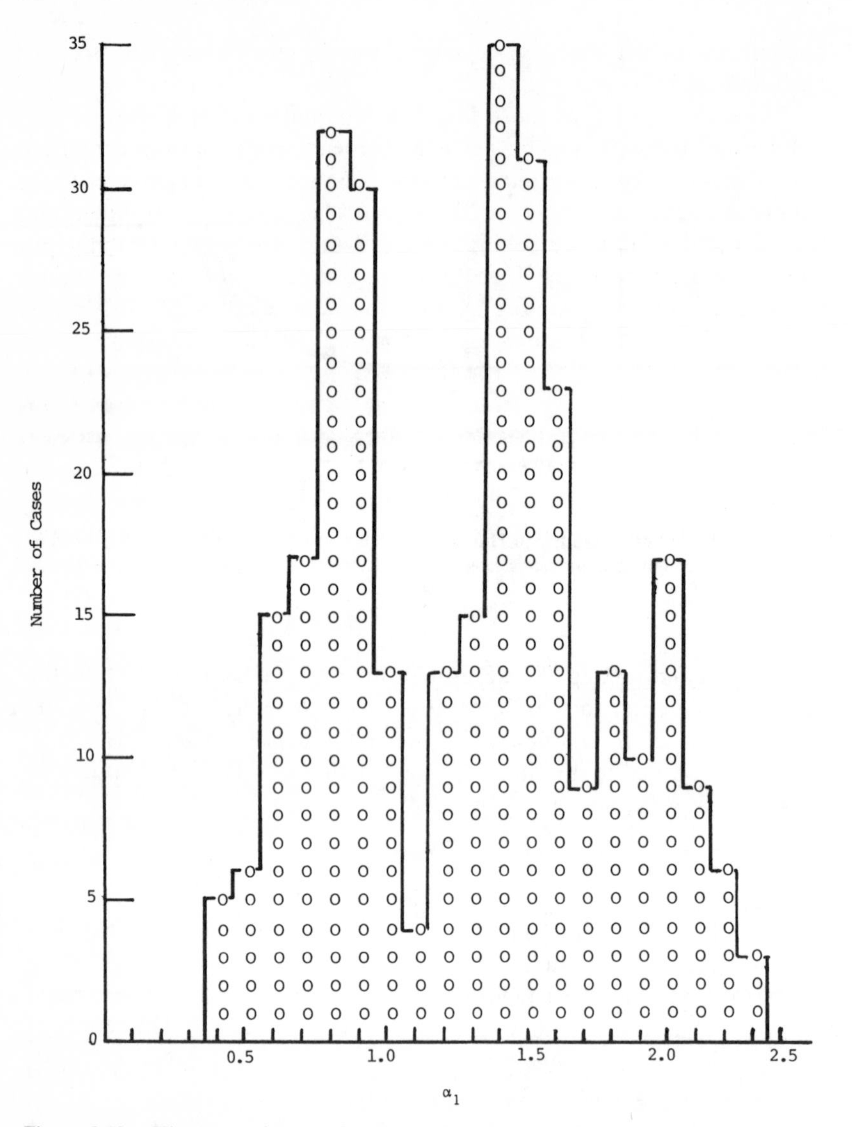

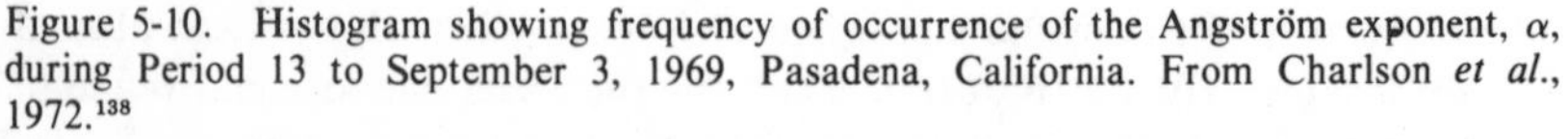
Figure 5-10. Histogram showing frequency of occurrence of the Angström exponent, α, during Period 13 to September 3, 1969, Pasadena, California. From Charlson *et al.*, 1972.[138]

For both normal wavelength dependence and Rayleigh scattering, blue scattered light against a dark background or red transmitted light (from the sun or a bright white object) is observed. Whether b_{sp} or b_{Rg} dominates is determined by the amount of particulate matter that is present. Anomalous wavelength dependence results in such rare phenomena as the blue sun or moon, usually attributable to very well aged aerosol.[614]

Blue hazes, such as those found in mountainous areas, may or may not be due to scattering by particles, depending on viewing conditions (e.g., dark or light background) and the distance from the observer to the background. Because $b_{Rg,550} = 0.15 \times 10^{-4}m^{-1}$, if $b_{sp} \sim 0$, mountains should not appear to be behind a haze if they are within 10 km. They will, however, appear hazy if the distance is much more than 100 km due to the omnipresent scattering by gas molecules. Conversely, if a mountain at such a distance is not visible at all, $b_{sp} \gg b_{Rg}$ and the haze is due to particles.

When viewing bright objects, such as the sun, the moon, sunlit snow-capped peaks, and cumulus clouds, hazes with $1 \leq \alpha \leq 2$ or sufficient optical depth cause reddening of the color.[538, 774] Because this color is remarkably similar to that observed through an optically thin layer of nitrogen dioxide,[363] its presence is no proof of the existence of nitrogen dioxide. To further complicate this issue, Husar and White[374] have shown that light scattered in the backward hemisphere calculated from typical measured size distribution is enriched in the red wavelengths, also causing the haze itself to appear reddened. In forward scatter ($0° \leq \phi \leq 90°$) this same haze appears white. Charlson *et al.*[138] showed that in perhaps 20% of the measured cases during August 1969 in Pasadena there was enough nitrogen dioxide to influence the coloration of the haze, and that in the remaining cases particles dominated the wavelength dependence of total extinction (b_{ext}). Figure 5-11 shows a plot of b_{sp} versus nitrogen dioxide in which the conditions dominated by particles and nitrogen dioxide, respectively, are shown. This figure applies only to the coloration of bright white objects viewed through haze.

b_{ap}—Light Absorption By Particles

Light absorption by particles is a well recognized but rarely measured quantity. Results to date with the methods of Lin *et al.*,[467] Fischer,[239, 240] and Lindberg and Laude[468] indicate that the imaginary part, $n_{imaginary}$, of the complex refractive index, $n_{complex}$ (which determines the light absorbing property of the particles), of a sample of urban particles is frequently in the range of 0.004 to 0.05, where $n_{complex} = n_{real} - in_{imaginary}$ and n_{real} = real part of refractive index $i = \sqrt{-1}$. Also, the absorption coefficient is typically up to 30% of the total extinction, i.e., $0 \leq b^{L}_{abs} \leq 0.3\ b_{ext}$. Measurement programs are currently underway to establish the generality of this result and of the value of the imaginary part of the refractive index for a variety of locations. Size dependence of the refractive index is an important factor due to the size dependence of chemical

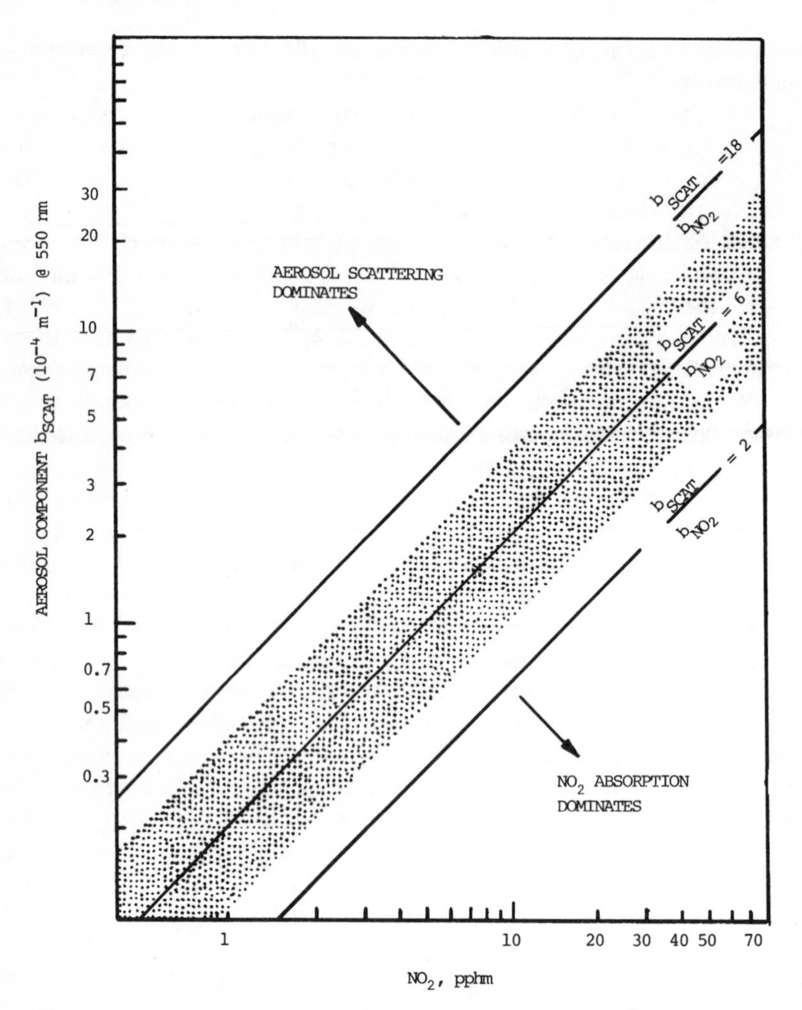

Figure 5-11. Light-scattering, b_{scat}, plotted against nitrogen dioxide (NO_2) in pphm, showing the ratio of $b_{scat}/b_{NO_2} \simeq 6$ at 546 nm for color effect of NO_2 to appear in transmitted light. Shaded region covers approximate 90% confidence intervals for the extra ordinate scales, mass concentration, and visible range (L_v). The lines $b_{scat}/b_{abs} = 2$ and 18 are limiting cases where NO_2 can be said to dominate and be totally unimportant, respectively. From Charlson *et al.*, 1972.[138]

composition. It makes the relationship between an average index of refraction and the integral property b_{ap} very tenuous.

RELATIONSHIP TO VISIBILITY

Having discussed the relationship of b_{sp} and b_{ap} to the other properties of the particles, it is logical to relate these correlations to visual range and associated quantities. According to Koschmieder's[423, 424] visibility theory, the following quantities are closely related:

Prevailing visibility or visual range, V(km)
Meteorologic range, L_v (km)
Light extinction coefficient, b_{ext} (m^{-1})

And in light of our previous discussion, these correlations should be related to particle mass concentration ($\mu g/m^3$), especially in the fine particle mode.

The Koschmieder theory relates the distance at which an ideal black object, viewed against the horizon sky, can barely be perceived in sunlight by a normal human observer. If the atmosphere is homogeneous and the illumination is uniform, this relationship is:

$$L_v = \frac{3.9}{b_{ext}}$$

where L_v is the meteorologic range. Definitions of prevailing visibility, visual range, and related terms are given in standard handbooks (see, for example, the *Glossary of Meteorology*[375]), but for our purposes they are nearly identical to meteorologic range.

There are published data with which these relationships can be checked, with the exception of simultaneous visibility observations and fine particle mass concentration. Only total and <5-μm particle mass concentrations have been measured in simultaneous field observations to date.

Horvath and Noll[364] conducted a study in Seattle relating total light-scattering, b'_{scat}, measured with an integrating nephelometer, and prevailing visibility as determined by two separate observers. Their results agreed with the theoretical expression of Koschmieder[423, 424] when only data for RH <65% were included. This limitation apparently resulted from the location of the nephelometer in a heated room, which caused reduced RH in the light-scattering measurements. At <65% RH, the correlations between b_{scat} and prevailing visibility were 0.89 and 0.91 with a coefficient in the Koschmieder expression of 3.5 ± 0.36 and 3.2 ± 0.25, respectively, for the two observers. This can be compared with the theoretical value of 3.9, indicating a slightly lower prevailing visibility than meteorologic range, assuming $b_{scat} = b_{ext}$. Since no ideal black targets were used (only trees, buildings, and other objects that reflect some wavelengths of light), these would have caused just such a deviation. Alternatively, 10% to 20% light absorption would yield this result.

Noll *et al.*[580] studied prevailing visibility, mass concentration, and soiling index in Oakland, California. They concluded that the prevailing visibility was inversely proportional to total mass concentration determined via 4-hr samples taken with 2.5 cm diam glass fiber filters.

The product of mass concentration and prevailing visibility was near 1.4 g/m². This compares well with a value of 1.8 g/m² determined with a nephelometer.[137] The correlation in Oakland between prevailing visibility and mass was approximately 0.9, indicating a good relationship between these variables.

Samuels *et al.*[662] have conducted the most extensive tests of the relationship of prevailing visibility to light-scattering and various mass concentration measures. They conclude that b_{scat} as measured with the integrating nephelometer is a good predictor of prevailing visibility and that the regression analysis is in basic agreement with Koschmieder's theory.

EFFECTS OF TURBIDITY ON SOLAR RADIATION

Besides visibility degradation, one of the most pronounced effects of man-made or man-caused particles is the diminution of the intensity of direct sunlight in urban regions. However, the main portion of scattered light is in the forward direction, and this reaches the surface of the earth along with the direct beam, where it is used in such processes as photosynthesis, heating of soil, or evaporation of water.[142] Most data on the optical properties of the atmosphere concern direct solar radiation, which must be weighed in terms of the actual fractions reaching the ground and being rejected upward. Clearly, data are needed on the actual amounts of radiation reaching the ground and not just that in the direct beam.

Extinction of the direct beam can be measured with the Volz sun photometer, pyrheliometer, and astronomical telescopes. The extinction coefficient is reported in various forms. The extinction vertically through the whole atmosphere may be given by the Beer-Lambert law in integrated form:

$$\frac{I}{I_0} = e^{-(\int_0^\infty b_{ext}\,dx)} \equiv e^{-\tau}$$

where τ $(\equiv \int_0^\infty b_{ext}\,dx)$ is the optical depth.

Another form of this expression results in a definition of decadic turbidity, B:

$$\frac{I(m)}{I_0} = 10^{-(\alpha_{ag}+\alpha_{Rg}+B)m}$$

where α_{ag} is the extinction due to light absorption by gases per unit air mass, m, and α_{Rg} is the extinction due to Rayleigh scatter. Another quantity is astronomical extinction, K:

$$\frac{I(m)}{I_0} = 10^{-\frac{K}{2.5}\cdot m}$$

where K is given in units of star magnitudes per air mass. Even though b_{ext}, α, B, and K are closely related, they are seldom compared and used together for studies of the effects of atmospheric particles.

Much of the data available for appraisal of extinction effects comes from the network of Volz sun photometers, which was reported by Flowers *et al.*[247] They interpolated the data into turbidity units, B. More recently, astronomical data have been used.[351] Four general features of the data are immediately apparent:

There is usually a turbid air mass centered over the eastern part of the United States. The turbidity (or extinction coefficient) in this air mass is about a factor of 2 higher than in the less turbid air found in the rural west[247] (see Figure 5-12; urban values are shown in parentheses).

There is a seasonal variation in the turbidity with a maximum in summer and minimum in winter, with exceptions in some urban locations where the opposite is found.[247, 648]

In and near cities there has been a trend of increasing turbidity,[523] increasing astronomical extinction,[350] and decreasing visibility[360] over the past half century.

In remote locations there is not yet a proven global trend over similar time scales.[214] Such a trend may exist, but the data quality and period of record are insufficient to detect it.

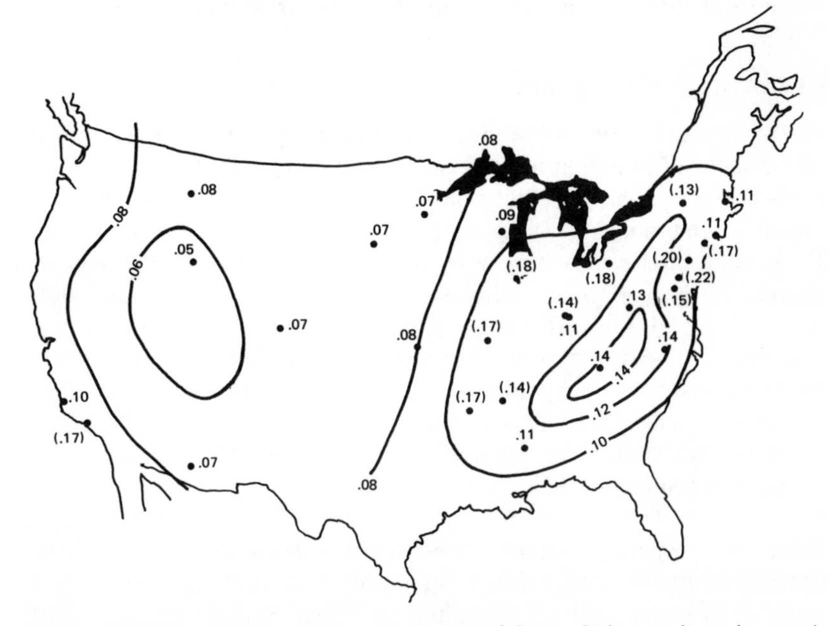

Figure 5-12. Mean annual turbidity over the United States. Urban station values are in parentheses. From Flowers *et al.*, 1969, p. 957.[247]

Losses of direct radiation of up to 50% have been measured in the rural Midwest in summer.[247] Of this decrease, assuming $b_{sp} \cong 0.2\ b_{ext}$ and $b_{bsp} \cong 0.1\ b_{ext}$, we can estimate that up to 20% of the radiation may be unavailable at the earth's surface. This decrease along with the atmospheric heating caused by b_{sp} might be important, for example, to agricultural production or to local or regional climate.

The few systematic studies of total (solar plus sky) radiation reaching the ground, which have been conducted simultaneously in cities and in rural areas, show reductions of this magnitude.[523] Much more extensive studies are needed to provide a more quantitative estimate of the possible reduction and fate of useable solar radiation over large rural areas.

Notable among these optical data are the long-term records of the Smithsonian Institution in Washington, D.C. and in Davos, Switzerland. These operations where started around the year 1900 as part of the determination of the solar constant. Almost by chance these operations continued and are among our only long-term records. The Volz photometer network, which was started in 1962, has grown steadily both in the United States and abroad.

In 1910, the late Charles Abbott of the Smithsonian Institution concluded that Mt. Wilson, California was not useful for solar photometry because of the hazes that emanated from the Los Angeles basin below. Hodge[350] noted that this effect has continued to develop there, virtually stopping photometric measurements at that observatory.

CLIMATE: DIRECT EFFECTS

In recent years there has been a great deal of speculation regarding the role of aerosol in determining regional or even global scale climate. The main concern is the possibility that aerosol might be responsible for the cooling trend noted globally in meteorologic data in recent years.[548] However, controversies have arisen largely because of insufficient data on relevant properties of particulate matter on both regional and global scales. Even if the particulate mass or number concentration were known, its climatologically important radiative effects would not be easily understood. Such radiative effects can be segregated into direct and cloud-related cases. The latter are due to the influence of the aerosol on cloud formation or on the optical properties of the clouds into which the aerosol is incorporated.

In the simplest case of direct radiative effect of aerosol without cloud involvement, disagreement exists on even the sign of temperature change to expect from secular changes in aerosol concentration on regional to global scales.[79, 140] Schneider[676] and Rasool and Schneider[629] suggested that aerosols would uniquely cool the earth's surface given one set of assumptions, which were disputed by Charlson and Pilat[141] and by Charlson *et al.*[139]

In spite of the lack of detailed data, it is clear that the regional and global effects of particulate matter due to scattering and sunlight absorption may not be disregarded. Bryson,[95] Robinson,[645] Kondratyev,[422] Budyko,[99] McCormick and Ludwig,[523] Joseph and Manes,[395] the Study of Critical Environmental Problems (SCEP),[512] the Study of Man's Impact on Climate (SMIC),[377] and other reports have all echoed the potential importance of this statement and the need for measurements and theoretical calculations. In many of these papers particulate matter is implicated via statistical correlation with past climate changes. The importance of volcanic debris is frequently mentioned. There is also interest in the current or future role of anthropogenic particles.

Various theoretical models have been used to aid in judging the importance of particles to radiative transfer. Rasool and Schneider,[629] Ensor *et al.*,[220] Atwater,[43] Chýlek and Coakley,[149] Yamamoto and Tanaka,[832] Mitchell,[548, 549] Wang and Domoto,[784] and others have explored a variety of optical calculations, different approaches to the radiative transfer, and atmospheric models over a range of complexity. Virtually all models depended on assumed values for the relevant optical constants of the atmospheric particles, notably complex refractive index as a function of particle size and the size distribution. The lack of data and the complexity of the climate problems render the results of such calculations extremely tenuous.

As an example of one of the simplest models,[548] the sensible heating, H, of both an optically thin aerosol layer and the underlying surface is given by:

$$H = S_0\, C(1 - A)e^{-(a+b)}[1 + A\,(1 - e^{-b})]$$
(sensible surface heating)
$$+ S_0\,(1 - e^{-a})\,[1 + Ae^{-b}e^{-(a+b)}]$$
(aerosol heating)

where S_0 is the incident solar flux, A is the surface albedo, a is the absorption optical depth, and b is the backscatter optical depth, defined as that fraction of the scattered light rejected into space. $C = B/(B + 1)$, where B is the ratio of sensible heating to latent (or evaporative) heating, known as the Bowen ratio.

The temperature rise in this system is proportional to H, and the quantities a and b define the role of aerosol; if we assume they are both small, then H depends on the ratio $a{:}b$. At $H = 0$, $a/b = (a/b)_0$, where

$$(a/b)_0 = \frac{C(1 - A)^2}{1 + A - C(1 - A)}$$

so that the critical ratio depends on both the albedo and the Bowen ratio. If $a/b > (a/b)_0$, then the aerosol would be expected to cause heating of the atmosphere; if $a/b < (a/b)_0$, cooling would be expected. In either case, the surface of the earth might cool due to a decrease of radiation.

A chief part of the controversy mentioned earlier arose from the fact that a/b, or its equivalent in other models, is unknown. It was estimated or calculated from assumed refractive indices or assumed size distributions and composition/size dependence. Depending on the assumptions, the value of the ratio $a:b$ ranges from values well above to well below $(a/b)_0$, giving rise to the questions of heating versus cooling due to particles. As Mitchell[549] concluded, ". . . it is of the utmost importance to measure the backscattering, absorption, and other optical properties of each type (each combination of chemical constituents) of aerosol layer commonly encountered in the atmosphere."

Among the factors that are important to climate studies are the spatial dependence of aerosol properties and the role of RH in determining these properties as humidity increases.

The horizontal spatial extent of turbid air masses typically spans 1,000 km, as expected from calculation of removal rates[647] and from turbidity data (Figure 5-12). The vertical distribution is important because there is a possibility that the particles exist below, between, or above clouds, which themselves have high albedos. Thus, heating is very much more probable for a given aerosol over clouds that over oceans or

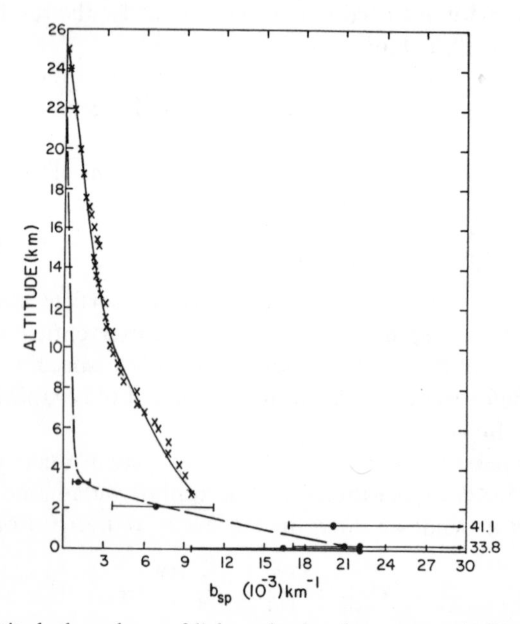

Figure 5-13. Altitude dependence of light extinction due to aerosols. Nephelometer data with 90% interval for b_{sp} at 500 nm (------) compared to Elterman's searchlight data (——) at 550 nm.[215] From Charlson *et al.*, 1974, p. 354.[142]

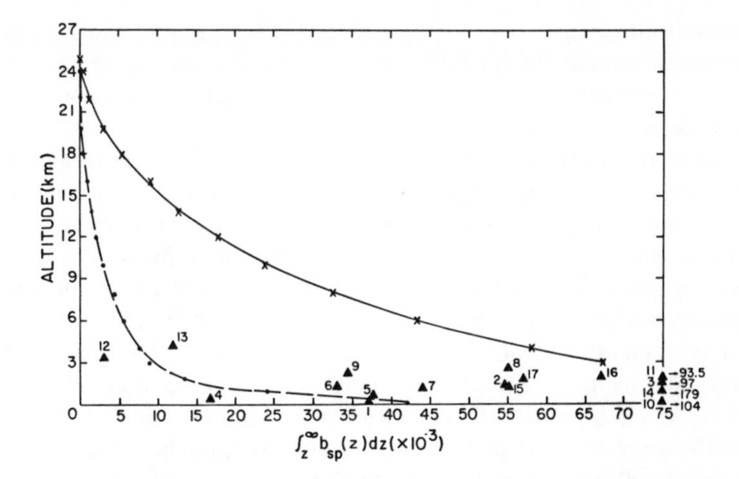

Figure 5-14. Mean aerosol optical depth versus altitude via astronomical telescope (data points) at 500 nm, the integral $\int_z^\infty b_{sp}(z)$ for the geometric mean nephelometer data in Figure 5-13 (------) and the same integral for the searchlight data (——).[215] The astronomical data compared here can be regarded as typical for remote locations. They were taken over varying time periods by many different observers. A value of T_{ozone} = 0.004 was assumed in the telescopic data. Observations are by number, with altitude (meters) in parentheses: (1) Royal, Cape of Good Hope (10); (2) Boyden, Republic of South Africa (1,387); (3) Radcliff, Republic of South Africa (1,542); (4) Mt. Bingor, Australia (460); (5) Mt. Stromol, Australia (768); (6) Siding Spring, Australia (1.164); (7) Mt. John, New Zealand (1,029); (8) Europ South, Chile (2,400); (9) Cerro Tololo, Chile (2,195) (10) Lo Houga, France (146); (11) Naini Tal, India (1,927); (12) Mauna Loa, Hawaii (3,370); (13) Mauna Kea, Hawaii (4,170); (14) Rattlesnake Ridge, Washington (1,080); (15) Lick, California (1,283); (16) Kitt Peak, Arizona (2,064); (17) McDonald, Texas (2,081). From Charlson *et al.*, 1974, p. 355.[142]

other low albedo surfaces. Data on the vertical distribution of particles outside of urban locations are fairly sparse. Figure 5-13 shows integrating nephelometer data at different altitudes and locations along with searchlight data from New Mexico.[142, 215] Figure 5-14 shows optical depth data from astronomical telescopes compared to those calculated from the searchlight data.[142] In both cases it appears that most of the optical effects of the particles occur below 3 km altitude. Because these data are limited in quantity, location, and period of record, they should be expanded.

The effect of humidity on b_{sp} has been measured, as pointed out above. The effects on a/b have not yet been measured, and their calculation is even more tentative than the estimation of a/b at low RH.

CLIMATE: CLOUD-RELATED EFFECTS

Over the past 15 years increasing evidence has indicated that major urban and industrial complexes are significantly modifying the downwind climate and, in particular, changing levels of precipitation. While there are certainly a number of operative mechanisms involved in the modifications, the concentration and chemical nature of the pollution aerosol play an important role.

This atmospheric interaction occurs because the transformation of water vapor into water droplets and ice crystals is a nucleation phenomenon requiring the presence of specific nuclei. The nuclei that are "weather active" usually constitute only a very small fraction of the total aerosol population. The nuclei involved in cloud droplet formation, cloud condensation nuclei (CCN), comprise approximately 1% of the total aerosol. Generally, the CCN outnumber the ice nuclei by a factor of 10^5 to 10^6. The relatively rare ice nuclei either freeze supercooled droplets or serve as sublimation sites for the direct deposition of vapor.

The concentration of these nuclei directly influences the concentration of cloud droplets and ice crystals. This in turn influences the cloud dynamics and the probability and amount of rainfall. Because, under normal circumstances, the concentrations of these nuclei are quite small in comparison to the total aerosol concentration, and because relatively small increases will produce significant effects, the concentration of weather-active nuclei is clearly a most sensitive environmental parameter.

Warm Cloud Processes

Cloud formation is due almost entirely to the expansion and subsequent cooling of ascending air. When the air is cooled to the saturation point (100% RH), condensation on the largest and most hygroscopic of the CCN accelerates. (Some of these particles may experience limited growth well below 100% RH.) The number of droplets that ultimately form in a given cloud depends on a complex feedback mechanism incorporating temperature, uplift velocity, the concentration of CCN, and the supersaturation required for each available CCN to act as a nucleating site. This mechanism generally limits the RH to less than 101% in clouds. Futhermore, it has been established that measurements of CCN concentration as a function of activation supersaturation and of estimates of the uplift velocity enable prediction of the approximate droplet concentration throughout the cloud.[746, 765]

Although a direct relationship exists between CCN and droplet concentrations, the relationship between these factors and precipitation probability is less well understood. Theory indicates that raindrops cannot be formed by condensation alone in a reasonable period.[231] Indeed, it shows that in the early stages of cloud development the assembly of droplets forms a rather size-stable array of nearly uniform 5- to 10-μm diam drops.[368, 762] This array becomes unstable if droplet-droplet coalescence occurs. This will happen only if a significant number of the droplets grow larger than a ~20 μm radius.[349] Warm clouds that fail to produce a reasonable number of large drops will remain stable and not produce precipitation.[719, 763]

Observation and theory suggest that the population of CCN can affect warm cloud stability, and hence precipitation probability, in several ways. First, if a large number of effective CCN (10^3 to 10^4 cm^3) have nearly the same activation supersaturation and size, then a cloud of high droplet concentration consisting of very small uniform droplets will result. Because the available water is spread among many droplets, the chance that any will grow large enough to initiate the coalescence process is relatively small, as is the probability of precipitation. Conversely, if there are relatively few CCN (10^1 to 10^2 cm^3), the available water shared among the resulting few droplets is far more likely to produce rainfall. Furthermore, if there are more than a few per liter of very large hygroscopic nuclei (larger than a few micrometers) in either polluted or unpolluted cloud masses, these nuclei may become large enough droplets to promote the coalescence mechanism regardless of the population of less effective CCN.[787, 831] This destabilizing technique provides the basis for most of the warm cloud modification schemes that have been tried to date.

Cold Cloud Processes

Ice particles appear in clouds once the droplets are cooled to below the freezing point. On the basis of ice nuclei measurements, the increase in ice crystal concentration is approximately an order of magnitude for every 4°C of cooling.[244] At about −40°C all of the remaining supercooled cloud droplets freeze spontaneously irrespective of the ice nuclei activity-temperature spectrum.

Unfortunately, the relationship between the measured concentrations of ice nuclei and of cloud ice particle concentrations is not understood. There is generally a large and highly variable difference between the two measurements, with usually more ice crystals than nuclei.[421, 560] In addition, the ice nucleus concentration has a much larger spatial and time variation than does the CCN concentration. Some short-term variations may be traced to local sources, but the general distribution of ice nuclei in the atmosphere is not clearly understood. Typical concentrations of ice nuclei measured at −20°C are from 0.1 to 1.0 per liter.

Once ice particles are present among supercooled droplets, the cloud stability is substantially decreased; the ice particles will always grow at the expense of the surrounding droplets. These growing ice crystals soon acquire a differential velocity compared to their surrounding droplets, which they collect by impaction. This provides a destabilizing mechanism similar to coalescence that, depending on ice nucleus concentration, can have a powerful influence on cloud precipitation probability. Most rainmaking projects proceed on the assumption that the natural concentration of ice nuclei is too low for optimally efficient precipitation and that the addition of ice nuclei will increase precipitation. However, this is not always the case. Rainmaking efforts have apparently decreased rainfall by providing too many ice nuclei and increasing the cloud stability in a way similar to the effect caused by too many CCN.[248] While there is considerable variation for different cloud types, a few ice nuclei per liter are probably optimal for precipitation.

Observed Effects

The addition of ice nuclei and CCN to a cloud affects the cloud's microphysical structure. This, in turn, may either increase or decrease the probability of precipitation. The probable effects and the approximate concentrations of nuclei to promote these effects are oulined in Table 5-2.

The literature indicates that the effects of urban areas on precipitation are often stronger than those of the most effectively planned weather modification projects. Landsberg[437] reported that many large urban areas show downwind precipitation increases of 5% to 15%. The now famous La Porte anomaly[132] appeared to show an increase of 31%. Moreover, a recent climatologic study[369] of eight major U.S. cities confirms Landsberg's results, showing precipitation increases of 9% to 17% with only one city showing no effect. Other studies downwind of nuclei sources have indicated, variously, no effect,[584] decreased precipitation of 25%,[786] and increases from 10% to 30%.[348]

Table 5-2. Expected effects on precipitation probability due to the addition of ice nuclei and cloud condensation nuclei

Precipitation probability	Ice nuclei (IN)	Cloud condensation nuclei (CCN)
Increase	The addition of sufficient ice nuclei active at the cloud temperature so that the total concentration becomes 1 to 10 per liter	The addition of sufficient giant hygroscopic nuclei so that the total concentration becomes 1 to 10 per liter
Decrease	Increase the ice nucleus concentration beyond a few hundred per liter	Increase the CCN concentration beyond a few hundred per cubic centimeter
Produce uncertain effects		The combination of high CCN concentrations together with optimal numbers of large hygroscopic particles

Unfortunately, despite ample evidence that there are many significant urban-industrial and agricultural sources of weather-active nuclei,[346-348, 420, 625, 711, 720, 751, 788] the link directly relating these sources with observed changes in cloud microphysics and precipitation remains tenuous. Much of this uncertainty has been due to the lack of a program to make simultaneous measurements of nuclei concentrations and inputs of heat and water vapor, cloud microphysical changes, and rainfall patterns.

The Metromex study,[133] conducted around St. Louis, was the first attempt to make these measurements. Semonin[690] indicates that there is evidence of a direct link between nuclei emission and precipitation. A few other observations also strengthen this link. In Australia, Warner[786] reported decreased rainfall downwind of agricultural burning of sugarcane refuse. This waste material, when formed, proved to be a copious source of CCN, which was traced in turn to substantial increases in cloud droplet concentrations in the area downwind. This case provides reasonable evidence linking a nuclei source with increased cloud stability and decreased precipitation.

Significant sources of CCN have also been associated with increased rainfall.[348] A study was made of many of the major sources of CCN in the northwestern United States and of the downwind winter rainfall patterns before and after the emitting industries were established. The strong sources showed a clear pattern of increased rainfall. Recent measurements[208] support the original hypothesis that the increases are not primarily due to the copious CCN emission, but to the simultaneous emission of moderate numbers of large (>1 μm) and very hygroscopic particles. These particles apparently form the large, destabilizing droplets necessary in the coalescence mechanism.

Because direct observations are lacking, evidence for inadvertent weather modification by the known urban-industrial ice nuclei sources must be argued largely from the similarity of those sources to planned programs of ground-based cloud-seeding. Some indirect observations have been made, such as Schaefer's[667] report of dramatic increases in ice crystal concentrations downwind of major urban-industrial areas.

Thus, a limited, but growing, body of evidence suggests that the urban-industrial aerosol significantly affects local downwind climate and cloud structure.

PRECIPITATION SCAVENGING AND RAIN CHEMISTRY

Precipitation scavenging is one of the major processes of removal of aerosol from the atmosphere. It is therefore one of the important links in the cycles of many elements. Scavenging can be divided into two major categories: washout, or removal below the cloud layer; and rainout, or removal within the clouds. For particles <1 μm diam, rainout is the dominant scavenging mechanism. For larger particles (>1 μm diam), washout is important; it competes with rainout and gravitational settling as the dominant removal mechanism. Because the removal efficiencies of washout and rainout are functions of particle size, the amount and frequency of precipitation events, both regionally and globally, have a pronounced effect on the size distribution of the regional and global "background" aerosols, respectively.

Of these two general types of scavenging, the below-cloud processes have been more extensively studied.[217] Impaction efficiencies integrated over both particle and droplet size have been related to a "washout coefficient," the characteristic decay time of aerosol concentration during precipitation. This treatment generally ignores other phoretic forces that dominate within the cloud. Because these phoretic processes are more difficult to quantify, the rainout mechanism has not been extensively studied. Some theoretical work on rainout has been done; however, the contributions of electrical interactions and ice nucleation require further investigation.[706]

The fact that atmospheric turbidity often decreases after precipitation does not necessarily imply that precipitation scavenging has cleaned the air. Flowers *et al.*[247] have shown that the turbidity change is more strongly related to the presence of unstable air (and thus greater mixing) associated with the precipitation than to precipitation scavenging. Thus, a majority of the aerosol has merely been diluted and not removed from the atmosphere.

The effect of precipitation scavenging of aerosols on the chemical composition of rainwater has received recent attention. It is complicated by the fact that some gases can contribute to the same chemical species in solution as the aerosol. Bielke and Georgii[71] have estimated that the aerosol contribution can be as low as 25%. In a well-mixed atmosphere, however, rainout undoubtedly predominates.[564]

The inorganic chemistry of precipitation has been extensively studied over the past 60 years in both Europe and North America. A good bibliographic review of the literature on the inorganic chemistry of precipitation has been published by Dana *et al.*[183] Most of the precipitation in these studies has been sampled at ground level over 1-month collection periods. Therefore, there have been relatively few, if any, attempts to relate the short-term (~<3-day) chemistry of precipitation to the equally short-term synoptic meteorology, or to determine the relative contribution of rainout and washout to the precipitation chemistry.

Perhaps the most reported chemical variable in rain is the hydrogen ion. Both hydrogen and sulfate ions have shown marked geographic variations and systematic changes in time, with both acidity and sulfate increasing most notably in Scandinavia. A Swedish report[78] concludes that increases in sulfate and acid are due to sulfur emissions associated with fossil fuel combustion in Germany, Holland, Belgium, and Great Britain, and that these increases have important biospheric effects. The most notable effects attributable to the increased acidity are decreases in the pH of lakes and rivers and the leaching of calcium from forest soils. The pH has become sufficiently low in affected lakes to cause some fish to cease production. The thin forest soils show important changes in nutrient balance. There is also evidence of decreases in growth rate of trees.

FOGS

As the RH in the lowest atmospheric layer approaches 100%, water vapor will condense on airborne particles, thereby forming fog. At a given RH, a certain fraction of the droplets thus formed grow until they reach an equilibrium radius, typically 1 to 3 μm, while other droplets continue to grow until they begin to fall, at 15 μm radius. The former are called unactivated, and the latter, activated droplets.

Both stability and visibility in fogs depend on the numbers of unactivated versus activated droplets. Large droplets tend to collide with smaller ones as they fall, thus providing a relatively efficient removal mechanism for water. Therefore, fog stability is enhanced by the presence of small, unactivated droplets and is decreased by the formation of large ones. Visibility varies inversely as the light-scattering coefficient, which is dominated by the smaller droplets even when many of the droplets are activated.[266] Thus, visibility is decreased in fogs containing large numbers of small droplets and is improved as a relatively large number of the droplets are activated.

The factors that determine the activation efficiencies of different kinds of small droplets are not precisely known. Laboratory experiments show that the application of organic surface films to the droplets at some intermediate stage of growth[418] or to the dry particles[459, 739] can suppress droplet growth, thus increasing stability and lowering visibility. The possible role of anthropogenic organic compounds in the atmosphere is unknown.

6

Effects of Inhaled Particles on Humans and Animals: Deposition, Retention, and Clearance

The major regions of the respiratory tract differ markedly in structure, size, function, and sensitivity or reactivity to deposited particles. Some regions also have different mechanisms for particle elimination. Thus, a complete determination of dose from an inhaled aerosol depends on the pattern of deposition, the retention times at the deposition sites and along the elimination pathways, and the physical, chemical, and biologic properties of the particles.

This chapter begins with a discussion of the factors that determine the amounts and patterns of particle deposition within the various regions comprising the respiratory tract. Following that is a discussion of the pathways and dynamics of particle translocation and elimination. Both discussions emphasize the normal patterns of particle behavior and transport that are applicable to most atmospheric aerosol of less than acute toxicity. This material provides a basic background for the sections on epidemiologic studies, on dose-response relationships, and on laboratory studies designed to produce toxic or functional effects on animals and humans.

As discussed in Chapter 1, most of the particles in the accumulation mode (0.1 to 2.0 μm diam) of atmospheric aerosol contain water and are hygroscopic. Thus, their size can change as they travel through the warm and humid atmosphere in the conductive airways. These droplet aerosols also contain dissolved materials that can rapidly diffuse within and through the fluid lining of the airways. On the other hand, inert and insoluble particles or components of liquid particles that are deposited within the conductive airways undergo passive transport determined by the motion of the mucus layer. Consequently, the fate of inert, insoluble particles during the bronchial clearance phase can be discussed more generally than that of rapidly soluble or highly irritating particles. To study the soluble particles that deposit in any part of the respiratory tract or the insoluble particles that are deposited within the nonciliated alveolar zone, the chemical composition of each aerosol must first be considered.

FUNCTIONAL ZONES FOR PARTICLE DEPOSITION

The respiratory tract (Figure 6-1) can be divided into zones because the insoluble particles that deposit therein contact or affect different cell populations and/or have substantially different retention times and clearance pathways. Each zone includes one or more anatomic regions. In each zone, the retention of soluble or reactive particles usually differs from that of inert, insoluble particles because they diffuse within the fluid layer at the surface, then through it into the underlying cells and circulating blood. The clearance kinetics for particles deposited in each region are discussed in the section on PARTICLE RETENTION (page 128).

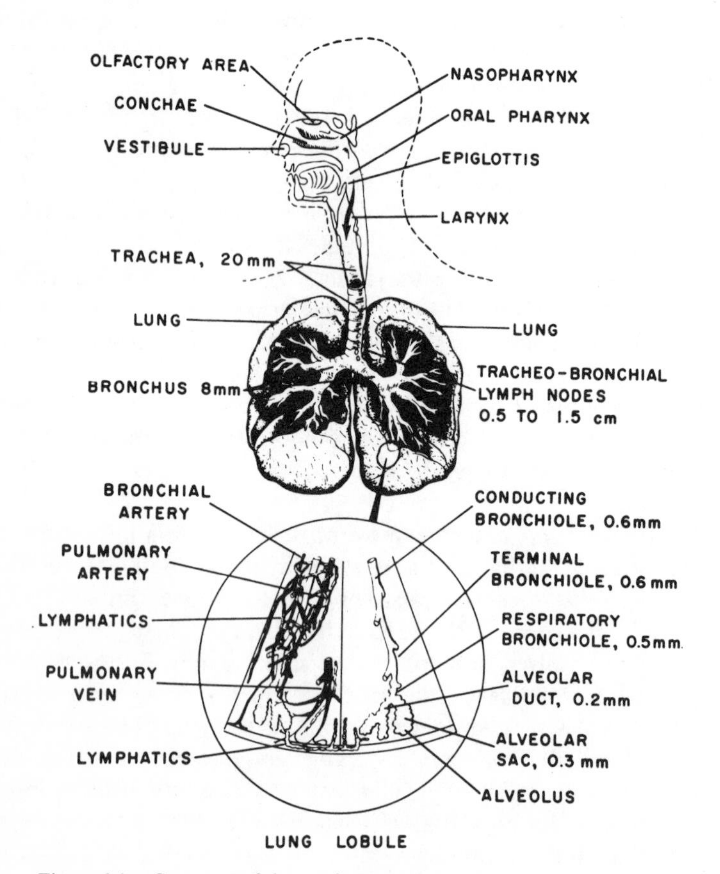

Figure 6-1. Structure of the respiratory tract.

Anterior Nares

Particles deposited in the unciliated anterior portion of the nose remain at the deposition sites for variable and usually indeterminate periods until they are removed mechanically by nose wiping, nose blowing, sneezing, etc. The effect of particle solubility on retention in this zone is not known.

Ciliated Nasal Passages

After leaving the nares, or nostrils, the inspired air passes through a web of nasal hairs, then flows through the narrow passages around the turbinates. It is warmed, moistened, and partially depleted of particles with aerodynamic diameters $>1.0\ \mu m$ by sedimentation and by impaction on the nasal hairs and on the passage walls at the bends in the air path. Particles $<0.1\ \mu m$ diam can be deposited in this zone by diffusion. The surfaces of the nasal passages are covered by mucus, most of which is propelled toward the pharynx by the beating of the cilia. Deposited insoluble particles are transported by the mucus; soluble particles may dissolve in it. Some mucus moves toward the anterior nares, carrying inhaled whole or dissolved particles into the zone of intermittent mechanical clearance.

Nasopharynx, Oral Passages, Larynx (Nonciliated Large Airways)

Particles inhaled through the nose and deposited in the nasopharynx or particles inhaled through the mouth and deposited in the mouth and oropharynx are swallowed within minutes. They pass through the esophagus to the gastrointestinal tract. Particles inhaled by either route can also be deposited on the larynx, where they are transported rapidly to the esophagus in or on the mucus coming up from the trachea.

Although the residence times for particles in these zones may be very short, local deposition is important. Concentration buildup on some of these surfaces can be very high and may exceed the capacity for clearance. This may account for cancers of the anterior nares in furniture workers[2, 551] and for laryngeal cancers in cigarette smokers.[342, 723]

Tracheobronchial Tree

The tracheobronchial or conductive airways have the appearance of an inverted tree, with the trachea analogous to the trunk and subdividing bronchi to the limbs. The airway diameter decreases distally, but because of the increasing number of tubes the total cross section for flow increases and the air velocity decreases. In the larger airways, particles too massive to follow the bends in the air path are deposited by impaction. At the low velocities in the smaller airways, particles are deposited by sedimentation and diffusion. Ciliated and secretory cells are found throughout the tracheobronchial tree. Inert nonsoluble particles deposited on normal ciliated airways are cleared within one day by transport on the moving mucus to the larynx.[13] Soluble particles are cleared much faster, presumably via bronchial blood flow.[836]

Alveolar Zone

Gas exchange occurs in the alveoli beyond the ciliated airways. The epithelium there is very thin, permitting soluble particles to enter the pulmonary blood within minutes. Insoluble particles deposited in this zone by sedimentation and diffusion are removed at a very slow rate, with clearance half-times measured in days, months, or years. The mechanisms for clearance of insoluble particles from the alveolar zone are only partly understood, and their relative importance is a subject of debate (see the section on ALVEOLAR CLEARANCE, page 140).

FACTORS AFFECTING PARTICLE DEPOSITION

Particles are deposited in the various zones of the respiratory tract by a variety of physical mechanisms. Deposition efficiency depends on the aerodynamic properties of the particles, the anatomy of the airways, and the geometric and temporal patterns of flow.

Deposition Mechanisms

There are five mechanisms by which significant particle deposition can occur within the respiratory tract: interception, impaction, sedimentation, diffusion, and electrostatic precipitation. Of these, impaction, sedimentation, and diffusion are the most important.

Interception Interception is usually significant only with fibrous particles. It takes place when the trajectory of a particle brings it so close to a surface that an edge of the particle contacts that surface; thus, the larger the particle, the greater the chances for interception. The probability of deposition by impaction and sedimentation is determined by aerodynamic diameter (D), which is the diameter of a unit density sphere having the same terminal settling velocity of the particle under study. Nonfibrous particles with $D > 5$ μm are usually deposited in large airways by impaction or sedimentation rather than by interception. In his studies of asbestos and other fibrous particles, Timbrell[755] found that D averaged ~3 times the fiber diameter for fibers with a length to diameter ratio >10:1. Consequently, a fiber with a length of 200 μm and a diameter of 1 μm would have an aerodynamic diameter of ~3 μm. Such fibers have a low probability of deposition in the conductive airways by impaction or sedimentation. However, the probability of their deposition by interception is higher because of the fiber length. Some 200-μm-long fibers have been observed in human lung samples.[756] Such straight fibers as amphibole asbestos are more likely to penetrate to the alveoli than the similarly sized, but curly, chrysotile asbestos because the straight fibers assume orientations more parallel to the flow streamlines.

Impaction Inhaled air follows a tortuous path through the nose or mouth, then into the branching airways in the lung. Each time the air changes direction, the momentum of particles tends to keep them on their preestablished trajectories. This can cause them to impact on airway surfaces. Impaction probability is determined by stop distance $d_s = \nu\tau$, where ν = air velocity and τ is the relaxation time. It also depends on the location of the particle within the airway. For example, in a bifurcating airway, deposition probability on inhalation is much higher for a particle traveling along the centerline of the parent tube than for particles moving nearer to the walls. The most likely deposition sites are at or near the carina of the bifurcation.

No adequate mathematical model exists for impaction deposition in the bronchial tree. The periodic air flow can change from turbulent in the trachea to laminar in deeper airways. In the larger airways, where most impaction occurs, the flow pattern is never fully developed because the Reynolds number† is high and the length of each segment is only about 3 times the diameter.[362, 792]

Sedimentation Gravitational sedimentation is an important mechanism for deposition in the smaller bronchi, the bronchioles, and the alveolar spaces where the airways are small and the air velocity is low.

Sedimentation becomes less effective than diffusion when the terminal settling velocity of the particles falls below ~0.001 cm/sec, which for unit density spheres is equivalent to a diameter of 0.5 μm.

Diffusion Submicrometer particles in air undergo a random motion caused by the impact of gas molecules surrounding the particles. This Brownian motion increases with decreasing particle size and becomes an effective mechanism for particle deposition in the lung as the root-mean-square displacement approaches the size of the air spaces.

Diffusional deposition is important in small airways and alveoli and at airway bifurcations for particles less than ~0.5 μm diam. The diffusional efficiency of radon and thoron daughters, whose particle sizes are molecular, can be high, especially in the head and in large airways like the trachea.

Electrostatic Precipitation Particles with high electric mobility can have an enhanced respiratory tract deposition even though no external field is applied across the chest. Deposition results from the image

---

† The Reynolds number (*Re*) is dimensionless. It is equal to the product $(DVp)/\mu$, where D is the diameter of the airway, V is the velocity of the air, p is the density of the air, and μ is the viscosity of the air.

charges induced on the surface of the airways by the charged particles. Test aerosols, produced from the evaporation of aqueous droplets, can have substantial mobility. The results of deposition studies using such aerosols without charge neutralization are accordingly suspect. Because most ambient aerosols have reached charge equilibrium, their charge levels are much lower. Thus, the deposition due to this mechanism is usually small compared to that due to the mechanisms discussed above.

Aerosol Factors

Particle size is always an important variable in regional deposition. There are a number of ways of expressing particle size. Usually, particle size is expressed in terms of actual or equivalent diameters, although some scientific literature uses particle radius. When particle size is measured by one parameter and expressed in terms of another or "equivalent" size, the basis for the conversion must be clearly established. Nonspherical particles are frequently characterized in terms of equivalent spheres, i.e., on the basis of equal volumes, equal masses, or aerodynamic drag. The aerodynamic diameter, discussed above, is used with increasing frequency. This measurement incorporates both particle density and drag.

Aerodynamic diameter is the best measurement to use in studies of particle deposition by impaction and sedimentation, which usually account for most of the deposition by mass in the head and lungs. Interception also depends on both the linear dimensions of the particle and its aerodynamic drag, which can affect the particle's orientation within the airway. On the other hand, diffusional displacement, which is the dominant mechanism for particles $<0.5\ \mu m$, depends only on particle size and not on density or shape.

The conversion of linear or projected area microscopic measurements of nonspherical particles to aerodynamic diameters requires assumptions about the relationship between projected area and volume, about density, and about aerodynamic shape factors. Sometimes these conversions have been made accurately, but more often they have not. An alternative is to measure aerodynamic size directly with an aerosol spectrometer.[425, 730, 755]

A complicating factor for water-soluble particles is the change in size that may occur in humid atmospheres. Furthermore, dry aerosols of materials like sodium chloride, sulfuric acid, and glycerol will take up water vapor and grow in size within the warm and nearly saturated atmosphere in the lungs.[539, 615] Because total and regional deposition depend on particle size, droplet growth could significantly change deposition pattern and efficiency.

Respiratory and Flow Factors

Increasing air velocity increases impaction deposition but decreases sedimentation and diffusion by decreasing residence time. The calculation of deposition at a constant velocity is relatively simple but cannot be applied to normal breathing, during which velocity is continuously variable and reverses direction twice in each cycle. During exhalation, the flow profiles within the airways differ from those during inhalation, thereby affecting particle deposition probabilities, especially for the impaction mechanism.

The velocity profiles in the large conducting airways are quite different from those normally encountered in fluid-filled conduits. Because the average length of a bronchial airway is only about 3 times its diameter,[362, 792] the flow therein never achieves a fully developed flow profile. Moreover, because each segment terminates in a bifurcation, the entering flow profile is asymmetrical, i.e., the maximum velocity is close to the carinal ridge. Such profiles have been demonstrated by Olson *et al.*[586] and Pedley *et al.*,[604] who also have shown that the asymmetrical entry results in a secondary swirling motion that is imposed on the bulk flow. Finally, the flow is cyclical and reverses many times per minute. At its peak rate, it may be turbulent in the trachea. Because the Reynolds number decreases with increasing lung depth, the flow is always laminar in the smaller conducting airways and viscous in the alveolar region.[186]

The flow is laminar in the more distal bronchioles, which contain most of the volume of the conductive airway system. In these distal segments of the anatomic dead space, the central core velocity is almost twice the average velocity. Even during very shallow breathing, a substantial fraction of the inhaled air penetrates beyond the anatomical dead space. Particles with appreciable sedimentation rates (>2 μm) or large diffusional displacements (<0.1 μm) are deposited efficiently in peripheral airways.

Tidal volume is an important respiratory parameter. The air inhaled at the start of each breath goes deeper into the lung and remains there longer than the air inhaled later in the breath. The deeper the air goes and the longer it stays, the greater its depletion of inhaled particles. During quiescent breathing, when the air velocity is low, mixing is minimal, and the tidal volume is only 2 to 3 times the dead space volume, a large portion of the inhaled particles can be exhaled.

Anatomical Factors

Intrasubject variations in airway anatomy affect particle deposition in several ways:

The diameter of the airway influences the displacement required by the particle before it contacts the airway surface.

The cross section of the airway determines the flow velocity for a given volumetric flow rate. Velocity affects particle deposition, as discussed under Respiratory Factors (above).

The variations in diameter and branching patterns along the bronchial tree affect the mixing characteristics between the tidal and reserve air in the lungs. Such convective mixing can be the single most important factor determining deposition efficiency of particles with aerodynamic diameters of less than ~2 μm.

There are also significant intersubject differences in respiratory tract anatomy. For example, the average dimension of alveolar zone air space has a substantial coefficient of variation when measured either post-mortem on lung sections or *in vivo* by aerosol persistence during breath holding. In the former case, Matsuba and Thurlbeck[513] reported a mean size and variation of 0.678 mm ± 0.236; in the latter, Lapp *et al.*[438] found values of 0.535 mm ± 0.211.

Physiologic Factors

The effective diameters of the conductive airways for airflow are defined by the surface of the mucous layer. In healthy subjects, where the mucous layer on the conductive airways is believed to be approximately 5 μm thick,[180] the reduction in air path cross section by the mucus is negligible. In individuals with bronchitis, the mucous layer can be much thicker; it can accumulate and partially or completely occlude some of the airway. Air flowing through partially occluded airways forms jets. This probably increases small airway particle deposition by impaction and turbulent diffusion.

An important function of the conductive airways is to warm and humidify the inspired air. Air inhaled through the nose is already saturated at body temperature before it reaches the pharynx. Air inhaled through the mouth may not reach saturation before the major or segmental bronchi. The warming and humidifying of the inspired air within the upper airways will cause hygroscopic particles to grow in size,[539, 615] thereby affecting their regional deposition.

Environmental Factors

There is relatively little information on the effects of natural environmental factors (e.g., temperature and humidity) on particle deposition. While low ambient humidity may dry the tracheal mucus during mouth breathing, it does not appear to affect the mucous transport of particles in the nose in normal nose-breathing humans, even at a relative humidity (RH) level as low as 10% at 23°C.[621] However, ambient humidity can greatly affect the size of many pollutant particles, thereby affecting their total and regional deposition efficiencies.

Most of the atmospheric aerosol in the accumulation mode contains water-soluble acids and/or salts that are formed from gaseous precursors. As discussed in Chapter 4, as much as 50% or more of the atmospheric aerosol mass may be composed of sulfuric acid, ammonium bisulfate, and/or ammonium sulfate. Sulfuric acid and ammonium bisulfate are aqueous solution droplets at all relative humidities from 30% to 100%, while ammonium sulfate undergoes a transition from the dry crystal to a solution droplet at an RH of ~80%.[143, 588] Sodium chloride, potassium chloride, and sodium bromide also undergo sudden transitions from dry crystals to aqueous droplets at ~70%, 75%, and 40% RH, respectively.[163, 588] However, when the humidity then decreases, these droplets retain water at humidities considerably lower than those at which they originally changed to their aqueous state.[143, 163, 588]

Effects of Pollutant Gases and Aerosols

When discussing the deposition on inhaled atmospheric aerosols, the influence of airborne cocontaminants on the lungs of the people inhaling the particles should be considered. Such inhaled irritants can affect the fate and toxicity of the inhaled particles by altering airway caliber, respiratory function, clearance function, and/or the function, survival, and distribution of the cells lining the airways.

A large portion of the overall pollution aerosol, and the dominant portion of the aerosol in the accumulation mode, is formed within the atmosphere from gaseous precursors associated with combustion sources. Thus, any atmosphere containing elevated sulfate and nitrate aerosol concentrations may also contain elevated sulfur dioxide and nitrogen dioxide gas concentrations. Elevated gaseous oxidant is associated with elevated hydrocarbon aerosols.

The effects of pollutant gases and aerosols on lung function are discussed in Chapter 7. Increased pulmonary flow resistance reflects the bronchoconstrictive effects of some pollutants. Any reduction in airway cross section in the larger bronchial airways results in increased flow velocities. This should increase particle deposition by impaction. Increased bronchial deposition of inhaled particles was observed in two healthy young men who had been exposed to sulfur dioxide. One subject, exposed for 7 min to 34 mg/m^3 (13 ppm), showed decreases (from 10% to 2%) in alveolar zone deposition of 4.6-μm aerodynamic diameter particles. The exposure also produced a marked proximal shift in the bronchial deposition pattern. The second subject, exposed for 6 min to 31 mg/m^3 (12 ppm), showed 18% to 4% decreases in alveolar deposition of 5.9 μm diam particles. Tests on two other healthy subjects at exposures of 13 and 24 mg/m^3 (5 and 9 ppm) did not produce any significant shift in regional deposition.[474]

Effects of Chronic Lung Disease

While some healthy cigarette smokers have increased bronchial deposition, the increase is small compared to that in individuals with clinically defined chronic bronchitis.[473] Greatly increased tracheobronchial particle deposition has also been seen in some asymptomatic asthmatics[473] who did not smoke.

Among 58 working coal miners, total respiratory tract deposition of 1 μm particles was significantly correlated with lung function measurements characterizing airway obstruction. The increased deposition was observed among smokers in the group. The presence of simple pneumoconiosis was not associated with the degree of aerosol deposition.[488]

EXPERIMENTAL DEPOSITION DATA

Total Deposition

Few measurements of regional particle deposition in humans have been attempted. In most studies, total deposition has been explored. For particles between ~0.1 and 2 μm aerodynamic diameter, deposition in the conductive airways is generally small compared to deposition in the alveolar regions. Thus, total deposition is close to that of alveolar deposition alone. Total deposition as a function of particle size and respiratory parameters has been measured experimentally by numerous investigators. Many reviews of deposition studies have called attention to the very large difference in the reported results.[185, 319, 470, 741, 749]

Much of the discrepancy can be attributed to uncontrolled experimental variables and poor laboratory technique. The major sources of error have been described by Davies.[186] Figure 6-2 shows data from studies that were performed with good techniques and precision. All were done with mouth breathing at respiration frequencies of from 12 to 16 breaths/min. Tidal volumes varied from 0.5 to 1.5 liters. All appear to show the same trend with a minimum of deposition at ~0.5 μm diam. All four studies in which di(2-ethylhexyl)sebacate (DES) was used appear to have somewhat lower absolute values.

In most studies involving more than one subject, there was considerable individual variation among the subjects. Davies *et al.*[187] showed that some of this variation could be eliminated by standardizing the expiratory reserve volume (ERV) and thereby the size of the air spaces. They found that deposition decreases as ERV increases. This was confirmed by Heyder *et al.*,[326, 327] who reported little intrasubject variation among six subjects when their deposition tests were performed at their normal ERV's.

The data of Heyder and his colleagues appear to represent minimum deposition for healthy men. Their test procedures were precisely controlled. There were no electrical charges on particles. With more natural aerosol and respiratory parameters, higher deposition efficiencies would

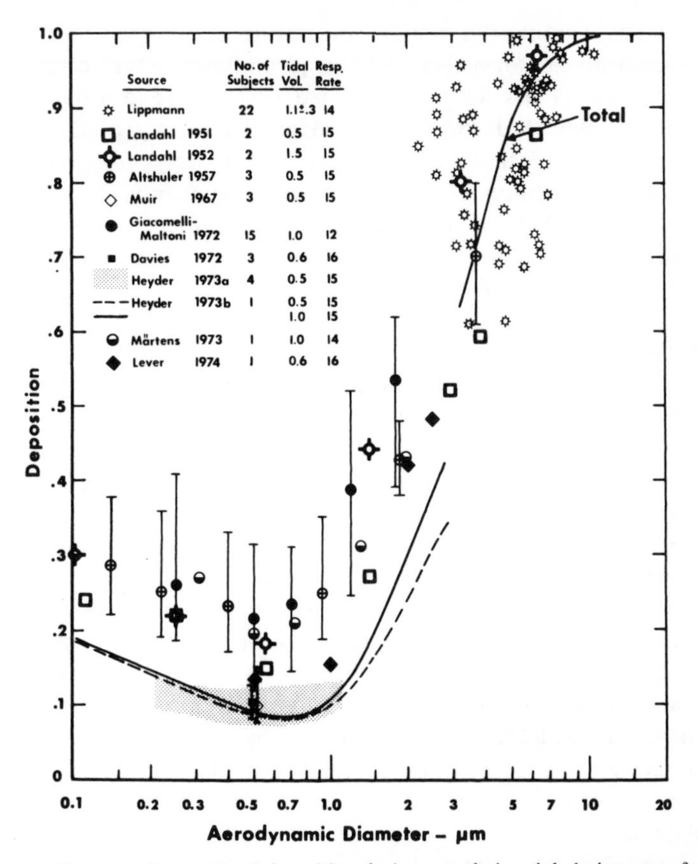

Figure 6-2. Total respiratory tract deposition during mouthpiece inhalations as a function of D (aerodynamic diameter in μm) except below 0.5 μm, where deposition is plotted versus linear diameter. Data of Lippmann[471] and Lippmann and Altshuler[474] are plotted as individual tests, with eye-fit average line. Other data on multiple subjects are shown with average and range of individual tests. Monodisperse test aerosols used were ferric oxide,[471] triphenyl phosphate,[17, 435, 436] carnuba wax,[269] polystyrene latex,[507] and di(2-ethylhexyl) sebacate.[187, 327, 381, 562]

be expected. The data of Landahl *et al.*,[435, 436] Altshuler *et al.*,[18] Giacomelli-Maltoni *et al.*,[269] and Märtens and Jacobi,[507] which are higher and agree quite well with each other, provide the best available data for total deposition in normal humans.

The deposition data in Figure 6-2 were based on the difference between inhaled and exhaled particle concentrations, except for Lippmann's data,[471] which are based on external *in vivo* measurements of γ-tagged particle retention. Most of his tests were performed with particles larger than 2.5 μm diam, while smaller particles were used in most of the other studies. Despite the minimal overlap in size range, the two sets of data appear to be consistent. The large amount of scatter among the individual data points for the larger particles is due to a quite variable deposition in the head and tracheobronchial tree (see Figures 6-3, 6-4, and 6-5). Figure 6-3 shows data for nonsmokers only. Cigarette smokers have a similar median behavior for head deposition, but even more scatter. Figure 6-5 shows that the median and upper limits of tracheobronchial deposition are higher for cigarette smokers than for nonsmokers, but the lower limit is about the same.

Head Deposition

Some inhaled particles are deposited within the air passages between the point of entry at the lips or nares and the larynx. The fraction deposited is highly variable; it depends on the route of entry, particle size, and flow rate. The nasal flow path is usually a more efficient particle filter than the oral, especially at low and moderate flow rates. Thus, those people who normally breathe through the mouth either part or all of the time may be expected to have more particles deposited in their lungs than those who always breathe through the nose. During exertion, the flow resistance of the nasal passages causes a shift to mouth breathing in almost all people.

There are very few data on head deposition during mouth breathing. The available data on nonsmoking healthy subjects (Figure 6-3) are based on external scintillation detector measurements of γ-tagged particles deposited in the head immediately after inhalation via a mouthpiece.[471]

Figure 6-4 shows head deposition data for monodisperse aerosols during inhalations via nose masks as a function of D^2F. The solid line is the International Commission on Radiological Protection (ICRP) Task Group model,[749] based on Pattle's data on one subject.[599] These data and the Hounam *et al.*[366] data on three subjects were based on the reduction of airborne particle concentration as constant flows were drawn in through the nose and out of the mouth while the subject held his breath. The data of Lippmann[469] on two subjects, Giacomelli-Maltoni[269] on 15 subjects, Märtens and Jacobi[507] on one subject, and Rudolf and Heyder[653] on four subjects were obtained during normal nasal inhalations, with exhalations via the mouth. Lippmann measured the amount of γ-tagged aerosol actually deposited within the head, while Giacomelli-Maltoni, Märtens and Jacobi, and Rudolf and Heyder estimated nasal passage deposition on the basis of the total deposition difference between paired studies on the same subject; i.e., mouth-in/mouth-out versus nose-in/mouth-out. Deposition in the oral passage is assumed to be negligible in this technique. Because oral passage deposition values are not negligible

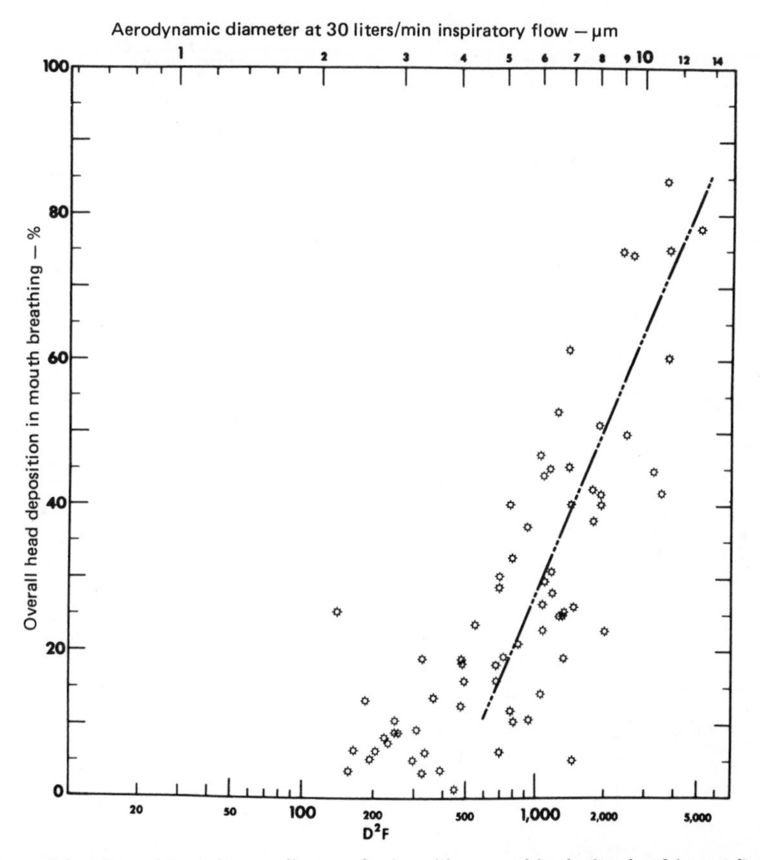

Figure 6-3. Deposition of monodisperse ferric oxide aerosol in the heads of (normal) nonsmoking human males during mouthpiece inhalations as a function of D^2F, where F is the average inspiratory flow in liters/min. An eye-fit line describes the median behavior for deposition between 10% and 80%. Total respiratory tract depositions in these tests are shown in Figure 6-2.

for particles $> \sim 2$ μm diam, these nasal deposition values may be too high.

Pattle used methylene blue particles between 1 and 9 μm aerodynamic diameter and constant flow rates of 10, 20, and 30 liters/min. Hounam used 1.8- to 8.0-μm diam particles and constant flows from 5 to 37 liters/min. Lippmann used 1.3- to 3.9-μm diam ferric oxide particles and inspiratory flows averaging ~30 liters/min. Giacomelli-Maltoni used 0.25- to 1.8-μm diam wax particles and average flows of about 24 liters/min. Märtens and Jacobi used 0.3- to 1.9-μm diam latex particles and an average flow rate of ~30 liters/min. Rudolf and Heyder used 0.5- to 3.0-μm diam di(2-ethylhexyl)sebacate particles and constant inspira-

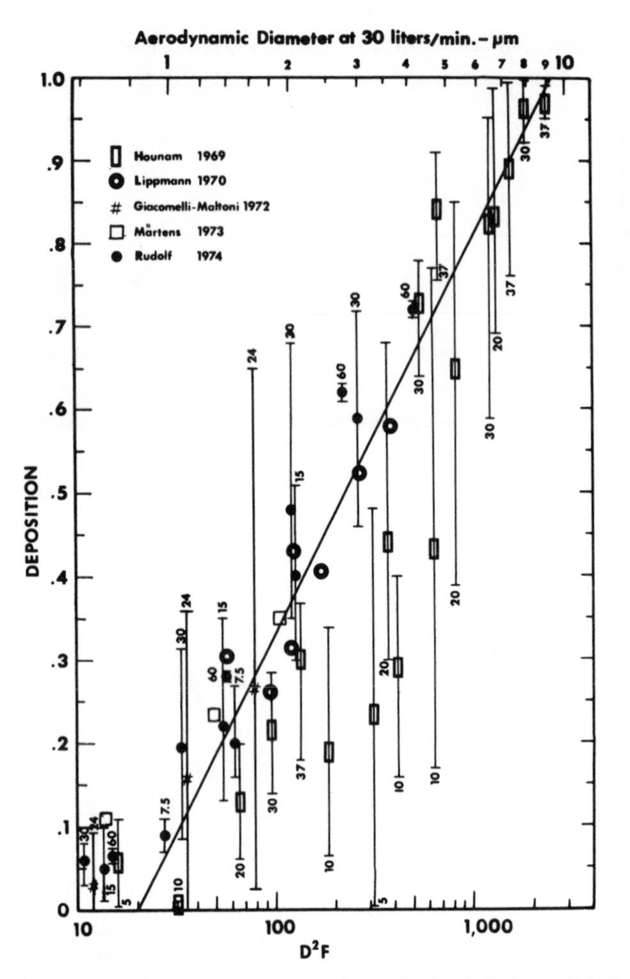

Figure 6-4. Deposition of monodisperse aerosols in the head during inhalation via the nose versus D^2F. The heavy solid line is the International Commission on Radiological Protection Task Group deposition model,[749] which is based on the data of Pattle.[599] For the Hounam *et al.*,[366] Giacomelli-Maltoni *et al.*,[269] and Rudolf and Heyder[653] data, the symbol shows the median value, and the bars show the range of the individual observations. The number at the end of the bar indicates the inspiratory flow rate in liters/min. The monodisperse aerosols used were methylene blue,[599] polystyrene latex,[366, 507] ferric oxide,[469] carnuba wax,[269] and di(2-ethylhexyl)sebacate.[653]

tory flows from 7.5 to 30.0 liters/min. Pattle found that the data from the three different flows could be normalized by plotting D^2F against retention. Hounam found that the best straight line fit to his data was obtained by plotting retention against $D^2F^{1.6}$. In Figure 6-4, his D^2F data at 37 and 30 liters/min lie close to the line, while those at 20, 10, and 5 fall increasingly below the line. The Rudolf and Heyder data show a similar, but lesser trend, with their 60 and 30 liter/min data tending to lie above the line while the 15.0 and 7.5 liter/min data straddle the line. A better straight line fit can be obtained through the Rudolf and Heyder data by plotting $D^2F^{1.3}$ versus retention.

In the Hounam *et al.*, Giacomelli-Maltoni, and Rudolf and Heyder data, most of the scatter is due to intersubject variability. Hounam *et al.*[366] also measured the pressure drop across the nose and mouth during each deposition study and plotted the data in terms of percent deposition versus D^2R, where R is the resistance across the nose and mouth in millimeters of water. The same data on this plot exhibited less scatter than the D^2F plot, with the greatest improvement at large values of percent deposition. The fact that particle deposition correlates better with D^2R than D^2F indicates that the air path dimensions are variable. Because the air paths are distensible, their dimensions can be expected to vary with flow rate in a given subject. Perhaps this accounts for the observation that a flow factor exponent higher than 1 helps to normalize the data.

Rudolf and Heyder[653] determined head deposition efficiency for aerosols exhaled through the nose to be essentially the same as that measured during inhalation for a flow rate of 7.5 liters/min. At 15 and 30 liters/min, the efficiency for 1- to 3-μm diam particles during exhalation was approximately 25% higher than during inhalation. However, at 60 liters/min, the exhalation efficiency was about 7% lower. Deposition efficiencies are similar during inhalation and exhalation; however, the amount deposited during exhalation is usually much lower because the aerosol concentration is depleted while passing through the nasal passages and then into and out of the lungs.

Deposition in the Tracheobronchial (T-B) Zone

The only measurements of tracheobronchial deposition in the literature are those of Lippmann and his colleagues[471, 472, 473] (Figure 6-5). T-B deposition for a given particle size varies greatly among individual nonsmokers, cigarette smokers, and patients with lung disease. The average T-B deposition is slightly elevated in smokers, and greatly elevated in the patients with lung disease. Healthy nonsmokers and nonbronchitic smokers exhibit individual particle size-versus-deposition relationships. T-B deposition includes deposition both by impaction in the larger airways and be sedimentation in the smaller airways. Impaction deposition predominates for large particles (>3 μm diam) and high flow rates (~ 20 liters/min). T-B deposition of smaller particles is slight. The small amount that occurs may be attributable to either sedimentation or impaction.

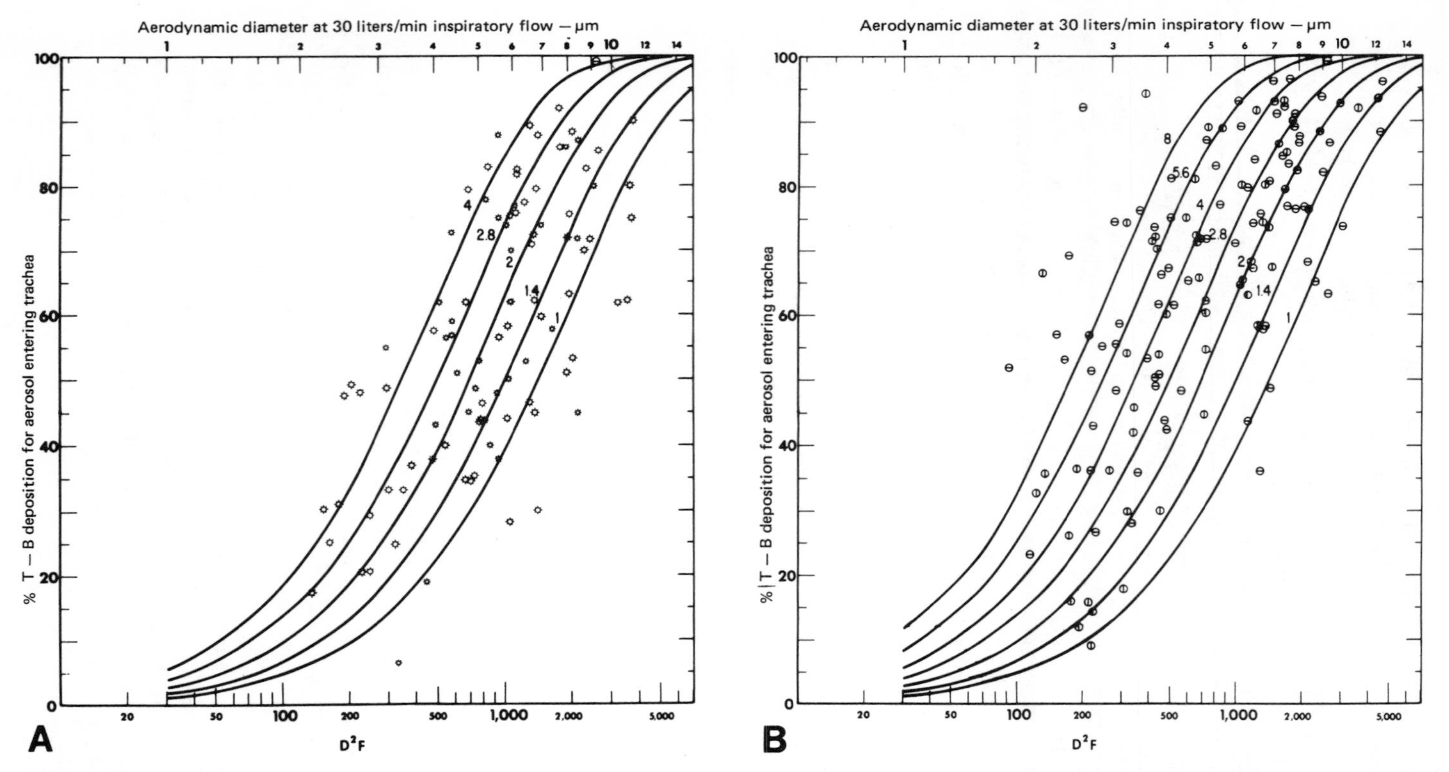

Figure 6-5. Deposition in the ciliated tracheobronchial (T-B) region during mouthpiece breathing, in percent of the aerosol entering the trachea. Panel A shows data for nonsmoking healthy human males, while Panel B contains data for cigarette smokers. The curves represent the change in T-B deposition as a function of D^2F for different values of the characteristic airway dimension parameter developed by Palmes and Lippmann.[595] A comparison of the two panels demonstrates that many cigarette smokers have increased T-B deposition. The nonsmokers' data are from the same tests for which total respiratory tract depositions are shown in Figure 6-2 and head deposition in Figure 6-3.

Deposition in the Alveolar Zone

Landahl *et al.*[435, 436] conducted mouth-breathing experiments with monodisperse aerosols of triphenyl phosphate (0.11 to 6.3 μm diam). The exhaled air was separated into four sequential components. The residual airborne particle burdens in each were presumed to have come from different lung regions. The results obtained were compared to predicted values for the same breathing patterns obtained from Landahl's[433] previous theoretical calculations. The agreement was sufficiently good for the authors to conclude that the predictions were verified.

Brown *et al.*[91] studied regional deposition during nose breathing using classified china clay aerosols with count medium diameters between 0.9 and 6.5 μm (σ_g = 1.25). They collected the exhaled air in seven sequential components, and used the carbon dioxide content of each fraction as a tracer to identify the region from which the exhaled air originated. The validity of these data depends on the accuracy of the association between the various exhaled air fractions and their presumed sources. The Brown data have been criticized[17] on the basis that their simple two-filter model was inadequate and that the rapid diffusion of carbon dioxide caused the measured carbon dioxide values to differ significantly from the corresponding concentration in the alveolar spaces. The resulting error in the volume partitioning caused an underestimation of the alveolar deposition.

Altshuler *et al.*[17] estimated regional deposition from mouth-breathing experiments on three subjects. They measured continuously, during individual breaths,[18] both the concentration of a monodisperse triphenyl phosphate aerosol and the respiratory flow. Using a tubular continuous filter bed model as a theoretical analog for the respiratory tract, regional deposition in the upper and lower tract components was calculated for various values of anatomic dead space. The upper tract penetration during inspiration, pause, and expiration was derived from the expired aerosol concentration corrected for aerosol mixing. The alveolar deposition estimates varied from subject to subject, and for each subject they varied with the volume of anatomic dead space. The particle size for maximum alveolar deposition was estimated to be >2 μm diam. The exact value could not be estimated, however, because only one particle of this size was used in the experiments.

The alveolar deposition values of Altshuler *et al.*, based on their best estimates of dead space, are plotted on Figure 6-6. This figure also shows alveolar deposition values obtained by Lippmann during his mouth-breathing inhalation tests on nonsmoking, healthy subjects. These data are based on external measurements of the retention of γ-tagged particles after the completion of bronchial clearance. The Lippmann and Altshuler data, in the region of particle size overlap, are in good agreement.

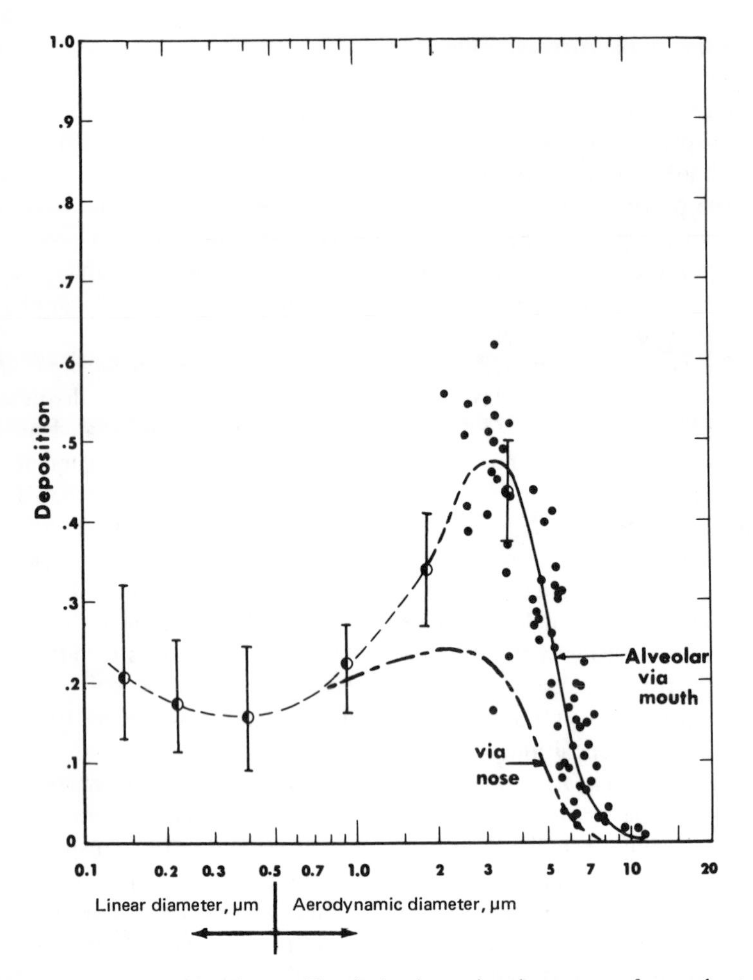

Figure 6-6. Deposition in the nonciliated alveolar region, by percent of aerosol entering the mouthpiece, as a function of diameter. Individual data points and eye-fit solid line are for the same ferric oxide aerosol tests plotted in Figures 6-2, 6-3, and the left panel of Figure 6-5. The dashed line is an eye-fit through the median best estimates of Altshuler *et al.*,[17] based on their 1957 studies[18] of three subjects whose average deposition range is indicated by the vertical lines. The lower curve is an estimate of alveolar deposition during nose breathing. It is based on the difference in head deposition shown in Figures 6-3 and 6-4.

Figure 6-6 also shows an estimate of the alveolar deposition when the aerosol is inhaled via the nose. It is based on the difference in head retention during nose breathing and mouth breathing from the straight line relations plotted on Figures 6-3 and 6-4. For mouth breathing, the particle size for maximum alveolar deposition is ~3 μm diam. Approximately half of the inhaled aerosol of this size deposits in this region. For nose breathing, there is a much less pronounced maximum of

~25% at 2.5 μm diam, with a nearly constant alveolar deposition averaging 20% for all sizes between 0.1 and 4.0 μm diam.

PREDICTIVE DEPOSITION MODELS

Mathematical models for predicting the regional deposition of aerosols have been developed by Findeisen[230] in 1935, Landahl in 1950[433] and 1963,[434] and by Beeckmans[60] in 1965. Findeisen's simplified anatomy, with nine sequential regions from the trachea to the alveoli, and his impaction and sedimentation deposition equations, were used in the ICRP Task Group's 1966 model.[749] For diffusional deposition, the Task Group used the Gormley-Kennedy[289] equations; for head deposition, it assumed entry through the nose with a deposition efficiency given by Pattle's empirical equation.[599] A comparison between the various predictions and the experimental data indicates that for both total and alveolar deposition, Landahl's model comes closest, but it overestimates alveolar deposition for particles with aerodynamic diameters larger than 3.5 μm.

The ICRP Task Group's 1966[749] model was adopted by ICRP Committee II in 1973, with numerical changes in some clearance constants. The 1966 Task Group report has been widely quoted and used within the health physics field. A major conclusion of this report was that the regional deposition within the respiratory tract can be estimated using a single aerosol parameter, the mass median aerodynamic diameter (MMAD) (Figure 6-7A). For a tidal volume of 1,450 ml, there are relatively small differences in estimated deposition over a very wide range of geometric standard deviations ($1.2 < \sigma_g < 4.5$). The Task Group report contains calculations of the mean deposition curves for 750 and 2,150 ml tidal volumes (Figure 6-7B).

None of the previously proposed models provide reliable estimates of aerosol deposition in healthy normal adults, because their predictions for total and alveolar deposition efficiencies differ from the best experimental data (Figures 6-2 and 6-6). Furthermore, they do not take into account the very large variability in deposition efficiencies among healthy subjects. The effect of intersubject variability on total respiratory tract deposition is illustrated in Figure 6-8. Here, Palmes and Lippmann[595] have used the experimental data of Giacomelli-Maltoni *et al.*[269] to construct curves showing deposition by percentiles of the overall population. Within this particle size range, deposition for the top 2% of the population is from 34% to 54% higher than for the median subject. Other variations in deposition are produced by cigarette smoke (Figure 6-5) and lung disease.[14, 474] There have been significant advances in the measurement of deposition in recent years; however, considerable effort is underway to improve theoretical understanding and predictive models.

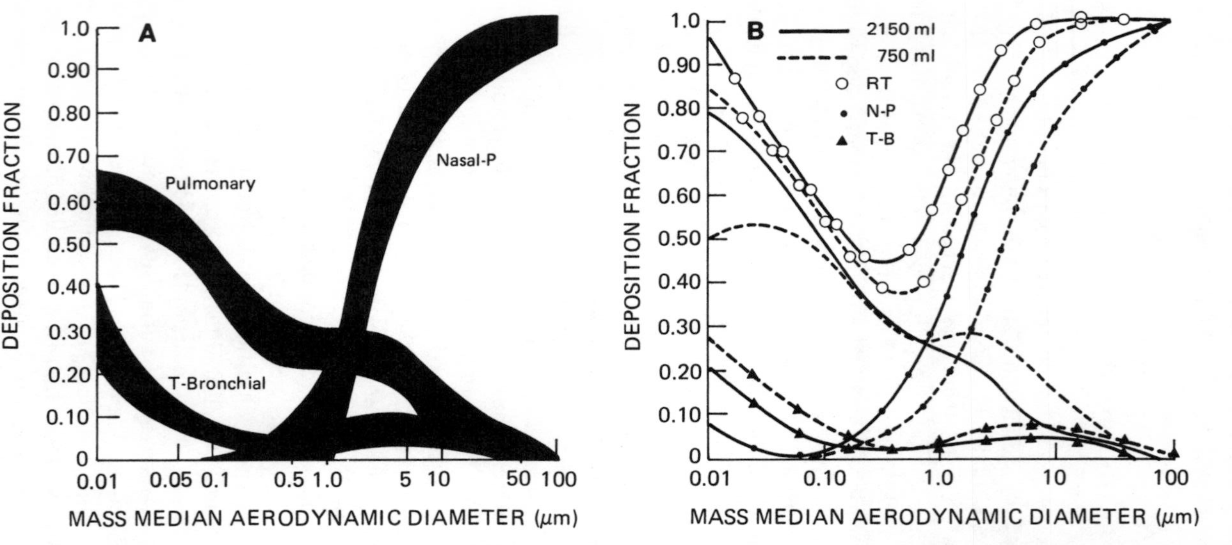

Figure 6-7. Regional deposition predictions based on model proposed by the International Commission on Radiological Protection Committee II Task Group on Lung Dynamics[749] indicating effect of variations in σ_g and flow rate. (A) Each of the shaded areas (envelopes) indicates the variable deposition for a given mass median (aerodynamic) diameter in each compartment when the distribution parameter σ_g varies from 1.2 to 4.5 and the tidal volume is 1,450 ml. (B) Two ventilatory states, 750 and 2,150 ml tidal volume (~11 and ~32 liter/min volumes, respectively), are used to indicate the order and direction of change in compartmental deposition that are induced by such factors. Reprinted with permission of Pergamon Press, Ltd., and Task Group Chairman. From *Health Physics* 12:173–207, 1966.[749]

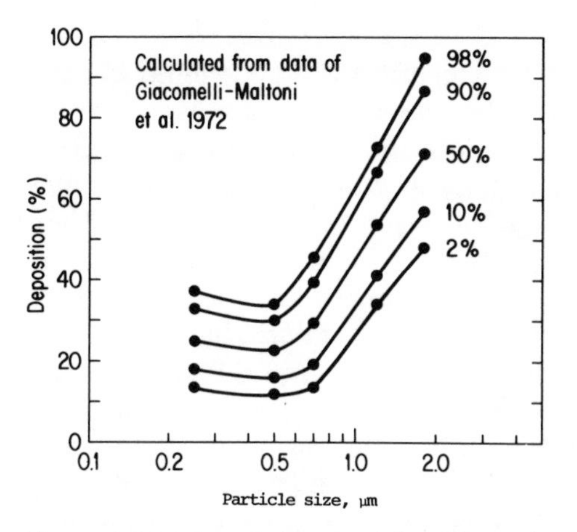

Figure 6-8. Range of expected total respiratory tract deposition values based on experimental data of Giacomelli-Maltoni *et al.*[269] Each curve represents the indicated percentile of the overall population. From Palmes and Lippmann,[595] with permission from Pergamon Press.

DEPOSITION OF AMBIENT ATMOSPHERIC AEROSOL

As discussed in Chapter 4, the mass or volume distribution of the ambient aerosol as a function of particle size generally has two distinct modes. One mode results from the formation of small particles in the atmosphere and their growth by coagulation. These particles vary from ~0.01 to 1.5 μm diam. The second mode is created by particles injected into the atmosphere by mechanical forces. Most of the mass of these particles is contributed by particles larger than ~3 μm diam. The upper size limit is quite variable, increasing as atmospheric turbulence increases.

Because total and regional deposition depend on particle size, changes in size due to droplet growth can significantly change deposition pattern and efficiency. However, for the atmospheric accumulation aerosol, it is not clear that major shifts will occur. When the humidity is low, the accumulation mode is generally 0.3 μm MMAD or less. Thus, even if the particle size were to increase by a factor of 3 to 5, the total deposition efficiency would remain close to 20% (Figure 6-6), and most of the deposition would still occur within the alveolar region. On the other hand, if a hygroscopic aerosol having a dry size of 1 μm diam or larger is released into the atmosphere, its growth in the atmosphere[143, 163, 588, 825, 826] or in the respiratory tract[539, 615] could greatly affect its regional deposition. The amount deposited in the head and T-B zones increases rapidly with increasing droplet size (see Figures 6-4 and 6-5).

Wherever possible, consideration should be given to the particle size distributions for specific toxic constituents of the ambient aerosol, such as airborne lead and particulate polycyclic organic matter, which were the subjects of two reports by the Committee on Medical and Biologic Effects of Environmental Pollutants of the National Research Council.[571, 572] Natusch *et al.*[575] measured the trace element concentrations in fly ash samples of different particle size ranges, with diameters from an upper limit of 74 μm to less than 1 μm. They reported that the percentage of arsenic, nickel, cadmium, and sulfur in the particles increased as particle size decreased. They attributed this to a volatilization-condensation process within the furnace and stack. The MMAD's of ambient airborne lead and aerosols of other metallic elements were measured in six U.S. cities by Lee *et al.*[457] Annual MMAD's ranged from a low of 0.4 to 0.7 μm for lead and vanadium to a high of 2.4 to 3.5 μm for iron. The other metals measured had median diameters of intermediate sizes. Similar measurements of vanadium and iron aerosol MMAD's have been reported by Gladney *et al.*[276] for Boston, which was not one of the six cities studied by Lee and his colleagues. This later study did not include lead, but did include many trace metals not measured by Lee *et al.* The small size of the particles containing lead has also been noted by Harrison *et al.*[318] and Bernstein *et al.*[70]

The size distributions of several polynuclear aromatic hydrocarbons were measured in and around Toronto, Canada by Pierce and Katz.[610] MMAD's ranged from ~0.5 μm near a rural freeway to ~3.0 μm at suburban York University. Measurements also indicated seasonal variations. For example, at the York University site, the average summer concentration of benzo[*a*]pyrene was 12.6 ± 0.7 μg/g particulate, the MMAD was 2.4 μm, and the σ_g was 2.2. The corresponding winter values were 17.1 ± 0.9 μg/g particulate, 1.5 μm MMAD, and 2.9 σ_g. The only other report on the particle size of a polycyclic aromatic hydrocarbon is that of Kertész-Sáringer *et al.*,[413] who found an MMAD of 0.3 μm for benzo[*a*]pyrene in Budapest, Hungary.

PARTICLE RETENTION

Particle retention is a time-dependent variable that is equal to the difference between the amount of aerosol deposited and the amount cleared. Rates of clearance vary greatly among the regions of the respiratory tract. This variation was a major basis for the characterization of the functional zones described above. In two of these zones, the ciliated nasal passages and the tracheobronchial tree, clearance in healthy individuals is completed in less than one day. In the alveolar zone, clearance proceeds by slower processes, most of which vary considerably with the composition of the particles (see section on ALVEOLAR CLEARANCE, page 140).

MUCOCILIARY CLEARANCE

The mucociliary system of the conducting airways collects, transports, and eventually eliminates inhaled particles. The ciliated epithelial lining of the nasal passages and tracheobronchial tree is covered by a layer of mucus, which is a mixture of material from several sources. The coordinated beating of the cilia, which extend to the respiratory bronchioles, results in the movement of this fluid from distal areas toward the pharynx, where it is swallowed.

Sources of Respiratory Tract Mucus

The respiratory tract mucus is derived from mucous glands, goblet cells, Clara cells,[151, 210] bronchial transudate, and fluids drawn from the respiratory bronchioles. The total volume secreted by the mucous glands is ~40 times that secreted by the goblet cells in the epithelium of the tracheobronchial tree of a healthy adult human.[637] Thus, these glands are thought to be the primary source of the viscous mucus.

The goblet cells are part of the respiratory tract epithelium; they are more numerous in the larger proximal airways than in the more distal branches.[773] Mucous glands are located predominantly in the submucosa of the cartilage-containing airways, and often lie between and external to the cartilage plates.[638] The glands contain both serous cells and mucous cells, whose secretions are discharged onto the ciliated epithelium through a common duct.[536]

Sensory and secretomotor nerve endings have been observed among the mucous and serous cells of human tracheobronchial glands.[65, 568, 710] Studies with laboratory animals[246] have provided physiological evidence that these glands are innervated and secrete under nervous control. Not only the quantity, but also the character of the secretions may be under such control; a change in the consistency of bronchial secretions from tenacious to thin was observed in humans following bronchopulmonary denervation.[87] Secretion seems to be stimulated cholinergically through a system mediated by the vagus nerve.[246] Atropine reduces the secretory response of the gland cells,[53, 742, 743] while parasympathomimetic stimulation causes both mucous and serous cells of the glands to secrete.[130, 742, 743, 834]

Goblet cells do not appear to discharge by autonomic nerve stimulation. Their secretion is not affected by atropine.[53] No difference in the character or number of goblet cells in denervated and normal tracheobronchial trees has been observed.[280] These cells seem to have a baseline of activity that may be altered by local stimulation, such as irritation.[246, 521]

In the bronchiolar airways there are a few goblet cells but no mucous glands. The Clara cells found in these airways are secretory.[174]

Ebert and Terracio[210] have recently summarized the work on Clara cells. In the airways, these cells probably produce a protein secretion that forms the sol layer in which the cilia beat.[415] Inflammation and various immunologic stimuli may result in the addition to these secretions of mast cells, plasma cells, and other cell types normally found under the tracheobronchial epithelium.[712]

The lining of the tracheobronchial tree of rats contains a small number of brush cells.[643] Similar cells have been found more rarely in human tracheobronchial epithelium.[642] The exact role of these cells is unknown. Some investigators have suggested that they are discharged goblet cells[710] or that they have a role in fluid reabsorption.[642]

Volume, Composition, and Rheology of Respiratory Tract Mucus

Estimates of daily volume production of tracheobronchial secretions in humans range from 100 cm^3, based on inferences from animal studies,[613] to 10 cm^3, based on actual work with tracheotomized adult humans.[758]

Pure secretions from the normal human bronchial tree are difficult to collect. Most studies to determine the biochemical composition of the secretions have been conducted using expectorated sputum, often from chronic bronchitis patients, who produce more sputum than normal subjects. However, Matthews *et al.*[515] have determined the chemical composition of tracheobronchial secretions from "normal" laryngectomized patients. The composition, per 100 g wet weight, was found to be as follows: water, 94.79; protein, 1.000; carbohydrate, 0.951; ash, 1.13; DNA, 0.028; and lipid, 0.840. Most likely, the DNA and lipids are derived from disrupted cells, such as degenerated bronchial epithelial cells.

The major macromolecular components of tracheobronchial secretions are glycoproteins. These long chain molecules consist of small polysaccharide units glycosidically linked to a protein core.[272] The polysaccharides constitute about 40% of the molecular weight of the glycoprotein. The basic polysaccharide unit in human bronchial glycoproteins is composed of galactose, *N*-acetylglucosamine, and *N*-acetylgalactosamine. This unit is attached to the polypeptide chain. Superimposed upon this basic structure are fucose and either *N*-acetylneuraminic acid, sulfate groups, or both.[46] The relative amounts of these three moieties serve as a basis for the classification of tracheobronchial glycoproteins into three specific groups: fucomucins, sulfomucins, and sialomucins. The fucomucins are neutral or very weakly acidic glycoproteins characterized by a richness in fucose. Both the sulfomucins and sialomucins are acidic glycoproteins, the former being rich in sulfate and the latter characterized by a high amount of *N*-acetylneuraminic acid. Histochemical analysis of mucous glands has resulted in the further separation of the sulfomucins and sialomucins into two groups each;[639]

these four types of acidic glycoproteins display a specific pattern of distribution within mucus-secreting cells.[430]

Other macromolecular constituents of tracheobronchial secretions are immunoglobulins (primarily IgA produced locally in the mucosa),[408, 511] lactoferrin, lysozyme, and kallikrein.[652]

Normally, tracheobronchial secretions (350 milliosmol) are moderately hyperosmotic to serum (290 milliosmol). An average electrolyte composition (in mg/1,000 g of water-soluble material) is sodium, 211; chloride, 157; potassium, 16.6; and calcium, 2.45.[617]

Tracheobronchial secretions exhibit anomalous flow characteristics and do not behave like simple fluids, such as water. Rather, mucus exhibits the rheological phenomenon of viscoelasticity, which implies both fluidlike and solidlike behavior. This viscoelastic behavior is probably caused by the constituent glycoprotein molecules.[476] Keiser-Nielsen[409] demonstrated that mucous glycoproteins exhibit elastic properties under rapid stretch conditions and viscous properties under slow stretch. These phenomena are characteristic of viscoelastic materials. Tracheobronchial secretions are very shear-sensitive. That is, a reduction of viscosity occurs with increasing shear rates. This type of behavior is known as shear rate thinning or thixotropy. It may be due to an alignment of the glycoprotein molecules under the shear force with the result that the viscous resistance between adjacent layers of molecules decreases.[594]

Respiratory Cilia

The beating cycle of each respiratory cilium consists of a fast, rigid effective stroke in one direction and a slower, limp recovery stroke in the opposite direction. The effective stroke in the mammalian respiratory tract takes from one-third to one-fifth of the time required for a complete cycle.[47, 339, 541] The beat frequency of mammalian respiratory tract cilia has been estimated to be between 700 to 1,200 beats/min.[34, 339, 541]

Rows of cilia do not beat synchronously. Rather, each cilium beats shortly after the cilium in front of it, and before the cilium behind it. Because of this metachronous rhythm, the overall ciliary activity creates an impression of a wave traveling over the epithelial surface in a direction opposite to that of the effective stroke of the cilia.

The exact mechanism controlling ciliary rhythm and coordination in mammals is not known. One postulated mechanism is local chemical control,[103] similar to the action of serotonin as a ciliary pacemaker in mollusks.[290] Some investigators have suggested coordination by mechanical coupling,[705] in which rhythmical contact between adjacent cilia maintains a constant beat frequency. A neuroid control mechanism has also been proposed.[704, 705] In this mechanism, a buildup of excitation in a pacemaker cilium determines the frequency of beat. The other cilia beat at constant intervals thereafter as a result of conducted impulses. However, this innervation of cilia in vertebrates has been demonstrated only in the frog palate.[290]

The energy for ciliary beating is supplied by the dephosphorylation of adenosine triphosphate (ATP). The bending of the cilia is caused by the sliding of filaments against one another.[664]

Mucociliary Interaction

In their studies of ciliary activity and nasal clearance in mammals, Lucas and Douglas[492] noticed that under some conditions, while the cilia were vigorously beating, the overlying secretory blanket did not move. They also observed that a particle in the fluid surrounding the cilia could be propelled beneath a stationary top fluid layer. They suggested that the total secretory blanket was actually composed of two parts: a layer of low density, serous-type fluid (sol), which floated on top of the watery fluid and rested on the tips of the cilia. In this model, flow results from the application of shearing forces on the top fluid layer. More recent studies involving a mechanical analogue[49, 540-542] of the mucociliary system indicated that optimum fluid flow occurs only when a viscoelastic (gel) fluid layer lies above the tips of the beating cilia.

The source of the fluid in the sol layer is unknown. Suggestions include serous cells,[492] intercellular fluid that oozes onto the epithelial surface,[492] alveoli,[414] brush cells,[526] bronchioles,[273] and Clara cells.[210, 415] Negus[576] believes that both the sol and gel layers may have the same sources and that the two distinct layers are formed when the pressure exerted by ciliary action on the gel mucus causes liquefaction of the layer in which the cilia beat.

Specific interplay between the cilia and the mucus blanket seems to be involved in effective mucus clearance and particulate transport because ciliary beating alone may not be enough. For example, although beating continues for long periods after the death of laboratory animals under various adverse conditions,[180] particulate transport by the mucociliary system becomes irregular and may even stop.[338]

The production of effective fluid movement requires the "coupling" of cilia to the overlying mucus layer. This may occur through the elastic properties of the gel layer.[475] It is effectively coupled with the sol layer, whose flow is induced by the beating cilia.[650] Computations from a mathematical model of the mammalian mucociliary system[73, 651] indicate that, although changes in the serous (sol) fluid viscosity would significantly affect transport rate, moderate changes in mucus (gel) viscosity would not.

Sadé *et al.*[658] have suggested that there is a minimum amount of mucus necessary for transport. Under conditions of mucus depletion, particulate clearance did not occur but the cilia continued to beat. This lack of particulate clearance was attributed to the failure of the mucous secretions to get above the tips of the cilia. The renewed particulate clearance, which occurred upon the addition of more mucus to the system, indicated the need for this "coupling" mucus to be present above the ciliary tips. Sadé and his colleagues tried various solutions as substitutes for mucus on the mucus-depleted, ciliated palates of frogs and toads, but no particulate transport occurred. However, in further investigations, they identified some synthetic materials that restore effective particle transport.[274, 416] The requirement for mucociliary transport is probably rheological, rather than chemical. The substances that were chemically dissimilar to mucus and also capable of sustaining transport were all lightly cross-linked macromolecular systems like the mucus glycoproteins.

A possible further explanation for the specificity of mucus has been proposed by Sadé and Eliezer.[657] The presence of very high molecular weight substances in the small interciliary spaces might impede the ciliary rhythm because of viscous forces. No viscous mucus in the interciliary spaces of the mucociliary system of the ear[657] was observed. Microscopic examination revealed that mucus passed from the goblet cells directly to the overlying mucus blanket in narrow streams.[657] This has also been observed at the mouths of mucous glands.[299]

Mucociliary Clearance Rates

The rate of mucus transport is dependent on both the motility of the respiratory cilia and the physical and chemical properties of the mucus. Various *in vitro* and *in vivo* techniques have been used to determine mucus transport rates in many species. The reported transport rates differ greatly. Much of this variability can be attributed to experimental artifacts. In *in vitro* preparations, it is unlikely that the thickness or composition of the mucus layer can be maintained at normal physiologic levels. Most *in vivo* animal preparations and some human studies are also suspect, because they involved the use of anesthetics and/or other drugs that could affect mucociliary transport.

In vitro preparations of the lining of the buccopharyngeal mucosa of frogs have shown mean transit rates ranging from ~21 to 40 mm/min.[80] Mean rates for other species are listed in Table 6-1. Iravani and Van As[381] found the mean mucus transport rates in *in vitro* preparations of rat airways to increase from ~0.4 mm/min in preterminal bronchioles to 11.5 mm/min in the upper part of the lobar bronchi. Using excised dog lungs, Asmundsson and Kilburn[41] found mucus flow rates to average 17.6 mm/min in the trachea, 8.3 in the lobar bronchi, 4.0 in the segmental bronchi, and 1.6 in distal airways. Andersen[34] reports that the mean tracheal transport rate in the rabbit is 2.5 mm/min, on the basis of both *in*

Table 6-1. Mean transport rates of *in vitro* tracheal preparations

Animal	Rate (mm/min)	Reference
Cat	36.1[a]	412
	10.0	48
Cattle	15.0	40
Chicken	33.1	412
Dog	36.1	412
Monkey	21.0	412
Rabbit	24.6	412
	3.1	66
Rat	20.9[a]	412

[a] The discrepancies in the rates for cats and rats are real and have not been adequately explained. Such differences are not necessarily surprising, based on the discussion in the text.

vitro and *in vivo* measurements; however, his *in vivo* measurements were complicated by the fact that the tracheal segment measured was separated from the more distal airways.

In vivo mucus transport rates in the lungs of various species have also been studied. Transport rates averaged 22 mm/min on the major longitudinal ridge of the bullfrog lung.[414] Barclay and Franklin[48] found the average transport rate to be 2.5 mm/min in the small bronchioles and 10 mm/min in the tracheas of rats. Table 6-2 compares *in vivo* transport rates for several species.

Morrow *et al.*[557] estimated that mean transport rates in humans were 14 mm/min in the trachea, 1 mm/min in the upper bronchial tree, and 0.4 to 0.6 mm/min in the lower tree. At these rates, bronchial clearance would be completed within approximately 3 hr. Mucus transport in the trachea of 42 unanesthetized, healthy nonsmoking adults was studied by Yeates *et al.*[837] Boluses of radioactive microspheres were transported in the trachea at discrete but constant rates. The coefficient rate of variation in an individual over the period of a few hours was 25%. There was a

Table 6-2. Mean *in vivo* transport rates in the trachea

Animal	Rate (mm/min)	Reference
Cat	15	127
Dog	14	505
Donkey, unanesthetized	12-15	12
Rat	10	48
	22	441
	13.5	180

large interindividual variation: the 95% limits of the geometric mean velocities were 0.8 and 10.8 mm/min (coefficient variation 75%). The geometric mean tracheal transport rate for the population studied was 3.6 mm/min. Repeat studies on 22 subjects showed that the variation within individuals was less than between individuals. Santa Cruz *et al.*[663] reported a mean tracheal transport rate in humans of 21.5 mm/min, but this value may be influenced by the prior application of xylocaine and/or the presence of a bronchoscope for visualization of the surface flow.

Although the *in vivo* transport rate in humans cannot be measured in smaller ciliated airways, the total duration of bronchial clearance can be measured. Albert *et al.*[14] reported that a 90% completion of bronchial clearance among nonsmoking healthy subjects occurred anywhere from 2.5 to 20.0 hr, with an intrasubject variability of 30%. Because particles of 2 μm diam are presumably deposited in more distal ciliated airways than 6 μm particles, they require considerably more clearance time. Therefore, the transport rates appear to decrease distally. Calculations by Morrow *et al.*[557] suggest the rates in terminal ciliated airways to be 100 μm/min, while calculations by Yeates[835] suggest 1 μm/min.

Physiologic Alterations in Mucociliary Transport Rates

Mucociliary transport is controlled by cellular mechanisms in the epithelium and by the sympathetic and parasympathetic nervous systems, as well as by environmental and other factors. Evidence indicates that goblet cells in cats are not under neural control but are stimulated by local irritation.[246] Sympathomimetic drugs have had varied effects on cilia beating rate and mucociliary transport in the animal models used.[42] However, in humans, terbutaline,[122] isoproterenol,[251] epinephrine,[251] and Thll65a (a sympathomimetic bronchodilator)[837] increase transport. Parasympathomimetic drugs also speed mucociliary clearance in humans.[121] Parasympatholytic drugs decrease mucus transport in animals[42] and, for several hours, in humans.[251, 601, 837] They probably also decrease ciliary beating.[42] Reversibility of atropine effects was demonstrated when isoproterenol aerosol was given subsequent to systemic atropine.[251] In studies where adrenergic drugs were administered as aerosols, a small but significant part of the clearance acceleration was attributable to the water in the droplets.[251]

Effects of Pollutants on Mucociliary Transport

Cigarette smoke Reported effects of smoking on mucociliary transport are confusing due to the wide variation of effects of tobacco smoke on animals.[42] In humans, the immediate response to the smoking of several cigarettes has been either an increase in tracheobronchial clearance or no effect.[14, 15, 119, 837] In the donkey, short-term exposures to fresh, whole smoke from two cigarettes accelerated clearance,[12] but exposures to the whole smoke from 10 or more cigarettes, or the filtered smoke from 20 or more cigarettes impaired clearance.[12] When an increase in transport occurs, it is greatest in the small airways,[15] probably because of the adrenergic stimulation caused by tobacco smoke.

Chronic smoking appears to have a more variable effect on mucociliary transport. This is due to the variations in dosage and to the different degrees of individual susceptibility to tobacco smoke. Studies of these effects have been quite diverse, and the number of subjects studied has been small. Yeates *et al.*[837] have shown that tracheal transport rates were within normal limits in nine smokers studied. Although rapid clearance occurs in some long-term smokers, impairment of large airway clearance has also been observed.[13, 77, 118, 485] Pavia *et al.*[600] and Thomson and Pavia[754] were unable to demonstrate any effect caused by long-term smoking. Cigarette smoking causes disease in the small airways. Albert *et al.*[13] showed that some heavy cigarette smokers had long overall bronchial clearance times that could be interpreted as slow clearance in small airways. Clearance observed in smokers 3 months after cessation of smoking was faster than during the smoking period.[120] Chronic high-level exposure to cigarette smoke has severely impaired bronchial clearance in the donkey[10] but recovery after cessation of exposure was almost complete within a few weeks.

Sulfur Oxides *In vivo* exposures to sulfur oxides at ambient concentrations are not likely to affect mucociliary clearance. At higher concentrations, more typical of some occupational exposures, effects have been observed. Wolff *et al.*[828] exposed nine nonsmokers to 5 ppm (13 mg/m^3) of sulfur dioxide for 3 hr after an aerosol exposure to albumin tagged with ^{99m}Tc, a gamma-emitting radionuclide. The tracheobronchial mucociliary clearance of the tagged aerosol was essentially the same as in control tests except for a transient speedup at 1 hr after the start of the sulfur dioxide exposure. Further tests by Wolff *et al.*[827] show that exercise accelerates bronchial clearance, and 13 mg/m^3 (5 ppm) of sulfur dioxide during exercise significantly speeds clearance beyond that produced by the exercise alone.

Higher concentrations of sulfur dioxide can slow bronchial clearance. Donkeys exposed for 30 min to sulfur dioxide via nasal catheters exhibited delayed bronchial clearance, severe coughing, and mucus discharge from the nose at concentrations exceeding 785 mg/m^3 (300 ppm).[718] Mean residence times following exposures to 139 to 785 mg/m^3 (53 to 300 ppm) were not significantly different from those at control test concentrations. The one test performed at a lower concentration produced an acceleration in bronchial clearance, which would be consistent with the clearance accelerations seen by Wolff *et al.*[827, 828] with 13 mg/m^3 (5 ppm) exposures. These results (acceleration at low concentrations and slowing at higher concentrations) are similar to those produced by cigarette smoke.

Fairchild *et al.*[223] showed that high concentrations of sulfuric acid mist, e.g., 15 mg/m^3 with a median diameter of 3.2 μm administered for 4 hr after tagged streptococcal aerosol exposures, reduced the rate of ciliary clearance from the lungs and noses of mice. At mist concentrations of 1.5 mg/m^3 with a median diameter of 0.6 μm, there were no significant effects.

Inert particles Camner *et al.*[114] have conducted the only study of the effect on human mucociliary clearance of an acute exposure to high concentrations of particulates. They studied 8 subjects exposed to 20 breaths of 11-μm diam carbon particles at a concentration of 50 g/liter. These exposures produced either no effect or only a slight increase in mucociliary transport, presumably in large airways.

Mucociliary Transport and Lung Disease

The tracheobronchial tree can be subdivided into four subregions: the larynx, the trachea, the large airways, and the small airways. Each subregion is susceptible to viral and bacterial infections as well as other disease processes that could either cause or be caused by perturbations in mucociliary clearance. Cancer of the larynx and large bronchi could be due to the greater surface deposition of particles, to long residence times in these regions, or to a combination of both. Small airways are probably more susceptible to chronic obstructive lung disease involving mucous gland hypertrophy, goblet cell hyperplasia, bronchoconstriction, edema, and airway closure. The relationships between mucociliary transport rates in the trachea and in the large and small airways, as well as the relationship of these rates to disease processes, remain unknown.

Numerous *in vitro* and *in vivo* studies indicate that the mucociliary system can transport particles of a wide range of sizes (from <1 μm to >500 μm diam) and types (charcoal, soot, lycopodium spores, cells, 680-μm diam teflon discs, ion exchange resins, and iron oxide, protein, polystyrene, and teflon spheres). Whether all these types of particles are removed as effectively in each of the various ciliated regions of the human respiratory tract has not been established.

The carinas of the larger airways appear to be particularly vulnerable to the accumulation of inhaled particles. Hilding[337] suggested that this is due to ciliastasis in these carinal areas. Schlesinger *et al.*[673, 674] and Bell and Friedlander[61] have shown that carinas are also sites of enhanced particle deposition. However, the relative roles of deposition pattern and mucociliary clearance rate in the ability of toxic substances to cause cellular damage or transformation has not been delineated.

Defects in passive mucociliary particle transport may be important in the pathogenesis of lung disease. In addition, particles with biologic activity, such as bacteria, can be inactivated by the immunologic and enzymatic activity of the fluid lining the lung, by alveolar macrophages, and by other free cells on the lung surface. Investigators studying tracheobronchial clearance in patients with chronic lung diseases[13, 117, 486, 493, 516, 744, 754, 778, 838] have variously reported transport to be faster than, similar to, and slower than transport in healthy subjects. The rapid disappearance of deposited aerosol in some patients was probably aided by more proximal deposition patterns and/or by coughing, by which mucus is eliminated from the diseased lungs.[117, 835, 838] Faster transport can also result from an increase in mucus secretion. Studies of subjects for whom control data are available indicate that poor mucociliary transport could result from infections[115, 116, 659] that cause damage to human ciliated epithelium in culture.[361] (In these acute studies, major hypertrophy of mucous glands could not be expected.) This resultant defective transport may be one of the reasons why patients with lung disease are more susceptible to insults from inhaled pollutants.

The importance of the wide range of tracheobronchial clearance rates found in apparently healthy subjects is not known. It has often been supposed that slow clearance in a healthy subject is detrimental. It may be beneficial if they have a greater reserve capacity for coping with increased secretions.

The mucociliary system can probably cope with a considerable increase in secretions above the normal baseline level, but the relationship between mucus volume transported and the surface transport rates is not known. Refluxing of mucus[835, 837, 838] could occur when mucus becomes too thick for the cilia to propel or when there is an excessive loss of cilia, which has been observed in bronchitic rats.[380] Refluxing mucus in the trachea can be cleared by coughing,[835, 838] in contrast to healthy nonsmoking adults in whom coughing appears to have little effect.[837] Coughing is not expected to have as much effect in small airways, although this has not been demonstrated.

The nature of the protective barrier of mucus that the airways present to inhaled toxicants is altered in disease. This has been shown in bronchitic rats by Dalhamn[180] and Iravani.[380] Comparisons of mucoid mucus from patients with hypersecretory conditions such as chronic bronchitis, bronchiectasis, and cystic fibrosis have produced few differences, either histochemically[431] or rheologically.[145, 224, 745] However, purulent mucus from diseased subjects had altered biochemical[515, 616] and rheological[194] properties.

The study of mucous glands and their hypertrophy[299, 639, 640] has been of major importance in understanding the pathogenesis of chronic bronchitis and chronic obstructive pulmonary disease. However, Reid[639] only measured gland-to-wall ratios and not the physiologically important parameters of gland volume/airway circumference and gland volume/lumen size. Some work using this latter parameter has been reported by Bedrossian *et al.*[59]

The proposed sequence of events leading to chronic obstructive pulmonary disease is similar to that suggested by Albert *et al.*[14] Acute exposures to dust increase mucus production and mucociliary transport. Continuation of these exposures leads to bronchial mucous gland hypertrophy and goblet cell hyperplasia. Mucus transport in small airways could remain normal or increase during this stage. As gland hypertrophy continues, the mucociliary transport system becomes inadequate in removing the excess secretions. This leads to chronic cough, accumulation of secretions, and increased susceptibility to inhaled particulates, noxious gases, and pathogenic organisms.

In summary, particulate atmospheric pollutants may be involved in chronic lung disease pathogenesis as causal factors in chronic bronchitis, as predisposing factors to acute bacterial and viral bronchitis, especially in children and cigarette smokers, and as aggravating factors for acute bronchial asthma and the terminal stages of oxygen deficiency (hypoxia) associated with chronic bronchitis and/or emphysema and its characteristic form of heart failure (cor pulmonale).

Indirect Indicators of Mucociliary Transport Abnormalities

In addition to the direct measurement of the effects of particles on mucociliary transport, there are other indices from which changes in mucociliary transport may be inferred; namely, sputum production, alterations in pulmonary function, and pathologic abnormalities.

Epidemiologic studies of a group of men showed that phlegm production decreased concomitantly with the reduction of airborne particulates despite little change in sulfur dioxide concentrations.[714] Another study indicated that concentrations of particulates, not sulfur dioxide, were related to decreased pulmonary function.[226] Lowe and his colleagues[490, 491, 785] found little such correlation. The detrimental effects of particulate sulfates on lung function are discussed in Chapter 7.

A number of studies indicate that inhalations of very high concentrations of inert dusts can result in increased airway resistance caused by bronchoconstriction.[161, 203, 675] Carbon dust stimulates both cough[810] and irritant receptors[689] in cats, thereby increasing vagal afferent activity. This activity is probably mediated by the nerves in the epithelium (Jeffery and Reid[385]) and may be responsible for the increased mucociliary transport mentioned above.

A few investigators have studied the gross pathology of the bronchial mucosa after inhalation of dusts. Negative findings do not necessarily mean that the mucosa is functioning normally, but these studies do indicate the kind of damage that dusts can cause to the mucociliary transport system.

A 6-week exposure of piglets to cornstarch or corn dust produced no "significant" lesions within the respiratory tract,[510] nor were bronchial lesions reported in monkeys exposed to fly ash for 18 months.[501] Schiller[670] exposed dogs and rats to coal dust. After 3 years, hypertrophy of the epithelium and goblet cells had developed, but no pneumonoconiotic lesions, connective tissue reactions, or cellular infiltrations were evident. Animals exposed for 6 years incurred atrophic epithelium partially denuded of cilia and formation of digitate appendices probably consisting of apocrine secretion. Intratracheal injections of silica dust in rats caused obliteration of the respiratory bronchioles.[298]

Hadfield[305] studied the nasal epithelium of workers in the furniture industry who had been exposed to wood dust. She reported adenocarcinoma at the ethmoid sinuses, squamous metaplasia of the mucous membrane covering the anterior ends of the middle turbinates, vasomotor rhinitis, and allergic nasal palpi.

ALVEOLAR CLEARANCE

Mechanisms

Once a material is deposited on the respiratory epithelium of the alveoli, both "absorptive" and "nonabsorptive" clearance processes ensue. Neither term explicitly describes the pulmonary transport that occurs. Most likely, both processes occur together or with temporal variations.

Respiratory absorption from a mechanistic point of view is poorly understood, but evidence exists for both passive and active transport mechanisms.[554] Collectively, these processes bring about an apparently unique and selective permeability for the respiratory epithelium, *vis-à-vis* other epithelia.[147] Permeation of the alveolar epithelium precedes both lymphatic transport and uptake into the blood. Thus, both pathways also require the penetration of an endothelial barrier, which is generally more variable and permeable than the epithelial surface.[553]

In the "nonabsorptive" category, materials may become "fixed" within the parenchymal tissue. Such depots may contain chemically or physically altered material or material that has been concentrated by cytologic and/or physiologic processes. These depots often appear to undergo very little clearance by nonabsorptive processes, but it is conceivable that both the normal turnover of cells and endocytosis affect their persistence.

The most widely accepted "nonabsorptive" clearance mechanism is the transport from the lungs of alveolar macrophages that have phagocytosed particles. Many investigators have studied the role of alveolar macrophages in dust clearance[84, 225, 242] as well as such other functions as their maintenance of alveolar sterility and their etiologic significance in pulmonary disease.[291] Methods used to evaluate the macrophage response quantitatively include "free-cell harvests" by endobronchial lavage[84, 428] and histologic examinations of alveolar tissue.[242] Despite the common acceptance of phagocytosis and subsequent cell removal as the dominant nonabsorptive clearance mechanism, it has, so far, defied quantification.

The clearance pathways for phagocytosed particles remain controversial. It is generally agreed that macrophages ingest particles and transport them proximally on the bronchial tree to the pharynx, where they are swallowed. There is considerable disagreement concerning which pathway is predominant between the alveoli and the bronchial tree. There are proponents for an interstitial route, while others favor a continuous, proximally moving surface film that draws the cells onto the ciliated surface at the terminal bronchioles.

The interstitial route was proposed by Brundelet[93] on the basis of microscopic sections of rat lungs containing dye particles. He suggested that particle-laden alveolar macrophages migrated onto ciliated airways through lymphoid collections at the bifurcations of bronchi and bronchioles. Green[292] reported similar observations for macrophages containing coal dust. He proposed that alveolar fluid, cells, and particles flow within "liquid veins" between alveoli. This fluid flow is driven by the variation in tension created by respiratory movements in the alveolar walls. This moves the fluid from the midzone of the lobule toward areas of least pressure adjacent to subpleural, paraseptal, and peribronchial lymphatics. Tucker *et al.*[761] reported that bare particles (those not ingested by macrophages) follow this clearance route in the first few hours following dye particle inhalations.

The surface route was proposed by Macklin,[503] favored by Hatch and Gross,[319] and supported by Ferin's studies.[225] Ferin exposed rats to titanium dioxide at low concentrations (15 and 100 $\mu g/m^3$) that were representative of air pollution. He studied lung sections of rats killed at 1, 8, and 25 days. Particle-laden macrophages were concentrated in alveolar ducts and at junctions of respiratory and terminal bronchioles. A very small fraction was found in lymphatics. The studies of Brundelet,[93] Green,[292] and Tucker *et al.*[761] involved much more massive exposures, which may have induced a change in the normal pathway.

Forms of endocytosis, such as pinocytosis, cannot be easily differentiated from true absorption, inasmuch as they both may result in transmembrane transport of material. Thus, any description of an "absorptive" process that cannot be precisely localized and made on a mechanistic basis must be considered tentative and uncertain as to both the number and types of transport mechanisms operating. Consequently, "absorptive" clearance of inhaled materials from the alveolar surface is largely based on, or inferred from, the ultimate disposition of the material, especially its appearance in the blood, other body organs, and the urine.

Bona fide absorptive mechanisms and pathways have been established for specific materials in the alveolar membrane. Specific information on alveolar absorption mechanisms comes mainly from the studies of Chinard,[147] Taylor *et al.*,[750] Liebow,[464] and Bensch and Dominguez.[64] Other relevant data, generally lacking mechanistic information, are available on specific materials.[556, 737] Many of these lung absorption studies and the investigations of alveolar clearance were the subjects of a recent review.[554] An important component of "absorptive" clearance is solubilization that either precedes or is a part of true absorption. In other words, absorption presumably depends on materials being either monomeric or, to a lesser extent, polymeric forms of small dimension.[554] The significance of solubilization as a principal clearance mechanism in the alveolar region has been brought into focus by the success of several simple solubility models.[533, 556] Solubility must presently be considered the most important "absorptive" process.

Kinetics

The clearance of particles from the alveolar region proceeds in several phases. These phases can usually be described by a series of exponents, each of which presumably corresponds to a different clearance mechanism.

Casarett[128] proposed that the earliest alveolar phase, with a half-time measured in weeks, is generally associated with phagocytic clearance, while a slower phase, with a half-time in months or years, is generally associated with solubility. The ICRP Task Group model[749] does not include the initial alveolar phase. Casarett attributes this omission to overreliance on data from studies in which the "fast" alveolar phase was absent because of the cytotoxicity of the dusts used. Jammet *et al.*[384] showed that for hematite dust, a clearance phase with a half-time of from 10 to 12 days is normally present in the cat, rat, and hamster, preceding a slower phase with a half-time exceeding 100 days. The 10- to 12-day phase disappeared when the animals were exposed to toxic amounts of plutonium,[578] silica dust, or carbon dust,[452] while the half-time for the slower phase remained relatively unaffected.

Numerous inhalation studies using insoluble radioactive aerosols have been performed on beagles, with long-term follow-up of the lung retention by external *in vivo* gamma counting.[45, 173, 241, 559] In most, but not all, of these studies, a fast phase with a half-time of 10 to 14 days was evident. The half-life for the slower phase of alveolar clearance, which appears to be related to *in vivo* solubility, is quite variable. Morrow *et al.*[559] used test aerosols of ferric (^{59}Fe) oxide, mercury (^{203}Hg) oxide, barium (^{131}Ba) sulfate, manganese (^{54}Mn) dioxide, and uranium dioxide. The biologic half-time for the slow alveolar clearance phases were 58, 33, 8, 34, and ~200 days, respectively. Bair and Dilley[45] found the slow phase of alveolar clearance of ferric (^{59}Fe) oxide to be greater than several hundred days, suggesting that the surface-volume ratios or other surface properties of the particles can have a major effect on dissolution in the lung.

Because alveolar retention of relatively insoluble particles is recognized to be important to the pathogenesis of chronic lung disease, it is surprising that the literature yields virtually no useful data on the rates or routes of alveolar particle clearance in man. This paucity of data is attributable to the laboratory sophistication required and the relatively high cost incurred in human studies. The only feasible way to perform such studies is to have the subjects inhale tagged test aerosols. The subsequent lung retention can then be monitored with external detectors. In addition, the types and varieties of particles that can be used in human studies are limited. For example, only demonstrably nontoxic particles can be intentionally inhaled.

The only laboratory studies of human alveolar clearance were conducted by Albert and Arnett[11] and Morrow *et al.*[557, 558] In the Albert and Arnett study, eight healthy human males inhaled neutron-activated metallic iron particles. Three subjects had sufficient residual activity after completion of bronchial clearance for continued measurement of retention. In a 32-year-old male nonsmoker and a 27-year-old male moderate smoker, the postbronchial clearance occurred in two phases—a "fast" phase lasting about 1 month and a much slower terminal phase. The faster phase was missing in a 38-year-old male two-pack-a-day cigarette smoker with a chronic cough. Although firm conclusions cannot be drawn from these limited data, they do suggest that the "fast" alveolar phase can be detected in humans, and that dust retention may be increased by cigarette smoking (beyond the retention of the smoke particulates themselves).

The only other studies of human alveolar clearance under controlled conditions are those of Morrow *et al.* In an initial study, four healthy human subjects inhaled a manganese (^{54}Mn) dioxide aerosol with a median diameter of 0.9 μm, a geometric standard deviation of 1.75, and a mass concentration of 4 mg/m^3.[558] The aerosol was inhaled for 20 to 30 min. The breathing pattern consisted of four normal inhalations alternated with one maximal inhalation. Measurements made more than 48 hr after the inhalation indicated a single clearance phase in all four individuals, with biologic half-times varying from 62 to 68 days.

In their other human studies, Morrow *et al.*[557] used several different aerosols and again reported alveolar clearance rates in terms of a single exponent. The half-times varied with the composition of the particles used. Half-times of 65, 62, and 35 days were found for manganese (^{54}Mn) dioxide, ferric oxide labeled with chromium-51, and chromium-51 labeled polystyrene, respectively. The measurements indicating the shorter half-time for the polystyrene particles were based on <14 days. Consequently, they may have been influenced by a rapid initial rate of alveolar clearance, in contrast to the iron and manganese oxides, which were followed for 45 to 120 days.

Scattered data on human lung retention of inhaled particles exist in reports of *in vivo* measurements in atomic energy industry workers who had been accidentally exposed to airborne nuclides.[241, 301, 654] While these data are interesting and useful for some evaluations, they must be interpreted with their major limitations in mind. Among these limitations are:

The time the exposure took place is either not known or it extended over an indeterminate period before it was discovered.

The chemical form and particle size distribution of the inhaled aerosol are usually not known, and the exposure may be to more than one aerosol and/or nuclide.

Because the exposure may be detected from routine bioassay or *in vivo* counting long after the exposure, it may be impossible to establish the initial amount deposited.

For Atmospheric Aerosols

The foregoing suggests that a wide range of alveolar clearance rates involving both absorptive and nonabsorptive processes apply to atmospheric particulate matter. The special character of atmospheric aerosols, especially their enormous surface area and large surface-free energy, might result in higher reaction and dissolution rates for a given chemical species than would be estimated from conventional laboratory aerosols or industrial dusts.

Occupational and laboratory studies of aerosols are limited in their ability to predict the biologic behavior of atmospheric aerosols. The generally small particle size and other characteristics of atmospheric aerosols are difficult to simulate in the laboratory. For example, the biologic effects of atmospheric dust may depend in large measure on synergistic actions with other airborne pollutants.[428] The generally accepted view of synergism extends beyond potentiation to include the role of toxic vector. Such gases as sulfur dioxide are probably either absorbed to the particulate surface or absorbed into the particles, and thereby transported into the alveolar regions, where they exist in high, localized concentrations. These localized high concentrations could not be produced without particles. Accordingly, sulfur dioxide sorption to particulate matter might effectively allow sulfur dioxide penetration into the alveolar regions at even nominal environmental concentrations of the gaseous pollutant. This possibility has been widely cited by those who believe atmospheric concentrations of sulfur dioxide and particulate matter are associated with adverse health effects.

7

Effects of Sulfur Dioxide and Aerosols, Alone and Combined, on Lung Function

Inhalation toxicology plays two important roles: to describe the structural, biochemical, and functional effects of air pollutants and to reveal the mechanisms underlying these effects. A diversity of experimental models are used, ranging from subcellular and cellular elements to intact animals or human subjects. There is a natural inclination to extrapolate the results from experiments on *in vitro* preparations or animals to predictions of ill effects in large populations. Such extrapolations have a strong element of uncertainty that should be recognized. The attempt at extrapolation becomes more convincing if the experimental results can be demonstrated in more than one mammalian species, especially in primates, and if the biologic systems under study have close analogs in humans.

The types of exposure used in inhalation toxicology may mimic either intense and brief or low-grade and prolonged episodes of pollution; both exposures are typical of urban centers. Often, experimental pollutant concentrations must exceed ambient concentrations to be effective. At least two factors may be responsible. One is that these models of air pollution are relatively simple approximations of the complex, highly variable, and incompletely defined pollution that is found outside the laboratory. Ill effects in communities may be due to the cumulative action of many pollutants or to the interaction between pollutants and other forms of environmental stress. Therefore, no single component need be excessively high. Secondly, the most certifiable effects of pollution are seen in persons already ill with cardiopulmonary disease. It is difficult to simulate such disease in animals, and there are ethical and legal restrictions to the use of patients with these diseases in laboratory experiments. When human volunteers are studied, the exposures must of necessity be relatively short and the effects must be benign and completely reversible. Such considerations do not limit research on animals, but higher-than-ambient concentrations must be administered to produce effects of sufficient severity for a successful investigation.

This chapter concerns the effects of inhaled aerosols on lung function. Particular emphasis is given to sulfur oxide aerosols that are found in general or community pollution and to studies involving realistic concentrations of them. Pollens and aerosolized drugs, such as histamine and methacholine chloride, which are potent bronchoconstrictors, and the great variety of aerosols found in occupational settings are not discussed. Two of the largest groups of ambient aerosols are the sulfates and nitrates. Each may be emitted directly in a variety of combustive processes or may arise by chemical reactions in the ambient atmosphere. Of the two groups of compounds, the sulfates have prompted greater experimental interest and activity (see the recent, comprehensive review on nitrogen oxides[573]). Sulfur dioxide, the gas chiefly responsible for the genesis of sulfate aerosols, will also be discussed.

After a brief discussion of the effects of sulfur dioxide on the respiratory system, the chapter proceeds to the effects of aerosols alone and with sulfur dioxide. The principal focus will be on mechanical and ventilatory functions. Following are some of the functional measurements that are commonly used. Additional methods are required to assess ventilation-perfusion relations and gas exchange.

Flow resistance (R_L) and compliance (C_L) are two mechanical attributes of the lung. R_L is determined chiefly by the caliber of the airways, while C_L is an index of the size and distensibility of the lung. In animals, R_L can be subdivided into central and peripheral components by implantation of catheters. Less direct methods can be used to assess the peripheral airways in human subjects and intact animals. Flow resistance and compliance can be measured over a range of breathing frequencies as a means of determining the evenness of distribution of inhaled air.

Forced Vital Capacity (FVC) is the maximal volume of air that can be rapidly expelled following full inspiration. $FVC_{1.0}$ is that portion of the FVC that is expelled during the first second of the procedure. The Maximal Expiratory Flow Volume (MEFV) curve is a continuous recording of the maximal expiratory flow rate plotted against changing lung volume. These ventilatory tests are influenced by the caliber of the airways, size, distensibility, and recoiling force of the lung and chest wall, muscular strength, and even motivation. Measurements of ventilatory function have been used from day to day to assess the effects of environmental changes, including air pollution.[447, 448] (Clearance from the respiratory system is discussed in Chapter 6.)

SULFUR DIOXIDE

Because of its ready solubility in tissue fluids, sulfur dioxide is removed almost entirely by the upper airways (the nose, mouth, and pharynx) during quiet breathing.[34, 715] Greater penetration of the gas occurs when inspiratory flow rate increases and gas enters the airways through the mouth.[255] consequently, exposure of the lower airways (beginning at the larynx) and alveoli to sulfur dioxide as well as other gases of equivalent solubility is exaggerated during exercise, when both the ventilatory rate and the degree of penetration increase. Bates and his colleagues[54, 56] have shown that exercise potentiates the functional effects of ozone in human volunteers. To date there have been no published reports of the effects of exercise on the response to sulfur dioxide or to irritant aerosols, although several groups of investigators have begun such studies. The vulnerability of children to air pollution[202, 236, 453, 577, 696, 697] may reflect in part their great activity while out-of-doors.

The site(s) of uptake of the inspired sulfur dioxide beyond the upper airways is not known. Of the amount of gas inspired, less than 1% is likely to reach the alveoli. A preliminary attempt at predicting the sites of uptake of sulfur dioxide within the lower respiratory tract, and of the consequent dosage to these sites, is shown in Figures 7-1 and 7-2. Note that sulfur dioxide might be expected to act principally on the central and intermediate airways.

Sulfur dioxide alters the mechanical function of the upper and lower airways in humans and in laboratory animals. Nasal flow resistance may increase in human subjects exposed to 2.6 mg/m^3 (1 ppm at 25°C, 1 atm) of sulfur dioxide for 1 to 3 hr;[35] the response is accelerated by higher concentrations.[715] Nasal mucus flow rate is slowed by several hours of exposure to 13.0 mg/m^3 (5 ppm) of sulfur dioxide,[35] an effect that could imply a reduction in resistance to infection locally. However, the principal effect in acute experiments is to increase pulmonary flow resistance or impair forced expiratory flow rate. The concentrations of gas required to elicit these responses in subjects at rest are several times higher than the primary ambient air standard of 0.36 mg/m^3 (0.14 ppm; 24-hr average). Although 2.6 mg/m^3 (1 ppm) may increase R_L within 10 min in an occasional sensitive individual, an exposure of about 13.0 mg/m^3 (5.0 ppm) is required to affect a significant percentage of healthy volunteers.[253] The forced expiratory flow rate is slowed after 1 to 3 hr of exposure to 2.6 mg/m^3 (1 ppm) of sulfur dioxide.[35] When intermittent light exercise is superimposed on the exposure, maximal expiratory flow rate may be affected by 2.0 mg/m^3 (0.75 ppm) of the gas.

The variability in response to sulfur dioxide among different species of animals and within a single species[29] can be considerable. Amdur[26] has reported that guinea pigs show a statistically significant increase in R_L (12.8% $p < 0.001$) during response to a mean concentration of 0.68 mg/m^3 (0.26 ppm) [range: 0.08 to 1.7 mg/m^3 (0.03 to 0.65 ppm)]. However, earlier data from the same laboratory reflect a scatter of responses to different gas concentrations that far exceeded this small average change.[29] Another group of investigators[526] found no consistent

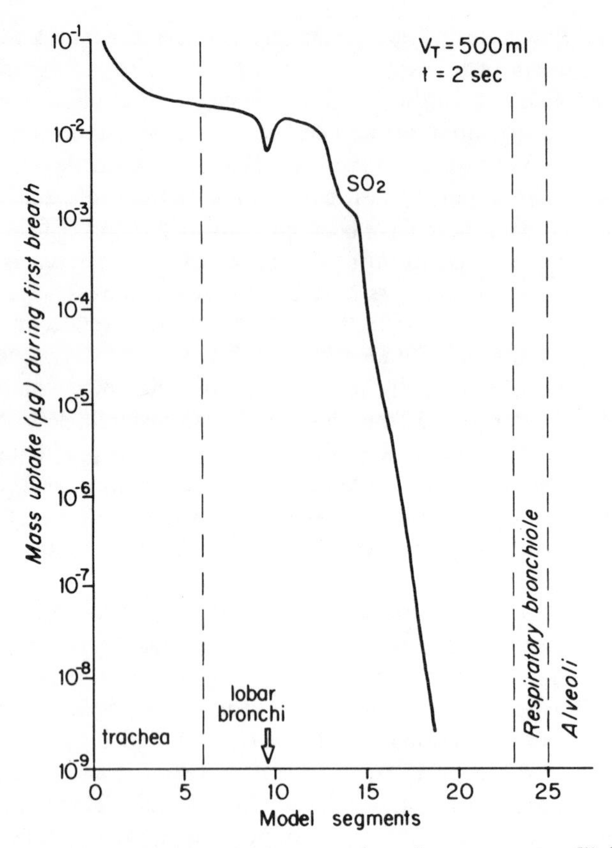

Figure 7-1. Weibel's generations of airways and respiratory structures[791] have been modified to accommodate the computer model for estimating uptake. The chemical and physical properties of water are used to represent the liquid layer lining the respiratory system. Not that the mass uptake of sulfur dioxide is virtually complete by the level of the intermediate airways.

change in R_L in lightly anesthetized guinea pigs exposed to 2.6 mg/m^3 (1 ppm) of sulfur dioxide for 2 hr; however, a significant increase in R_L resulted in response to an identical concentration of sulfur dioxide combined with sodium chloride aerosol.

For reasons that are not understood, the effect of sulfur dioxide on function may vary considerably in an individual from one exposure to the next. In several mammalian species the effect of sulfur dioxide on R_L is limited in time. The effect may begin to recede in humans after about 10 min exposure (10.4 to 44.2 mg/m^3)[252] and even sooner in anesthetized dogs (156 to 559 mg/m^3)[254] and cats (65 to 101 mg/m^3).[164] In contrast, the change in R_L in guinea pigs may be more or less steady or continue to rise for several hours over a wide range of concentrations (5.2 to 879.0

mg/m³).[23, 24] Sulfur dioxide appears to have a more sustained effect on nasal mucus flow rate[35] than on R_L.[252]

A notable feature of the experiments involving short-term exposures to sulfur dioxide is the exaggerated response among a few individuals for which there is no obvious cause. Amdur[27] has reviewed the evidence for a nonallergic "hypersensitive" response to sulfur dioxide in human subjects and test animals. The studies that she considered relied on measurements of flow resistance. She defined sensitive guinea pigs as those showing an increase in R_L more than 3 times greater than the average change for the group, and sensitive human subjects as those who responded with an increase in flow resistance to a concentration of sulfur dioxide that caused no changes in others in the group or those who showed a major increase at a concentration that caused only minor changes in the rest of the group. Roughly 10% of the humans and animals tested, all of whom

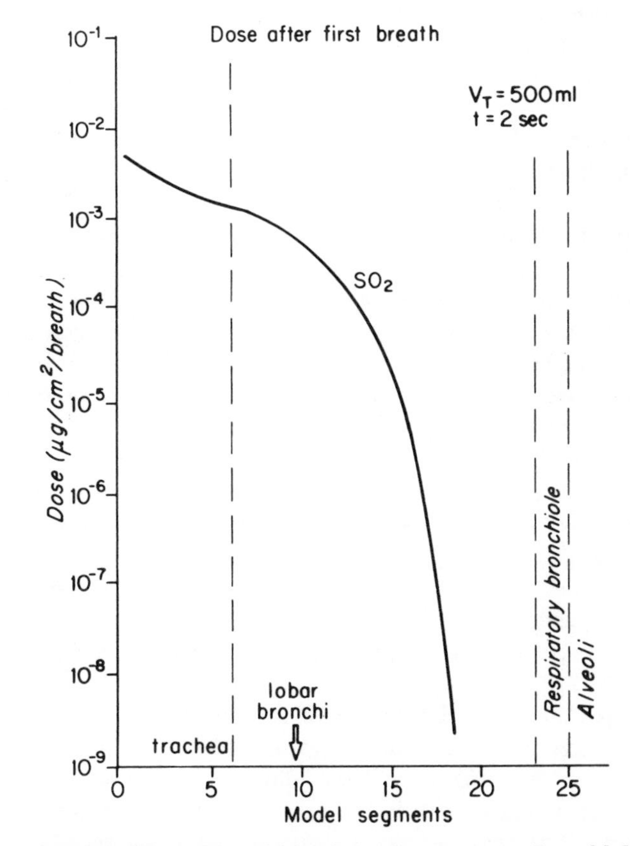

Figure 7-2. Sulfur dioxide uptake model for the respiratory system. From McJilton *et al.*, 1972.[527]

were ostensibly healthy, satisfied these criteria. The mechanism for the hypersensitivity is unknown.

Whether or not such individuals are more prone than the general population to illness arising from chronic or repeated exposure to sulfur dioxide, or perhaps to other airborne pollutants as well, is a challenging question.

The change in R_L in humans and in cats is reflex in origin and can be abolished by atropine or section of the vagal nerves.[61, 566] The more persistent change in R_L in the guinea pig suggests a different mechanism of effect. For example, slow-reacting substance (SRS), histamine, serotonin, or prostaglandin might be released from damaged tissues and act directly on the smooth muscle of the airways, causing constriction. Other possible mechanisms for narrowing the airways include swelling of the bronchial wall and excessive or retained secretions.

An intriguing question is whether or not sulfur dioxide and other inhaled irritants can alter mechanical behavior of the lung indirectly. Two such pathways have been postulated. One is a nasobronchial reflex in which stimulation of the nasal mucosa chemically, mechanically, or immunologically is said to induce bronchoconstriction. The action of pollen in asthmatic patients was cited as an example of this pathway by S. Ingelstedt at the Symposium on the Nose and Adjoining Cavities at the Armed Forces Institute of Pathology in 1969.[257] The afferent pathway for this reflex would be the fifth cranial nerve. It is arguable whether or not there is a nasobronchial reflex in healthy individuals leading to bronchoconstriction. Andersen *et al.*[35] have suggested that the nasobronchial reflex may account for some of the functional changes observed in healthy volunteers exposed to sulfur dioxide. Bronchodilation rather than bronchoconstriction has been observed in cats[567, 757] and rabbits[16] following irritation of nasal mucosa. The second, indirect pathway, for which there is less persuasive evidence, involves the excretion of sulfur dioxide as a gas into the lung, presumably from the pulmonary circulation.[256] Sulfur dioxide originally enters the blood following absorption in the upper airways.

Additional functional parameters, some directly related to the mechanical properties of the lung, have been used to assess the irritancy of sulfur dioxide. In 1953, healthy volunteers who were briefly exposed to 2.6 to 21.0 mg/m^3 (1 to 8 ppm) were reported to respond with rapid, shallow breathing and elevated pulse rates[30]; similar findings had been reported earlier from the same laboratory following exposure to sulfuric acid mist.[31] These effects of sulfur dioxide have not been confirmed by other investigators.[253, 444] It is worth noting that all of these studies[30, 31, 253, 444] were performed with the subjects breathing through fixed mouthpieces. The mouthpiece itself is likely to make breathing a self-conscious act so that the ventilatory pattern during the measurement often differs from that present with natural breathing. Any potential effect of an irritant gas or aerosol on the ventilatory pattern is also subject to interference or modification. Techniques that measure respiratory frequency or tidal volume without recourse to a mouthpiece are more suitable for this type of study, particularly when the subject is at rest. In guinea pigs, the increase in R_L caused by sulfur dioxide may be accompanied by slower, deeper breathing, an adjustment that acts to moderate the increase in work of breathing.[29]

Exposure to 2.6 mg/m^3 (1 ppm) of sulfur dioxide may impose slight reductions in forced expiratory flow rates after only 15 min.[709] Other investigators have noted reductions only after several hours of exposure.[35] The effects are magnified by higher concentrations and longer exposures and may persist for about 24 hr.[35] The patency of the peripheral airways, as reflected by the measurement of "closing volume," appeared to be unaffected.[35]

In 1973, Bates and Hazucha[56] studied pulmonary function effects in human volunteers. Sulfur dioxide [0.98 mg/m^3 (0.37 ppm)] administered with ozone [0.73 mg/m^3 (0.37 ppm)] produced exaggerated effects compared with the effects of the individual gases. Sulfur dioxide alone produced no response; ozone alone caused only slight functional changes. This important observation has spurred a continuing series of experimental studies in animals as well as humans. The mechanism of the exaggerated response is uncertain, but is likely to have involved the oxidation of sulfur dioxide by ozone to an ultrafine aerosol of sulfuric acid.[62, 112, 170]

Chronic studies: Sulfur dioxide has been administered at 0.34 to 15.0 mg/m^3 (0.13 to 5.72 ppm) to guinea pigs for 1 year,[8] and at 0.37 to 3.35 mg/m^3 (0.14 to 1.28 ppm) to cynomolgus macaques for 78 weeks without apparent functional or structural impairments.[7] One group of monkeys exhibited little functional impairment during 30 weeks of continuous exposure to 12.4 mg/m^3 (4.69 ppm) of sulfur dioxide. At that point, the monkeys were accidently exposed for 1 hr to a concentration of sulfur dioxide estimated to range between 524 mg/m^3 (200 ppm) and 1,310 mg/m^3 (1,000 ppm). For the remainder of the 78 weeks they were maintained on clean air but continued to show persistent functional impairment. This group alone had histologic evidence of tissue damage. The injury was extensive, involving the tracheobronchial tree, peripheral airways, and parenchyma.

It would appear that chronic exposure to low concentrations of sulfur dioxide alone (<13 mg/m^3) is not associated with cumulative functional or structural impairment of the lungs, particularly if the animal remains at rest. Intermittent spiking concentrations, which may occur in occupational settings, may represent a greater risk. Perhaps a more revealing long-term study would involve the combination of sulfur dioxide with other pollutants with which the gas may interact, and the testing of other aspects of function, including mucociliary clearance and small airway caliber.

AEROSOLS

Chemically inert as well as chemically active aerosols may cause changes in the mechanical behavior of the lung. Some aerosols are used therapeutically to produce bronchodilation and improve function. Aerosols tend to deposit at or near bifurcations of the airways, probably because of the changes in flow patterns at these sites.[61] Significantly, epithelial nerve endings also tend to concentrate near bifurcations,[565] a factor that increases the functional impact of the particles. The nerve endings of the more central airways are especially receptive to mechanical stimuli that lead to reflex coughing and bronchoconstriction. Evidence suggests that the peripheral nerve endings are more responsive to chemical stimuli, which may result in increased breathing rates and reduced pulmonary compliance.[565]

Many proven or postulated factors influence the type and magnitude of the response to aerosols. Some are inherent in the organism, others in the aerosol itself. The biologic factors that may modify responses, while widely recognized, have been assessed only superficially. These include the species, age, nutrition, level of activity, state of health, and level of consciousness. In animals exposed to aerosols while anesthetized, response may be influenced by the type and depth of anesthesia. In Chapter 6 there is a discussion of ventilatory (tidal volume, respiratory frequency, flow velocity) and morphometric factors that influence deposition. Among the relevant aerosol properties are mass concentration, aerodynamic size, molecular composition, pH, and solubility. Whether the soluble aerosol is in a dry or aqueous state prior to inhalation is important if it is administered with a soluble gas (see GAS-AEROSOL INTERACTIONS, below). In general, as the site of deposition of a relatively insoluble particle becomes more distal, the time required for its clearance increases. Some particles penetrate the alveolar wall and lodge in the interstitium or are disseminated through the lymphatic and circulatory systems.

Evidence suggests that many potentially toxic chemicals reside chiefly in the accumulation mode (see Chapter 1) of ambient aerosols. They include lead, cadmium, antimony, arsenic, nickel, zinc, and benzo[*a*]pyrene (a carcinogen), as well as sulfate and nitrate ions.[456, 575] The trace elements contained in soluble aerosols may ultimately affect extrapulmonary tissues irrespective of deposition site. A number of the aerosols that are found in general air pollution have been administered to laboratory animals. These include sulfates with a wide range of pH (e.g., ammonium sulfate and sulfuric acid); no reports on the effects of nitrate aerosols have appeared yet.[555] Much of this work has been done by Amdur and her associates.[24, 26, 28, 32] Their studies, conducted exclusively on guinea pigs, focused on changes in the mechanical behavior of the lungs, both R_L and C_L. The principal effect of the different aerosols was to increase R_L. The extent to which the response was due to reflex constriction of the trachea and bronchi, or to edema, excessive secretions, and the action of (humeral) smooth muscle constrictors released from local tissue, was not determined. C_L, reflecting distensibility of the lungs, was reduced.

Often, these changes in C_L and R_L were intimately related in time; they appeared, increased in severity, and remitted almost simultaneously. Such parallel timing suggests the following train of events. The increase in R_L resulted in an abnormal or uneven distribution of inspired air so that portions of the lungs became poorly ventilated. As a consequence, the "effective" size of the lungs, reflected in the measurement of C_L, was reduced. C_L alone may be reduced if the principal site of effect is in the small airways or alveolar region. These mechanical defects, besides increasing the work of breathing, may be severe enough to impair the normal transfer of oxygen across the alveolar-capillary membrane. With the exception of one study on human volunteers,[581] there has been no rigorous testing of gas exchange or blood gas concentrations in experiments using aerosols.

Particles up to a few micrometers in diameter generally cause greater functional and structural changes in the lung than do larger particles, probably because they are more likely to be deposited in the lower airways and alveoli (see Chapter 6). Probably the most noteworthy study of the effect of size on the magnitude of response was conducted with zinc ammonium sulfate.[28] For particles ranging from 0.3 to 1.4 μm diam, there was an inverse relation between the size of the particle and the change in R_L (see Figure 7-3). The mass concentration was approximately 1 mg/m^3. At least two factors may have contributed to the result:

The site or magnitude of deposition differed with size in a manner favoring the greatest change in R_L for the smallest particle. Paradoxically, the largest aerosol (1.4 μm diam prior to inhalation) would be expected to undergo the greatest total deposition (see Chapter 6). The solubility of zinc ammonium sulfate probably caused it to grow severalfold after inhalation. Such growth would probably influence the magnitude and site of deposition.[574]

The surface area/volume ratio of the particle, which is inversely related to its diameter, may somehow influence the response. Certainly, the rate at which moderately to slightly soluble particles enter solution (having deposited in the respiratory system) should increase as the diameter is reduced.

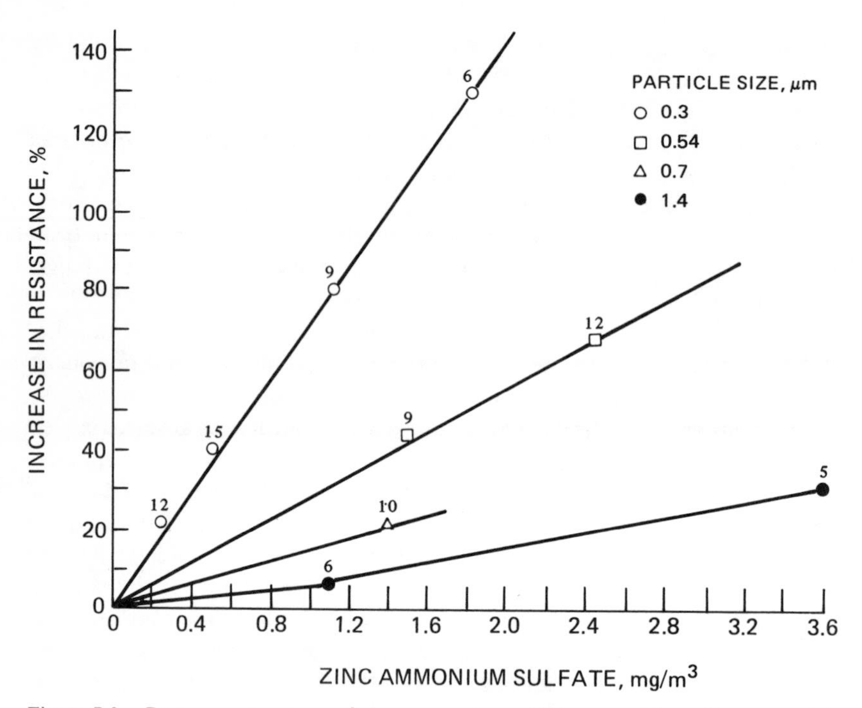

Figure 7-3. Dose-response curve of zinc ammonium sulfate aerosol for different particle sizes. Numerals beside each point indicate number of animals. From Amdur and Corn, 1963.[28]

Zinc ammonium sulfate was identified as a component of the severe pollution that struck Donora, Pennsylvania in 1948. Ammonium sulfate, a more common component of regional and urban pollution, and zinc sulfate are less irritating to the lung than zinc ammonium sulfate.[28]

Acute,[25, 26, 31] subacute,[108, 109] and chronic[6] exposure studies indicated that sulfuric acid causes a greater increase in R_L in guinea pigs than zinc ammonium sulfate at approximately equivalent particle sizes and concentrations.[25] Whether the greater toxicity of sulfuric acid is related to its acidity is unknown. No rigorous study of the role that pH plays in determining the irritant potency of sulfate aerosols has yet been made. In a series of studies, Amdur[26] compared the relative effects of a variety of sulfate aerosols (sulfuric acid, zinc ammonium sulfate, ferric sulfate, zinc sulfate, ammonium sulfate, cupric sulfate, ferrous sulfate, and manganous sulfate). The estimated pH of these compounds does not correlate well with their reported effects.

The size of the sulfuric acid aerosol does influence the response: particles 0.3 μm diam cause a greater increase in R_L than particles 2.6 μm diam. Indeed, within limits, size appears more critical than concentration in determining the response. Figure 7-4 shows that about

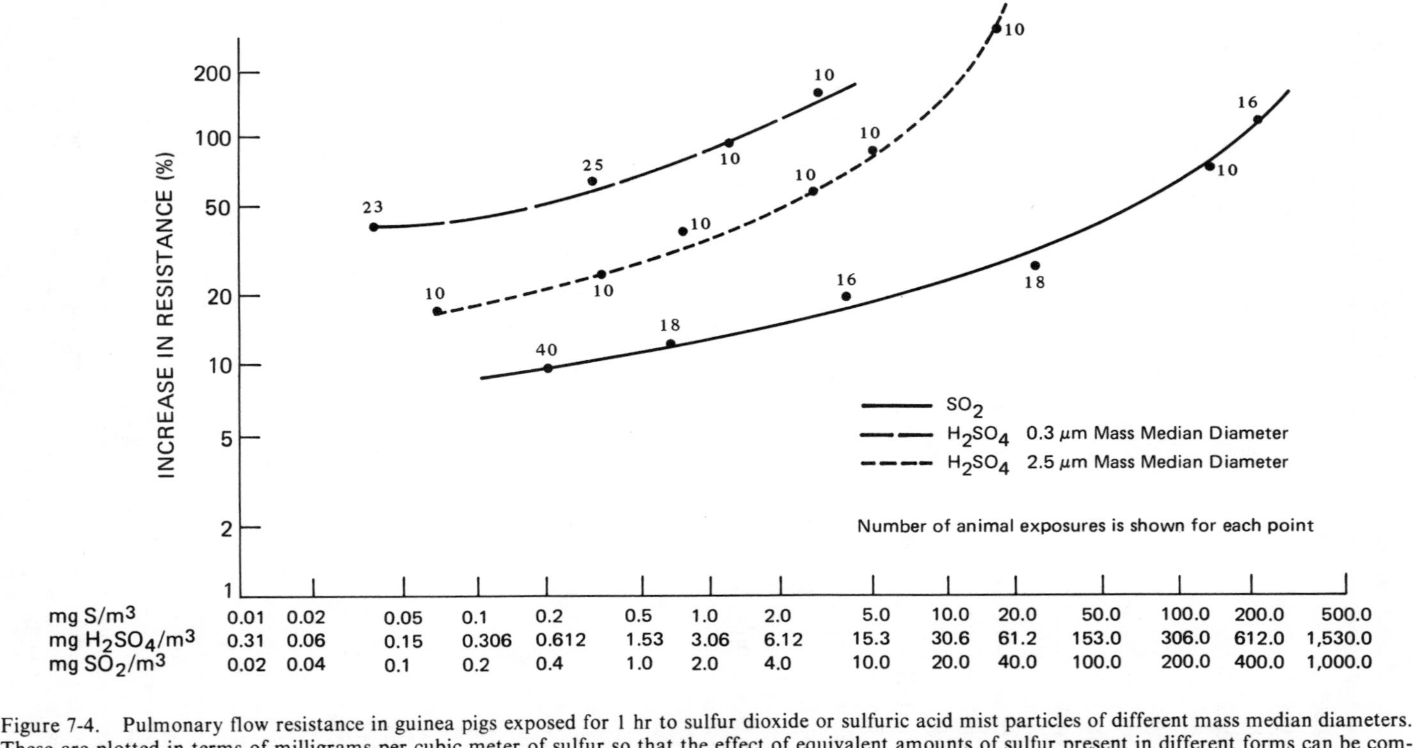

Figure 7-4. Pulmonary flow resistance in guinea pigs exposed for 1 hr to sulfur dioxide or sulfuric acid mist particles of different mass median diameters. These are plotted in terms of milligrams per cubic meter of sulfur so that the effect of equivalent amounts of sulfur present in different forms can be compared. Adapted from Amdur, 1973.[26]

0.1 mg/m^3 concentration of the smaller aerosol caused as much narrowing of the airways as did over 10 times the concentration of the larger aerosol. Ambient air may in the future contain sulfuric acid aerosols with two distinct mass median diameters, one between 0.4 to 0.7 μm and the other about one-tenth as large, i.e. 0.04 μm. Studies near St. Louis[144] indicate that the larger aerosol probably results from the oxidation of sulfur dioxide and the aging of the aerosol in the atmosphere. The smaller aerosol is generated by catalytic converters on cars and persists in this ultrafine state for a relatively short time before coalescence occurs. Whatever hazard may be posed by these ultrafine particles is probably confined to areas close to traffic.

The reaction chamber must be carefully controlled if such experiments are to provide useful information. Sulfuric acid is hygroscopic; its aerosol size is determined by the ambient relative humidity (RH). Growth in the respiratory system occurs within milliseconds. This growth rate is influenced by the original particle size; i.e., the larger the particle, the longer the time required to reach equilibrium at the RH of the respiratory tract (Figure 7-5). To reach equilibrium at 98% RH, a crystal of sodium chloride with a diameter of 0.3 μm would require about 0.1 sec; a 1 μm crystal would require about 1.0 sec.

If a droplet of sulfuric acid were to impinge on the respiratory surface before hydration were complete, its pH would be correspondingly

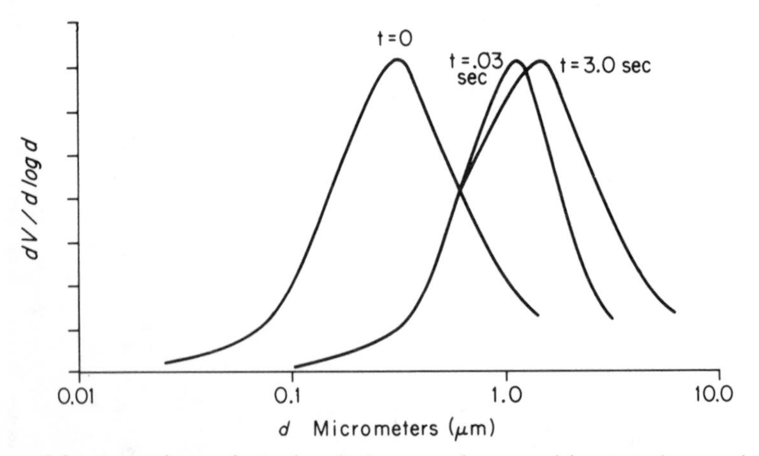

Figure 7-5. Approximate change in submicrometer log-normal hygroscopic aerosol distribution upon exposure to 99% RH. Example is for NaCl aerosol; H_2SO_4 would cause similar behavior. dV/d log d refers to the change in volume of the aerosol per unit log change in diameter. The area under each curve represents total aerosol volume. Smaller particles grow proportionately more than larger particles. After 0.3 sec, growth of the smaller particles is complete (ascending portions of curves t = 0.3 sec and 3.0 sec are identical), while the larger particles continue to grow. Provided by D. S. Covert, Ph.D., Departments of Civil Engineering and Environmental Health, University of Washington.

lower than at full hydration, a factor that might influence the biologic response. The pH of the particles is affected by hydration. It is lowest prior to inhalation and rises toward neutrality as the particles take on water. If a particle impinges on a mucosal surface before growth is complete, the pH at the instant of impact will be relatively low. Hence, in terms of the type and magnitude of biologic response that might be elicited by sulfuric acid or other hygroscopic aerosols, there is a complex interrelation between the size, RH, pH, and deposition site (see Chapter 6).

Changes in breathing pattern and mechanical status of the respiratory system will add to this complexity. As a result of the transition from slow to rapid and from nasal to oral breathing that occurs with exercise, a particle entering the airways with a specific aerodynamic size may impinge at a different site after having undergone a different degree of hydration and a different change in pH. A recent hypothesis[440] suggests that the ammonia excreted from the lung may convert sulfuric acid aerosol to ammonium sulfate while it is still airborne within the respiratory system, thereby increasing the pH of the aerosol considerably. Such an interaction would protect the lung. Limited knowledge precludes assessment of the contributions of these factors to the functional changes.

Because the guinea pig has been used so extensively in these studies, a brief comment about a few distinctive structural features of its lung is in order. The walls of the bronchi and especially of the terminal airways are thicker than in other laboratory animals or humans, and there is an abundance of smooth muscle in the pleura.[529, 545] In larger laboratory animals and in healthy young human adults, the peripheral airways are responsible for only a small fraction of the total R_L; in guinea pigs, this fraction may be much larger. Recent evidence indicates that, in humans, narrowing of the peripheral airways is chiefly responsible for the increase in both aging[579, 700] and emphysema.[700] The prominence of the bronchiolar wall in infancy suggests that peripheral flow resistance may also be relatively high at this stage of life. Infants appear to be more prone to "obstructive bronchiolitis" than adults. It is conceivable that the guinea pig's lung may be an appropriate analog of the infant's lung, with respect to the bronchiolar segment.

For this reason, narrowing of the distal airways may go undetected (these airways have been termed a "silent" zone[532]) unless other tests sensitive to the distribution of ventilation are employed. Such tests have been used to demonstrate functional impairment of the peripheral airways in circumstances as diverse as cigarette smoking,[522] coal mining,[439] and exposure to ozone.[56] In guinea pigs the smooth muscle layer of the peripheral airways, especially of the bronchioles, is thick.[529] It is therefore possible that the bronchiolar or peripheral flow resistance in these lungs is correspondingly large. The extent to which the peripheral airways of the guinea pigs were implicated in the changes in R_L produced by aerosols or irritant gases is not known. This question is important because evidence suggests that damage to these airways can be a forerunner of chronic pulmonary disease. By the same token, there is a compelling reason for studying species of animals, including primates, having a distribution of central and peripheral flow resistance more typical of the adult human lung.

Research with human subjects has focused on chemically inert dusts and powders, as well as sulfuric acid. Both healthy persons and patients with chronic ailments of the lung have been tested. In two of these studies[203, 489] the particles, which were <1 μm, caused an increase in airway flow resistance after brief administration. Abnormal intrapulmonary mixing and gas-trapping also occurred.[489] In a study using a polydispersed aerosol of calcium carbonate with particles from 0.2 to 5.0 μm diam, a variety of functional defects were elicited in both healthy subjects and patients with lung disease.[581] Principal among the defects were an uneven distribution of ventilation and impaired gas exchange, measured by an increase in the difference between alveolar and arterial oxygen tensions. The healthy subjects generally required longer exposure to elicit a response,[489] while recovery may have been delayed among asthmatic patients. None of these studies reported the concentration of the aerosols.[581]

In one study, sulfuric acid aerosol was perceived as a tickling or scratching sensation in the throat at a mean concentration of 0.72 mg/m^3 (range: 0.6 to 0.85 mg/m^3).[107] In an earlier study, the threshold of detection by odor, taste, or irritation was 1 mg/m^3; however, concentrations as low as 0.35 to 0.5 mg/m^3 caused an increase in rate of breathing and reductions in tidal volume and peak inspiratory and expiratory flow rates (see previous discussion under SULFUR DIOXIDE). Those changes occurred within minutes of the onset of exposure; the mean diameter of the particles was 1.0 μm. Higher concentrations of sulfuric acid, ranging from 3 to 39 mg/m^3 (1 μm MMD at 62% RH), have been shown to increase flow resistance, the effect being exaggerated at an RH of 91%.[701] It is worth noting that the concentration of sulfuric acid found in community air is in the order of 10 μg/m^3.[143]

Chronic Exposure

Fly ash and sulfuric acid mist have been administered to animals for periods in excess of 1 year. Monkeys exposed to 0.16 and 0.46 mg/m^3 of fly ash for 18 months incurred no functional impairment. Slight fibrosis of the lung was noted only at the higher concentration.[501] Sulfuric acid mist has been administered to both monkeys and guinea pigs in a range of concentrations and sizes.[6] The guinea pigs were unaffected. In monkeys, generally in response to higher concentrations of sulfuric acid, the distribution of the inspired air within the lung became uneven or abnormal, and the rate of breathing was increased. Respiratory mechanical behavior did not change. Histologic changes were frequent, consisting principally of epithelial hyperplasia and thickening of the bronchiolar walls. These structural changes probably contributed to the uneven alveolar ventilation. Because the mass concentrations were not constant in these experiments, it is difficult to draw conclusions about the relative effects of particles of different sizes.

In another study, four beagles were exposed daily for 620 days to 0.9 mg/m^3 of sulfuric acid, about 90% of the particles being less than 0.5 μm diam.[463] The dogs were reported to have developed a variety of functional defects by the end of the period, including reduction in the single-breath carbon monoxide diffusing capacity, in C_L, and in residual volume and total lung capacity, and an increase in R_L. Another set of animals exposed to 48.9 mg/m^3 (26 ppm) of nitrogen dioxide for 191 days prior to receiving the sulfuric acid was reported to have been less responsive to the sulfuric acid. Unfortunately, these conclusions are difficult to weigh because the information provided about the functional techniques and the presentation of the results was inadequate.

Guinea pigs have been exposed to sulfuric acid aerosols for periods of 5 days,[108, 109] and from 18 to 40 days,[752] and the effects were assessed by histologic examination. The 5-day exposures to concentrations of 1 and 2 mg/m^3, particle size not specified, produced lesions of the airways and parenchyma, including edema, that persisted several weeks.[108, 109] During the exposures of from 18 to 40 days,[752] three particle sizes, 0.6, 0.9, and 4.0 μm diam, were administered in concentrations ranging from 1 to 4 mg/m^3. The inflammatory response was confined to the larynx and lower airways. The greatest damage was elicited by the 0.9 μm diam aerosol.

Research on the toxicity of aerosols has not infrequently led to an overinterpretation of technically weak or insufficient data. In some studies, whose objective was to determine the importance of particle size in producing an effect, the aerosols being compared were polydispersed while differing in average size by only a few tenths of a micrometer, or they were administered in unequal concentrations. Criteria documents of the Department of Health, Education, and Welfare, although commendable in their goals, often cite such work uncritically.

GAS-AEROSOL INTERACTIONS

When an irritant gas is combined with an aerosol, the resulting toxicity may exceed that attributable to the two agents administered separately. The aerosol administered alone may have no physiologic effects or could be an irritant. The heightened, exaggerated response is referred to as "synergism." Principal among the pollutant gases that have shown the potential for synergism are sulfur dioxide and formaldehyde.[28] Two explanations of the phenomenon have been proposed. One is physical; that is, the gas is adsorbed by the aerosol and thus transported to a more sensitive or vulnerable site. This mechanism is thought to apply to highly soluble gases like ammonia and sulfur dioxide. Administered alone, each gas would be removed from the airstream almost entirely in the upper airways. The aerosol intervenes as a carrier, thereby increasing the penetration of the gas to the lower airways. LaBelle *et al.*,[429] through empirical observations on mice, and Goetz,[281] in a theoretical paper, gave credence to this hypothesis. The second explanation is chemical; that is, the two or more elements react to form a new molecular compound that is more toxic than the parent gas or particle. As an illustration, Amdur and Underhill[32] and Amdur[25] propose that the synergism resulting from the administration of sulfur dioxide and a submicrometer aerosol of sodium chloride to guinea pigs is attributable to the formation of sulfuric acid, which exaggerates the change in R_L produced by sulfur dioxide.

Experiments concerning synergism have elicited great interest because the phenomenon provides a basis whereby low concentrations of common ambient pollutants may be rendered more damaging to the lungs. Having observed the effects of mixing sulfur dioxide with a number of aerosols, Amdur and Underhill[32] concluded that synergism occurred only if the particles were soluble in water. The degree of synergism appeared related to the solubility of sulfur dioxide in the solutions of salts used to generate the aerosols. Accordingly, the relative effects from mixtures of sulfur dioxide and aerosols, measured in terms of changes in R_L, were greatest with ammonium thiocyanate, intermediate with potassium chloride, and least with sodium chloride. Synergism occurred sooner when sulfur dioxide was combined with soluble catalytic salts of manganese, vanadium, and iron (ferrous). This observation was cited as evidence that sulfur dioxide had been oxidized to sulfuric acid. By contrast, synergism was not produced when the gas was mixed with insoluble aerosols including carbon, activated charcoal, fly ash, ferric oxide, and manganese dioxide.[32]

Amdur and Underhill hypothesized that the interaction between gas and aerosol occurred within the respiratory tract, after the aerosol became a droplet. In all probability, most of the aerosols used in these experiments were "dry" prior to inhalation so that any absorption of, or interaction with, the gas in the reaction chamber would have been negligible. The requirement that the reaction occur within the airways harbors two conceptual difficulties. One is that while the formation of aerosol droplets in the upper airways may be quite rapid, perhaps requiring only milliseconds at body temperature and high RH,[588] sulfur dioxide is simultaneously absorbed at a rapid rate by the surrounding mucosal surfaces.[35, 255, 715] The surface area provided by an aerosol having a mass concentration of 1 mg/m^3 is negligible compared with that of the competing surrounding nasal passages. (Comparative surface areas: nasal mucosa = about 160 cm^2; aerosol, 1 mg/m^3 = 3×10^{-5} cm^2.) Therefore, exorbitant concentrations of the aerosol and gas may be required to counteract the imbalance in competing absorptive surfaces if a significant amount is to be incorporated in the droplet and carried to the lower airways. This may account for the observation that synergism between sulfur dioxide at 5.3 to 700.0 mg/m^3 (2 to 265 ppm) and the sodium chloride aerosol was greater when the dry aerosol was adminstered at a concentration of 10 mg/m^3 than at 4 mg/m^3,[24] and that other investigators found no synergism when the dry concentration of the gas was only 2.6 mg/m^3 (1 ppm) and that of the aerosol only 1 mg/m^3.[526] The second conceptual difficulty is that the catalytic oxidation of sulfur dioxide to sulfuric acid by the droplet aerosols of manganese, vanadium, or iron proceeds too slowly to have occurred to a significant degree within the respiratory system.

Another hypothesis or plausible mechanism for synergism places the chemical transformation of the gas and the production of the "irritant" aerosol in the ambient atmosphere prior to inhalation (see Chapter 3). This hypothesis has been tested by administering sulfur dioxide 2.6 mg/m^3 (1 ppm) and sodium chloride aerosol (1 mg/m^3) to guinea pigs at both low and high RH.[526] Sodium chloride aerosol is deliquescent; it is a dry crystal below 68% RH and rapidly becomes a droplet at higher RH (Figure 7-6). Below 40% RH, this combination caused no change in R_L; above 90% RH, there was a significant increase in R_L consistent with bronchoconstriction (Figure 7-7). The droplet absorbed sulfur dioxide, as evidenced by the reduced concentration of gaseous sulfur dioxide in the reaction chamber leading to the animal. The pH of the droplet fell below 4, and sulfurous acid, but not sulfuric acid, was found. It was not determined whether the irritant effect was caused by the physically dissolved sulfur dioxide, increased hydrogen concentration, the bisulfite ion, the sulfite ion, or by some chemical species produced in the tissue liquids following deposition of the aerosol. Alarie *et al.*[9] reported that sodium metabisulfite was irritating to the upper respiratory tract of mice, whereas sodium sulfite had no effect. In his experiment the release of sulfur dioxide from the metabisulfite could have been responsible for the sensory irritation. At a lower ambient temperature, the amount of sulfur dioxide dissolving in the droplet should increase. It would be interesting to test this additional "environmental" factor for exaggeration of synergism. There is ample epidemiologic evidence to indicate that the effects of air pollution on health are exaggerated in cold, damp weather (see Chapter 8).

A mixture of approximately 3.8 mg/m^3 (2 ppm) nitrogen dioxide and 330 $\pm$ 110 μg/m^3 sodium chloride aerosol has been administered at $<$ 75% RH to rats and monkeys for 14 months without causing

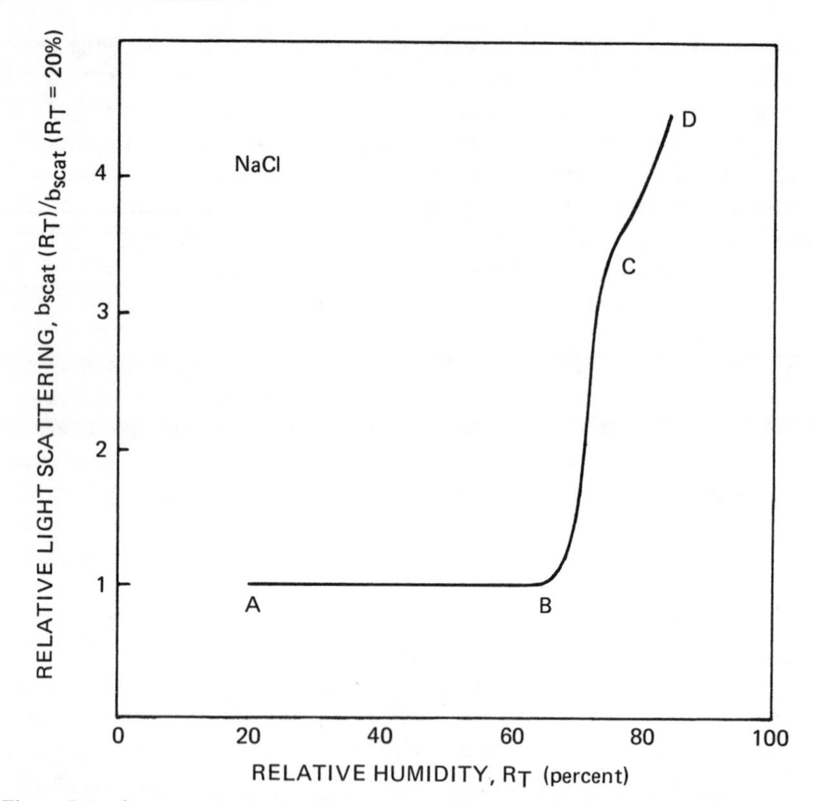

Figure 7-6. b_{scat} = extinction coefficient due to light scattering by both particles and gas molecules. R_T = relative humidity. Deliquescence occurs at a relative humidity of about 70%. From Covert, 1971.[166]

structural changes in the lungs greater than those caused by the same concentration of the gas alone.[265] At this RH the likelihood of a significant interaction between nitrogen dioxide and the aerosol would be small. The sodium chloride aerosol administered alone was innocuous.

Attempts to demonstrate synergism in humans have yielded conflicting results. Japanese investigators[569, 759, 760] reported synergistic responses with mixtures of sulfur dioxide [2.6 to 159.0 mg/m^3 (1 to 60 ppm)] and either sodium chloride (5 and 6 mg/m^3) or hydrogen peroxide (0.01 to 0.1 mg/m^3 and 0.8 to 1.4 mg/m^3) aerosols. Synergism was reported between sulfur dioxide and sodium chloride for particles 0.95 μm, but not for 0.22 μm in count median diameter.[569] Mixtures of nitrogen dioxide [11.3 to 75.3 mg/m^3 (6 to 40 ppm)] and the aerosol were also synergistic.[569]

These studies[569, 759, 760] relied on a method of rapid, repetitive interruption of gas flow at the mouth to measure flow resistance.[155] In at least one study,[759] this interruption method may have been used improperly, judging from an illustration provided by the author. The brief period of interruption should produce a relatively stable or "static" airway pressure, but Figure 2 of the referenced article shows that airway pressure changed rapidly during that period. It is not possible to judge whether the same methodologic problem was present in the other two studies.[569, 760]

Several groups of investigators in the United States have failed to elicit synergism between sulfur dioxide and sodium chloride aerosol.[106, 252, 709] The aerosol in one study had the same average size and concentration as that used successfully in guinea pigs.[252] None of these reports provided information about ambient RH. An experiment on anesthetized

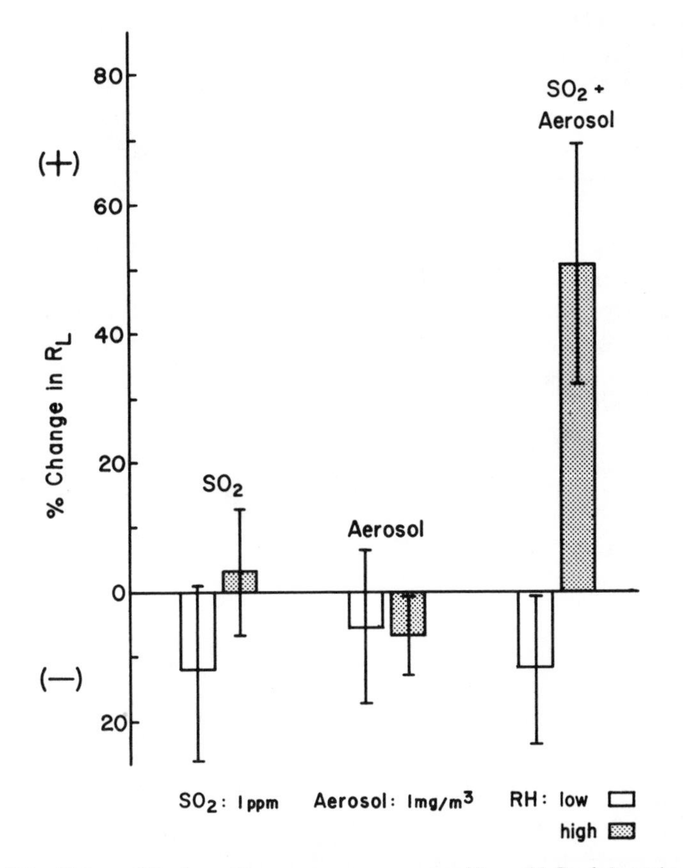

Figure 7-7. Values of R_L from 12 exposures were averaged (cm $-$ H_2O ml^{-1} sec^{-1}; means $\pm$ SE) for each of the six exposure modes (sulfur dioxide; in combination with sodium chloride aerosol, 1 mg/m^3). The results of the first and second exposures were not significantly different and were therefore combined. R_L was measured at approximately 4-min intervals. The increase in R_L for mode 6 exceeds the other changes by the following: compared to mode 2, $p < 0.05$; compared to modes 1, 3, 4, and 5, $p < 0.01$. From McJilton *et al.*, 1973.[526] Copyright 1973 by the American Association for the Advancement of Science. Reprinted with permission.

cats also gave no evidence of synergism with mixtures of sulfur dioxide and sodium chloride aerosol.[164]

CONCLUSIONS

These findings support the growing conviction that sulfur dioxide alone, in realistic ambient concentrations, is not especially hazardous to health. Instead, what appears to be of concern are respirable aerosols, many of which may result from oxidative reactions involving sulfur dioxide. These reactions and the generation of the aerosols occur to a large degree in the ambient atmosphere (see Chapter 2). For this reason it would appear necessary to control the emission of sulfur dioxide if the aerosols are to be held to acceptable levels.

Among the factors shown to influence the toxic potential of sulfate aerosols, and presumably other aerosols as well, are their mass concentration (little attention has been paid to molar concentration), molecular composition, and perhaps acidity. Whether sulfite and bisulfite ions contribute to toxicity is uncertain.

While there is an abundance of information on the toxicity of nitrogen dioxide in animals[172] and some epidemiologic evidence suggesting that nitrogen oxides may affect health,[694] there appear to have been no toxicologic studies involving nitrate aerosols.

8

Epidemiologic Studies on the Effects of Airborne Particles on Human Health†

Most epidemiologic studies of air pollution have been concerned primarily with such particles as mineral ash, carbonaceous material, sulfur oxides, and hydrocarbons that arise from the burning of fossil fuel (see Chapter 1). In the United States most data have been taken from high-volume samplers, which collect a fair number of larger particles outside the respirable range. In Europe, the British samplers used most often collect particles in the respirable size range ($<5\ \mu m$). It is therefore difficult to compare measurements made in Europe with those made in the United States. Furthermore, because the British method, which uses reflected light, is calibrated against a standard British smoke,* it cannot be used in the United States, where the smoke is different, nor in areas where there may be considerable white or colored particles as, for example, around a kraft mill.

Unless particle size is adequately considered, trends in pollution measured by a high-volume sampler may be quite misleading. For example, by removing the larger, nonrespirable particles, the mass concentration may be reduced without changing the respirable fraction. This may have little or no effect on health. In most studies, sulfur dioxide is the only sulfur compound that has been measured. The reactions that the gas undergoes in the atmosphere, particularly its oxidation to sulfuric acid, conversion to sulfates, and interaction with other gases, such as ozone and particulates, are described in Chapter 4. The sulfur dioxide, suspended particulate, or smoke concentrations often referred to in

† Portions of this chapter have been adapted from a paper presented by Dr. Ian T. T. Higgins, Chairman of the Committee on Airborne Particles, to the Society of Automotive Engineers (SAE) as part of its continuing education program. Subsequently, that paper was published with others in *Medical Aspects of Air Pollution: A Continuing Engineering Education Course*,[334] which was developed by the SAE Engineering Education Activity. Permission to use this material has been obtained from the Society of Automotive Engineers.

* In Britain, smoke samples are collected on filter paper and the darkness of the resulting stains is assessed photoelectrically. Readings are interpreted as if they had been derived from stains made by a standard urban smoke, using a formula giving the weight of equivalent standard smoke per unit volume of air.

epidemiologic studies should be considered as indices of pollution and not as specific pollutants causing certain effects. Pollution that arises from the presence of specific particles, such as lead, beryllium, or asbestos, or from viable particles, such as pollens, is not considered in this chapter. In addition to the ambient air pollution, pollution may also arise from occupational exposures and from tobacco smoking, especially cigarette smoking. It is often difficult to assess the relative contribution to ill health of these three forms of pollution.

ACUTE EPISODES

The acute episodes of high pollution, such as those that occurred in the Meuse Valley, Belgium;[238] Donora, Pennsylvania;[678] London;[483] New York City;[296, 297] Osaka, Japan;[790] and elsewhere, provide the most convincing and dramatic evidence for an effect of air pollution on health. It has been documented for over a century that smog can kill. The report of the British Ministry of Health on the London fog of 1952[546] records lethal fogs in London going back as far as 1873 and subsequent similar fogs in London and other British cities during the 1950s and 1960s.[83, 683-685, 811, 812] In New York City,[278, 294, 295, 519] episodes of high pollution have also led to increased mortality but these do not appear to have resulted in as many excess deaths as in London, possibly because the pollutant concentrations were lower or less continuously sustained in New York City than in London. In Britain, deaths arising from fogs were most often attributed to bronchitis, but also to coronary disease, myocardial degeneration, pneumonia, and other respiratory diseases. In New York City, deaths were attributed to influenza, pneumonia, vascular lesions of the nervous system, and cardiac diseases.[278] In both cities, morbidity as well as mortality increased during periods of sustained pollution, though again, the increase in morbidity in New York City has sometimes been less clearly associated with air pollution than has been usually the case in London.

In New York City, increasing pollution has sometimes been clearly linked to increasing symptoms. In a study conducted during the Thanksgiving holiday fog in 1966, Becker and his colleagues[58] showed that as pollution increased complaints of cough and phlegm, wheezing, breathlessness, and eye irritation also increased. Those with preexisting heart or lung disease and those aged 45 and over appear to have been predominantly affected. Mortality has been related to concentrations of routinely measured pollutants. In London, excess deaths resulted when smoke concentration exceeded 2.0 mg/m^3[286] and sulfur dioxide rose above 1.04 mg/m^3 (0.4 ppm).[102] In New York City excess deaths occurred when sulfur dioxide rose above 1.3 mg/m^3 (0.5 ppm) and particles were recorded at 6 or more coefficient-of-haze units (COHS).[278]

In Britain, fogs usually occurred in cold weather. Farr reported an influence of cold on mortality in 1841, and this has been confirmed many times since.[632] Scott and his colleagues[686] concluded that pollution caused deaths within 3 days but that cold alone acted more slowly, causing deaths after about 9 days. They concurred with Russell[656] that it was not possible to distinguish the respective shares of fog and low temperature on the genesis of mortality. The relative contribution of low temperature, humidity, sulfur dioxide, and smoke to mortality in London and a rural area in Britain was studied by Boyd,[81] who found that death rates from bronchitis and pneumonia were more closely related to cold and humidity than to fog frequency.

There has never been much doubt that these exceptional episodes of high pollution could cause illness and death. But there was much more uncertainty 20 years ago about the effects of lower, more sustained concentrations of pollution and their long-term effects. Research during the past 20 years has related these lower concentrations to various health indices and has provided some working dose-response relationships.

FOLLOW-UP OF DONORA

That persons with chronic illness are unduly susceptible to pollution was supported by a 10-year review of a sample of inhabitants of Donora, Pennsylvania. Immediately after the 1948 disaster, a representative sample of the town was studied by the United States Public Health Service. Ten years later Ciocco and Thompson[150] followed up this sample. They found that those affected by the fog had a higher mortality and morbidity rate than those who were not. But most of the excess occurred in persons who had chronic illness before the fog. This suggests that the fog's main effect was exacerbation of existing disease in sensitive people rather than initiation of illness. However, the 1948 mortality rates were slightly but consistently higher in persons who had been affected by the fog but had no prior chronic illness. The authors also compared the prevalence of heart disease, asthma, arthritis, and rheumatism in survivors of the 1948 episode with that in two neighboring Pennsylvania communities that had not been affected by the fog. Because they found little difference between the communities, the authors concluded that the fog had had little effect on the prevalence of these diseases.

VARIATION IN MORTALITY, MORBIDITY, AND LUNG FUNCTION OVER TIME

Daily Deaths and Illnesses

The relationships of daily measurements of black suspended matter (e.g., soot) and sulfur dioxide to daily deaths in London were first reported by

Martin and Bradley in 1960.[509] They attempted to assess the effect on mortality of relatively minor fog incidents and to establish a dose-response curve. During the early years of their study there was a significant positive correlation between black suspended matter and total deaths, a slightly lower but still significant association between sulfur dioxide and deaths, and a negative association between visibility and deaths. Bronchitis deaths showed lower correlations than total deaths with these indices of pollution; pneumonia deaths showed no significant association. Subsequently, the study was extended to include morbidity based on certificates of incapacity issued by physicians. These data indicated similar correlations. Since about 1963, however, this study has produced little evidence of any effect of pollution on mortality or morbidity. This has been attributed to the marked decline in pollution in London since 1958. From 1958 to 1963, average annual concentrations of black suspended matter fell from 0.30 to 0.06 mg/m^3 and sulfur dioxide from 0.25 to 0.175 mg/m^3 (~0.10 to ~0.07 ppm). The decline in both pollutants has continued, but unfortunately no further analyses from this study[508] have been published after 1964. In London, the more effective reduction of particulate matter than of sulfur dioxide has sometimes resulted in high 24-hr sulfur dioxide concentrations without correspondingly high particulate concentrations. Apparently, such peaks of sulfur dioxide have not been associated with excess mortality or morbidity.

In New York City, McCarroll and Bradley[519] first showed that excess mortality occurred during relatively minor episodes of pollution. Four groups of workers explored the effects of day-to-day variation in pollution in a more systematic manner. Glasser and Greenburg[277] related daily deaths to concentrations of sulfur dioxide and particulates measured by the smoke shade method‡ as well as temperature and certain other weather variables from 1960 to 1964. They noted a pronounced increase in mortality when daily sulfur dioxide concentrations exceeded 0.52 mg/m^3 (0.2 ppm). These authors estimated that the difference in the average number of deaths between days when sulfur dioxide was $\leq$0.52 mg/m^3 (0.2 ppm) and days when it was $\geq$1.04 mg/m^3 (0.4 ppm) was 10 to 20 deaths per day. Hodgson's[352] studies, from 1962 to 1965, indicated that daily deaths from respiratory and cardiac disease were highly correlated with indices of pollution, such as particle concentration. Presumably, their smoke shade measurements indicated a higher correlation than measurements of sulfur dioxide concentrations. Mortality from other diseases was not significantly related to levels of air pollution. Hodgson estimated that 73% of the day-to-day variation in

‡ In this method, the darkness of the filter stain, as measured by a reflectometer, is used to calculate concentrations of smoke or other dark suspended matter usually produced by combustion. It is rarely used in the United States at this time.

cardiac or respiratory deaths could be explained by variations in air pollution or temperature. This seems remarkably high. It is questionable whether or not adequate allowance was made for such other factors as influenza epidemics or seasonal influences.

In their 5-year study from 1962 to 1966, Buechley and his colleagues[100] gave particular attention to temperature, season, day of the week, influenza epidemics, and holidays. Heat waves, season, and influenza epidemics were the main determinants of mortality; other factors, such as day of the week, were also significant. After allowing for these, mortality still correlated with pollution and COHS correlated as highly with daily deaths as did sulfur dioxide concentrations. Deaths were 1.5% less than expected on the 232 days when sulfur dioxide concentrations were <0.0003 mg/m^3 (0.0001 ppm) and 2% greater than expected on the 260 days when sulfur dioxide concentrations exceeded 0.5 mg/m^3 (0.19 ppm). These authors present a dose-response curve that perhaps can best be interpreted as showing no threshold of sulfur dioxide concentration for mortality.

Finally, the initial study of Schimmel and his colleagues, which covered the 6-year period from 1963 to 1968, was extended 4 years to 1972.[671, 672] Unlike Buechley and his coworkers, who studied the entire New York Metropolitan area, Schimmel *et al.* limited their study to deaths in New York City. They included temperature and season in their analysis, but less thoroughly than Buechley *et al.* The highest estimate of deaths attributable to pollution was obtained using crude (uncorrected) pollution levels. Lower estimates were obtained when these were corrected. The authors calculated that the average daily excess deaths ranged from 18.2 to 36.7 (median 28.6) depending on the degree to which the effect of air pollution is standardized for temperature and other variables. The intermediate figure represents 12% of the more than 0.5 million deaths that occurred in the area between 1963 and 1968. From 1968 to 1972 mortality rates remained similar to those from 1963 to 1968 despite a large reduction in sulfur dioxide concentrations in the city. This finding suggests that sulfur dioxide itself cannot be the sole cause of excess mortality. Possibly mortality is more closely related to other pollutants, for example sulfates, which unlike sulfur dioxide, may not have declined. Alternatively, demographic changes in the population may have offset any reduction in mortality that might have been expected from the reduction in sulfur dioxide.

Buechley, Schimmel, and their colleagues based their studies on similar material, but used very different analytic methods. Their estimates of the effects of pollution on mortality were similar. They concluded that sulfur dioxide pollution might be responsible for about 3% of the deaths. Further study of these data is needed. Schimmel and Greenburg suggested that these studies should include analysis by age and sex, additional environmental monitoring stations, and analyses of more pollutants, notably sulfates.

Longitudinal Observations on Morbidity

Studies of day-to-day variation in mortality based, as they must be, on analysis of death certificates provide little information on the relevant characteristics of persons afflicted by pollution. To obtain these data, a group of people should be closely observed. From 1961 to 1966 Fletcher and Angel, with their respective colleagues,[39, 243] regularly observed over 1,000 men aged 30 to 59 living in North London. During the winter of 1962 to 1963, they monitored a subsample of 87 men more intensively. The investigators discovered that the incidence and prevalence of respiratory illnesses in this subsample were related to both smoke and sulfur dioxide concentrations in the district in which the men lived. Incidence of illnesses was related equally to smoke and sulfur dioxide concentrations; prevalence was, however, more closely related to smoke concentration. This group showed no significant correlation of respiratory illnesses with low temperature. The weekly smoke concentrations in London at that time were approximately 0.3 mg/m^3 (0.16 ppm) with peaks from 0.8 to 1.0 mg/m^3 (0.31 to 0.38 ppm); sulfur dioxide concentrations ranged from 0.4 mg/m^3 (0.15 ppm) to peaks of $\geq$1.2 mg/m^3 (0.46 ppm).

During the 6-year study, there was a consistent decline in the volume of morning sputum produced by all the subjects. This could have been due to the concurrent reduction in air pollution in London but it might also be attributable to the simultaneous reduction in the tar content of cigarettes.

During this same period (1960 to 1965) Lawther and his colleagues[451] demonstrated a progressive increase in peak expiratory flow rates by testing four trained subjects regularly. For 2 years, McCarroll and his coworkers[520] studied 1,860 persons living close to a monitoring station on Manhattan Island. These subjects reported on 25 symptoms daily and provided much more information weekly through interviews. The complaints of 1,090 persons indicated that cough and eye irritation were positively correlated with changes in the concentrations of sulfur dioxide and particulates. Eye irritation was immediate whereas the maximum effect on cough occurred 1 or 2 days after the exposure.

The possibility that respiratory disease might enhance susceptibility to air pollution prompted Waller, Lawther, and their colleagues[445, 450, 451, 781, 782] to study chest clinic patients with chronic bronchitis and emphysema. Each patient recorded daily whether his chest felt better than, the same as, or worse than on the previous day. The investigators calculated a mean score for all patients each day. Daily symptoms and pollution indices correlated highly when pollutant concentrations were exceptionally high.

During the winter of 1959–1960 Waller and Lawther observed approximately 1,000 patients. Correlations between symptoms and concentrations of smoke and sulfur dioxide were again high and positive. Five years later (winter, 1964–1965), when the observations were repeated following a considerable decline in the smoke concentration in London, the correlations were lower;[450] 2 years after this (winter, 1967–1968) the most susceptible patients in 1964–1965 showed even lower correlations. Patients appeared most sensitive to changes in pollution during early winter, suggesting some degree of acclimatization. The minimum daily concentrations that appeared to lead to a significant deterioration in chest condition were ~0.5 mg/m^3 (0.19 ppm) of sulfur dioxide and 0.25 mg/m^3 of smoke.

Waller and Lawther made no attempt to study lung function changes in their patients, although they observed daily a few healthy and bronchitic subjects (Lawther *et al.*, 1974).[449] Ventilatory lung function was studied, however, by Emerson in 1969 and 1970.[216] He observed 18 patients with chronic obstructive lung disease for 12 to 82 weeks. His weekly measurements of forced expiratory volume in 1 sec ($FEV_{1.0}$) and the midexpiratory flow rate (MEFR) were correlated with the average concentrations of smoke and sulfur dioxide on the day of testing and the previous 4 days. In no case did he observe a significant correlation of lung function with smoke concentration. A significant negative correlation of $FEV_{1.0}$ with sulfur dioxide occurred in one patient and of MEFR with sulfur dioxide in two patients. On the other hand, $FEV_{1.0}$ correlated negatively with temperature in six of the 18 patients and positively with relative humidity (RH) in four of them. At that time the average annual smoke concentration in London was 0.044 mg/m^3 with a maximum of 0.241 mg/m^3 and the average annual sulfur dioxide concentration was 0.193 mg/m^3 (0.074 ppm) with a maximum of 0.722 mg/m^3 (0.278 ppm).

Emerson concluded that these pollutant concentrations were too low to affect the lung function of patients with chronic obstructive lung disease. Unfortunately, the concurrent sulfate concentrations were not measured. They presumably varied with sulfur dioxide concentrations. This assumption suggests that sulfates did not affect ventilatory function. In London, Lawther has recorded sulfate concentrations of 0.015 mg/m^3 for 24 hr with an hourly maximum of 0.678 mg/m^3.

Two studies in Chicago have related symptoms in bronchitic patients to environmental factors. During 1963 and 1964 Burrows and his coworkers[105] asked a group of 115 patients to record their symptoms in a diary. Daily symptom severity correlated closely with low temperature and high sulfur dioxide concentrations. There were also significant correlations with nitrogen oxides, snowfall, and several other meteorologic and air pollution measurements. Most of these associations, however, were due to the similar seasonal pattern of respiratory symptoms and environmental conditions. When temperature and season were constant, only hydrocarbon concentration showed an independent positive correlation with symptoms.

Carnow and his colleagues[125] used a similar approach in their study of patients in Chicago. Observations of 561 patients enrolled in a bronchopulmonary disease registry indicated an increase in the frequency of acute chest illnesses with increasing sulfur dioxide concentration in men aged 55 and older with advanced bronchitis. When sulfur dioxide concentrations were 0.78 mg/m^3 (0.3 ppm) or higher, the percentage of days of chest illness was twice as high as when the concentrations were 0 to 0.104 mg/m^3 (0 to 0.04 ppm). In contrast to the younger men, those men aged 55 and over, with less severe bronchitis, and women showed no relation between frequency of chest illnesses and sulfur dioxide concentrations. Some of the apparent difference between these two studies may have been due to an inadequate allowance by Carnow *et al.* for meteorologic factors, notably temperature. In neither study were particulate concentrations measured. They were presumably high, since the average annual concentration in Chicago at that time, according to the National Air Sampling Network (NASN), was about 0.150 mg/m^3 (0.058 ppm). Although the implication of an increase in chest illness with sulfur dioxide concentration $\geq$0.78 mg/m^3 (0.3 ppm) is compatible with other findings, the suggestion that sulfur dioxide may influence respiratory symptoms at lower concentrations should be treated with reserve.

Changes in lung function attributable to environmental factors were reported by Shephard *et al.*[691, 692] and Spicer *et al.*[717] Shephard and his colleagues studied 10 patients with advanced respiratory impairment for 3 months. They noted negative correlations between several aspects of lung function and measurements of absolute humidity and suspended particulate concentrations. The changes were attributed to bronchospasm that might occur when particulate pollution reached 8 COHS or more. Spicer and his colleagues reported day-to-day changes in lung function in small numbers of healthy subjects and patients with chronic respiratory disease. They also monitored these changes in function over a longer period in a larger group of healthy subjects. This indicated a single cycle of function changes from October to May with lowest levels in February and March. While changes in airway resistance were correlated with low temperature, they were not obviously related to concentrations of either total suspended particulate or sulfur dioxide. Nevertheless, these authors concluded that variations appeared to occur in the group concurrently, suggesting the influence of some common environmental factor or factors. Bates and his coworkers[55, 57] repeatedly examined Canadian veterans with chronic bronchitis in Winnepeg, Montreal, Halifax, and Toronto. They tested many aspects of lung function. The cities vary in pollution; Winnepeg was by far the cleanest with respect to industrial dust fall and sulfur dioxide. Bates found that veterans in Winnepeg were less severely affected than those in the other three cities; their lung function was better and, over a 4-year period, showed no significant decline. Whether these effects were due to differences in pollution or to other causes, such as selection of patients, is uncertain.

Sick Leave in Relation to Pollution

Absence from work due to illness is often a useful index of air pollution effects. Dohan and Taylor[197] and Dohan[196] reported a high correlation ($r = 0.964$) between the mean concentrations of suspended particulate sulfates in the air of five cities and the incidence of respiratory illnesses lasting 7 days or more among women RCA employees from 1957 to 1960. No correlation was observed with other pollutants, and only respiratory diseases were correlated with particulate sulfate concentrations. The effect of sulfates appeared to be greatest in years of epidemic influenza. Much of this correlation may have resulted because the effects of season and climate were ignored. Because both of these factors have been correlated with pollution, they should be carefully considered in analysis.

Ipsen *et al.*[379] studied weekly morbidity rates at the RCA Camden plant for the 3 years from 1960 to 1963, the Curtis Publishing Company in Philadelphia for the 2 years from 1960 to 1962, and the Bell Telephone Company in Philadelphia for the 2 years from 1961 to 1963. Morbidity rates were higher during the colder, windy months. Air pollutants in the ranges measured during the study did not contribute to respiratory morbidity. Extensive analyses correlating respiratory diseases with meteorologic and pollution variables were conducted. Mean monthly values of 24-hr samples of sulfate concentrations varied from 0.013 to 0.018 mg/m^3. Ipsen *et al.* noted a different pattern in the relationship of respiratory disease to climate compared with the relationship of pollution to climate. They believed, therefore, that it should have been possible to show an effect of air pollution on respiratory disease, if such an effect existed, but they were not able to do so. Concentrations of suspended particulate matter, sulfur dioxide, sulfates, COHS, and nitrogen dioxide were positively correlated with both the incidence and prevalence of respiratory disease, though with considerable scatter. But this appeared to be explained by the common influence of season on pollution and respiratory disease. Because concentrations of all five pollutants changed simultaneously, separation of individual effects was impossible.

Sterling and his colleagues[724, 725] studied hospital admissions in relation to pollution using records from the Blue Cross and Blue Shield of Southern California. Admissions were categorized as relevant, highly relevant, or not relevant to air pollution. After allowing for the confounding effect of day of the week on hospital admission, relevant admissions and deviation of stay were significantly correlated with concentrations of sulfur dioxide, nitrogen dioxide, oxidants, ozone, and particulate matter, but not with temperature or humidity. The correlation coefficients were, however, very low even if statistically significant. Taken in conjunction with the extremely low concentrations of particulates and sulfur dioxide that were reported in this study, it is difficult to accept them as representing cause and effect.

Geographic Comparisons

An obvious way of measuring the effects of air pollution is to compare people living in areas with different pollutant concentrations. Unfortunately, such comparisons may not be successful. People who live in areas with different air pollution usually differ in many other respects that could also influence their disease frequency. The degree to which such confounding factors have been taken into account varies greatly among studies. These factors should always be carefully considered when drawing conclusions about the effects of pollution.

International Differences

The large differences in bronchitis mortality among countries documented by the World Health Organization (WHO) and others during the 1950s were often partially attributed to differences in pollution. Much of the international variation is best explained by differences in diagnostic practice and certification of cause of death; however, a number of careful morbidity comparisons have supported the existence of real differences in the frequency of chronic respiratory disease among countries, some of which may be due to differences in air pollution.[148, 228, 358, 359, 552, 585, 634] Mork[552] found a higher frequency of respiratory symptoms and lower average peak flow rates in male transport workers, aged 40 to 59, in Bergen, Norway, than in similar subjects in London, England. He showed that these differences could not be explained by smoking habits or socioeconomic factors, but suggested they resulted from the exposures to air pollution of the two groups. Subsequently, Holland *et al.*[358, 359] compared post office van drivers and telephone workers in London and in three English country towns with outside telephone workers in three areas of the United States. All groups were aged 40 to 59. A higher prevalence of respiratory symptoms, a larger volume of morning sputum, and lower average ventilatory lung function were found in the English workers than in their American counterparts. Because the differences could not be explained by differences in sitting height or smoking habits, the authors attributed them to the variations in suspended particulate and sulfur dioxide concentrations.

Some difficulties encountered in international comparisons are indicated by Ferris and Anderson.[228] They compared the prevalence of respiratory symptoms and ventilatory lung function in representative samples of the inhabitants of two towns that differed in degree of air pollution: a one in 10 sample of adults living in Berlin, New Hampshire in 1961,[227] and one in seven sample of Chilliwack, British Columbia two years later.[37] Comparable measurement methods were used in each survey by the same observers, thereby eliminating a major source of variation. At the time of these studies, total suspended particulate concentrations were approximately 0.180 mg/m^3 and sulfation rates were 0.731 mg of sulfites (SO_3)/100 cm^2/day in Berlin, whereas Chilliwack was free of pollution. The prevalence of respiratory disease in nonsmoking men was essentially the same in both areas, but a slightly lower prevalence in nonsmoking women was recorded in Chilliwack. Lung function values were consistently slightly higher than expected in both men and women in Chilliwack. The differences were not explained by anthropometric, social, or economic factors. Although they were attributed to a small effect of air pollution, ethnic differences could not be excluded.

The findings in Berlin, New Hampshire were compared[636] with those from a study by the British College of General Practitioners[157] of a representative sample of persons in Britain. A similar questionnaire had been used in both studies. The concentrations of particulates and sulfur dioxide were lower in Berlin than in most British cities. The prevalence of persistent sputum production was similar in both countries, but the incidence of repeated chest illnesses and breathlessness was higher in Britain. After allowing for cigarette smoking, the prevalence of these symptoms appeared to be similar in Berlin and in rural areas of Britain.

DIFFERENCES WITHIN COUNTRIES

Urban/Rural

For many years British mortality statistics have shown that death rates from bronchitis are approximately twice as high in urban as compared with rural dwellers. The gradient is much the same for both men and women. Differences in socioeconomic status, occupational exposures, or smoking habits cannot explain this. In the United States the gulf between respiratory disease mortality rates in urban and rural areas is smaller for men, and is nonexistent for women.[504] These observations have stimulated a great deal of research.[98, 181, 182, 504, 606, 733]

Early studies of air pollution in Britain related mortality from bronchitis, other respiratory diseases, and certain forms of cancer to sulfation rates, sulfate (as dissolved impurity in deposit gauges), smoke concentrations, or dust fall.[102, 104, 181, 182, 325, 606, 733, 734] From 1950 to 1952 Pemberton and Goldberg[606] correlated bronchitis death rates in men and women aged 45 and over with sulfation rates in 35 county boroughs in

England and Wales. They found statistically significant positive correlations in men but not in women. There also appeared to be a much weaker association of bronchitis mortality with particulate concentrations. Daly[181, 182] supported these findings. He also developed an index of pollution based on domestic and industrial fuel consumption per acre, which he showed to be highly correlated with bronchitis death rates and to a lesser extent with a number of other respiratory diseases. Stocks[733, 734] reported high correlations of mortality from bronchitis and certain types of cancer with dust fall and smoke concentrations in a large number of different areas of England and Wales after allowing for population density.

None of these studies included smoking habits, which were not then widely recognized as overwhelmingly important to the pathogenesis of respiratory disease. However, Buck and Brown,[98] in a somewhat comparable analysis, incorporated a crude adjustment for regional differences in cigarette smoking. Allowance for socioeconomic factors and smoking differences did not abolish the close correlation between air pollution and mortality from bronchitis. The relative contributions of smoking and urbanization to bronchitis mortality were clearly shown in Northern Ireland by Dean.[192] However, the extent to which the much higher death rates in central Belfast can be attributed to air pollution is uncertain.

Gorham[287, 288] paid particular attention to sulfates when he correlated mortality rates from bronchitis and pneumonia with various pollution indices in 53 country and metropolitan boroughs in England, Wales, and Scotland from 1950 to 1954. He studied monthly tar deposit, ash, sulfate per unit area in urban areas, and pH of precipitation. Bronchitis mortality correlated positively with sulfate concentrations and negatively with pH of precipitation. Partial correlations revealed that the pH was significant but not the sulfate concentrations. Conversely, pneumonia mortality was related to sulfate concentrations but not to pH. Gorham interpreted his findings to indicate that the acid droplets deposited in the major bronchi produced bronchitis, whereas the sulfate material, which was carried deeper into the lung, resulted in pneumonia.

Using multiple regression, Lave and Seskin[442, 443] reanalyzed mortality data from Britain. Bronchitis death rates were fairly well correlated with pollution indices (deposit index, concentration of suspended particles, sulfation rate, and population density). Much lower correlations were reported with socioeconomic variables. These authors also studied 114 United States Standard Metropolitan Statistical Areas relating 1960 rates of mortality from all causes and infant mortality to air pollution and other factors. Total death rates varied with total suspended particulates and with sulfate. The correlations of infant mortality were slightly lower.

The limitations of mortality studies makes one appreciate the need for morbidity studies. There have been many such studies, based either on sick leave records or on the results of specially designed surveys. A pronounced urban-rural gradient of morbidity, such as that already shown for mortality, resulted from a survey of a representative sample of men and women patients of nearly 100 general practitioners.[157] In 1961–1962, this was confirmed by the British Ministry of Pensions and National Insurance[547] in its investigation of the incidence of incapacity from bronchitis and other illnesses in representative samples throughout Britain. Bronchitis incidence in relation to particulates and sulfur dioxide were analyzed regionally. Incapacity due to bronchitis significantly correlated with winter concentrations of smoke and sulfur dioxide in each of four 10-year age groups (Figure 8-1). The interpretation of these data is somewhat uncertain, since similar correlations between these pollutant concentrations and illnesses were attributed to arthritis and rheumatism. Socioeconomic differences may also have contributed to the associations.

Many surveys of respiratory disease have been conducted in different countries. Some were designed primarily to measure an effect of polluted air on the respiratory tract. Others, though primarily concerned with assessing the impact of such factors as occupational exposures or smoking habits, also permit some conclusions about the effects of air pollution.

Fairbairn and Reid[222, 633] studied postal workers who, as a uniform group, received much the same pay wherever they worked, and, once established, tended to remain in the same area. An effect of pollution, after allowing for the confounding effect of socioeconomic factors, should therefore be possible. Sick leave, premature retirement, and death due to bronchitis or pneumonia were closely related to thick and presumably polluted fog but not to domestic overcrowding or population density. Their pollution index, based on visibility, had been developed by Wilkins.[812] The pattern of disability among the postal workers was similar to that of mortality in the whole British population.

Cornwall and Raffle,[165] in a comparable study of London transport workers, also showed that absence rates due to bronchitis were closely correlated with thick, polluted fog in London and that the rates for men working in the more rural, peripheral areas were lower than those working in central London. In neither study were smoking habits considered. But in a further study of a sample of men and women aged 35 and over, Lambert and Reid[432] showed that urban-rural differences in respiratory symptom prevalence could not be explained by smoking differences. Nearly 10,000 completed questionnaires, believed to represent 74% of a defined population aged 35 to 64, were analyzed. The prevalence of respiratory symptoms increased with increasing air pollution. This increase was confined mainly to smokers. Pollution appeared to

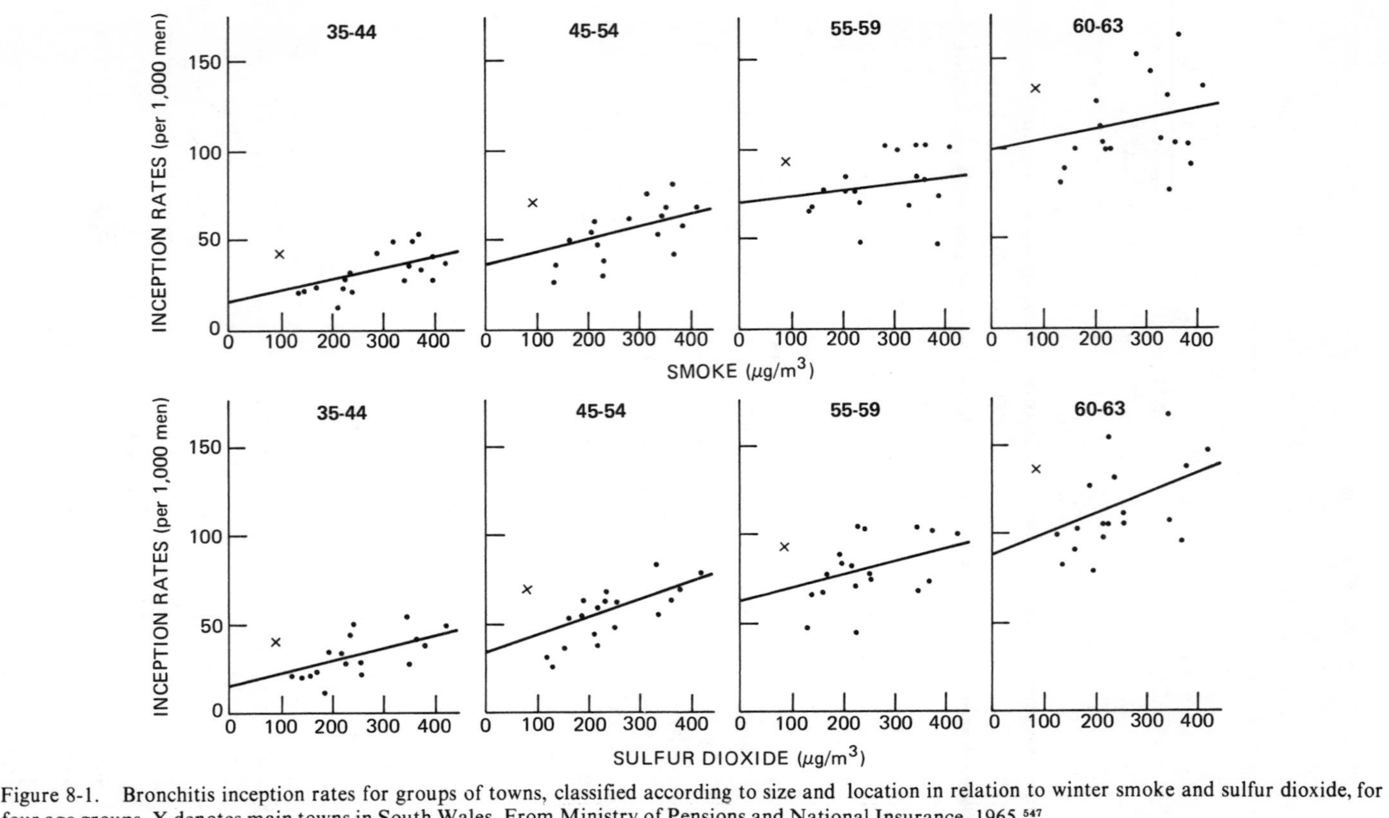

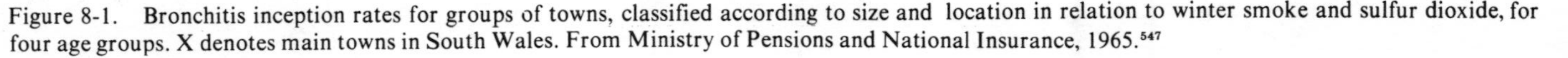
Figure 8-1. Bronchitis inception rates for groups of towns, classified according to size and location in relation to winter smoke and sulfur dioxide, for four age groups. X denotes main towns in South Wales. From Ministry of Pensions and National Insurance, 1965.[547]

have very little effect in nonsmokers. The prevalence of symptoms increased progressively from the lowest pollutant concentrations (<0.1 mg/m^3/year for either smoke or sulfur dioxide) to the highest (≥0.2 mg/m^3 for each pollutant).

More detailed studies on postal workers were conducted by Holland and Reid.[358] They compared 293 men aged 40 to 59 working in London with nearly 500 men of the same age working in three country towns in southern England. Standard methods (respiratory symptom questionnaire, morning sputum volume, and simple ventilatory lung function) were used. A higher prevalence of respiratory symptoms and lower average ventilatory lung function were found in the London men. There were differences in each smoking category but little real difference between nonsmokers or former smokers. The authors concluded that excessive effects in the London subjects were probably due to the greater degree of pollution there as well as possible interaction with smoking.

In the United States, similar methods were used in a series of studies of Bell telephone workers[193, 359] who formed a comparable occupation group of similar age. The effect of smoking on respiratory symptom prevalence and ventilatory lung function as observed in England was also demonstrable in the United States study. The smoking-adjusted prevalence of severe symptoms was higher and the mean $FEV_{1.0}$ was lower in the United Kingdom than in the United States. After considering differences in ambient temperature and respiratory infection rates, the authors concluded that the most likely cause of the differences were the concentrations and types of air pollution.

In a study of male telephone workers in San Francisco and Los Angeles, the prevalence of respiratory symptoms was found to be higher in the Los Angeles subjects. This was attributed primarily to differences in smoking habits between the two groups because the prevalence of persistent cough and phlegm appeared to be caused by cigarette smoking, while shortness of breath, illnesses, and level of lung function depended more on age.

Comstock *et al.*[160] reexamined the East Coast telephone workers originally studied by Holland and his colleagues after 5 to 6 years. At the same time (1967) similar examinations were given to the same type of employees in Tokyo, Japan. In all three surveys,[160] respiratory symptoms increased with age and with the number of cigarettes smoked per day. After allowance for age and smoking, there was no significant association of the findings with residence, from birth to time of study, nor with current place of employment. This study is one of a few that found no association between air pollution and respiratory disease. The reason for the discrepancy between this and most other studies is not clear. Selective migration of persons with respiratory symptoms from high to low pollution areas[411] seemed unlikely because there was no positive correlation of such symptoms with past residence or birthplace.

A comparison of respiratory disease and lung function in two Pennsylvania towns with contrasting levels of air pollution was made by the United States Public Health Service.[620] In Seward, the more polluted town, average dust fall was 3.2 times and sulfur dioxide 6.2 times that of New Florence. The only difference in lung function was a higher average airways resistance in Seward. Since there were other differences between the inhabitants of the two towns in addition to their exposure to air pollution, the higher airways resistance in Seward cannot be attributed solely to higher air pollution.

One of the most extensive studies of air pollution was conducted by the United States Public Health Service in Nashville, Tennessee.[841-843] Pollution was monitored through a network of 123 stations throughout the city. Dust fall, soiling index (COHS), sulfation rates, and sulfur dioxide were all measured. Mortality from respiratory diseases varied inversely with socioeconomic class. The large middle-income class incurred increased respiratory disease mortality as sulfation rates and COHS increased. Most of these deaths must have been due to influenza and pneumonia because mortality from bronchitis and emphysema actually declined as pollution increased. Morbidity from all causes in all age groups and from cardiovascular diseases in men and women over 54 increased with most indices of pollution. An association of increased asthma attack rates with pollution was also demonstrated in Nashville.

Pollution levels have also been related to mortality and morbidity in Buffalo, New York,[823, 824] where particulates and sulfation rates were extensively monitored from 1961 to 1963. From 1959 to 1961 mortality from all causes in men and women over 49 and mortality from chronic respiratory disease in men aged 50 to 69 were both positively correlated with suspended particulate concentrations (Table 8-1). No significant relationship was found between mortality from all causes and sulfation rates. There was, however, a positive association between sulfation and the death rates from chronic respiratory disease in each of the two lowest socioeconomic groups of men aged 30 to 69. In these mortality studies, smoking habits and occupation were not considered.

In a further study of respiratory symptoms in relation to pollution,[822] conducted on a random sample of white women in Buffalo, smoking habits were included. Among nonsmokers aged 45 and over, cough and phlegm were positively correlated with suspended particulate concentrations. Among smokers, however, the findings were confounded by an interaction between respiratory symptoms and residential mobility. Among smokers who did not change residence, cough and sputum were positively correlated with particulate concentrations measured near their home; but among women who had moved within the past 5 years, there

Table 8-1. Mean annual deaths attributed to asthma, bronchitis, or emphysema in white males, aged 50–69 years, per 100,000 population, Buffalo, New York, 1959–1961[a]

Median family income for each census tract, $	Mean annual particulate concentration, $\mu g/m^3$			
	<80	80	100	135+
3,005–5,007	—	126	271	392
5,175–6,004	136	154	172	199
6,013–6,614	—	74	110	128
6,618–7,347	70	80	177	—
7,431–11,792	79	109	0[b]	—

[a] From Winkelstein *et al.*, 1967.[823]

[b] Rate based on less than five deaths.

was a negative association. Socioeconomic factors could not explain these differences. There was no association between sulfur dioxide concentrations and prevalence of respiratory symptoms.

One of the most interesting studies of the effects of pollution in different parts of one city was conducted in Genoa, Italy, by Petrilli *et al.*[609] Genoa is a rapidly growing industrial city situated between the Appennines and the Mediterranean. Its topography causes the phenomena of strikingly different climates in various parts of the city. From 1954 to 1964, pollution was assessed at 19 sites. The investigators monitored sulfation, sulfur dioxide, carbon monoxide, lead, dust fall, suspended matter (estimated gravimetrically and by particle size), and 3:4 benzpyrene. Mean values for available data from 1962 to 1964 showed an increase of ~20% from 1954 to 1961. Petrilli *et al.* studied respiratory symptoms in nonsmoking women over 64 who had not worked in industry but who had resided for a long period in the same area. They also analyzed the results by floor level of residence. There were strong associations between the frequency of respiratory disease and the concentration of sulfur dioxide.

In seven districts where air pollution was measured continuously, 1961 and 1962 morbidity rates for bronchitis and other respiratory diseases varied with average annual sulfur dioxide concentrations. Unfortunately, inadequate analysis of this material prevents one from judging the effectiveness with which the authors standardized for socioeconomic factors or utilization of health facilities.

Studies of children Many studies of air pollution have been conducted on children because there are no occupational exposures to be considered and because cigarette smokers under 12 years old are rare. Children may also be unduly susceptible to pollution, as was suggested in the London fog of 1952.[483] Furthermore, as Reid[635] has suggested, bronchitis may begin in infancy or childhood and lung development during these early years may be adversely affected by environmental assaults. Reid has also indicated that international patterns in mortality from bronchitis are apparent even at young ages. He has drawn attention to a reversal of the urban-rural gradient of bronchitis mortality in the 5- to 14-year-old group. He attributes this to the fact that the country child first encounters infection when starting school. This occurs at a later age than that of the urban child's first exposure to infection.

An early comparison of schoolchildren living in highly polluted Sheffield, England,[779] with children living in the unpolluted Vale of Glamorgan, showed that the prevalence of ear disease, sinus opacity, repeated attacks of chest disease, and impaired ventilatory lung function was higher in the city than in the country.

These findings were subsequently confirmed by Lunn and his colleagues, who provided some evidence of the pollutant concentrations that produced effects. In their first study,[497] there was a progressive increase in respiratory morbidity from the lowest to the highest pollutant concentrations. The lowest average winter mean concentrations of smoke were 0.097 mg/m^3 and of sulfur dioxide, 0.123 mg/m^3 (0.005 ppm). Mean ventilatory lung function was significantly lower in children living in the highest pollution areas with daily concentrations of smoke at 0.301 mg/m^3 and sulfur dioxide at 0.275 mg/m^3 (0.106 ppm). Four years later,[498] during a follow-up study, the lowest concentrations of smoke and sulfur dioxide were 0.048 mg/m^3 (0.018 ppm) and 0.094 mg/m^3 (0.036 ppm). Other areas ranged from 0.041 to 0.169 mg/m^3 of smoke and 0.166 to 0.253 mg/m^3 (0.064 to 0.097 ppm) of sulfur dioxide. During this study, no significant differences in disease morbidity or lung function were found.

Additional quantitative estimates of the effects of air pollutants were provided by Douglas and Waller[202] in their study of health and development. They observed a group of children born within a 1-week period in March 1946 from birth until 1961. Their illness experience was related to the area of their residence and, therefore, probably to exposures to air pollution. The results showed that lower respiratory infections were consistently related to pollution but that upper respiratory infections were not (Table 8-2). Both the frequency and severity of such infections increased with the amount of pollution. The lowest concentrations of smoke and sulfur dioxide were 0.67 and 0.90 mg/m^3 (0.346 ppm), respectively. Higher illness rates were noted in all higher pollution classes.

The importance of childhood residence as a factor in respiratory illness was further supported by Rosenbaum.[649] He showed that the incidence of respiratory disease in servicemen correlated positively with previous residence in an industrial area.

A number of large-scale studies in Britain have evaluated the effects of pollution, past respiratory disease, and various social factors on

Table 8-2. Frequency of lower respiratory tract infections of children in Britain by pollution levels (%)[a]

Lower respiratory tract infections	Mean annual pollution levels ($\mu g/m^3$)			
	Very low	Low	Moderate	High
Smoke:	67	132	190	205
SO_2:	90	133	190	251
First attack in first 9 months	7.2	11.4	16.5	17.1
At least one attack in first 2 years	19.4	24.2	30.0	34.1
More than one attack in first 2 years:	4.3	7.9	11.2	12.9
Boys	5.7	8.1	10.9	16.2
Girls	2.9	7.7	12.1	9.7
Middle class	3.0	4.0	7.7	9.3
Manual working class	5.1	10.8	13.9	15.4
Admission to hospital in first 5 years:				
Lower respiratory infection	1.1	2.3	2.6	3.1
Bronchitis	0.0	0.9	1.0	1.4
Pneumonia	1.1	1.4	1.6	1.8

[a] From Douglas and Waller, 1966.[202]

respiratory symptoms and lung function in children. Holland and his colleagues[357] studied over 10,000 schoolchildren in four areas of Kent, two predominantly urban and two rural. They found peak expiratory flow rates to be related to area of residence, socioeconomic status, family size, and history of pneumonia, bronchitis, or asthma. These factors appeared to act independently; their effects were additive.

Colley and Holland[158] attempted to assess the varying influences of smoking, places of residence and work, overcrowding, family size, and genetic factors in the etiology of chronic respiratory disease by measuring respiratory symptoms and lung function in all family members. A preliminary analysis of 2,342 families living in two areas of North London indicated that the frequency of such symptoms as winter cough in mothers and children differed between the two areas. In the fathers, however, differences due to socioeconomic factors, occupation, or smoking seemed to mask any geographic differences.

Colley and Reid[159] studied respiratory disease frequency in urban and rural areas of England and Wales. They observed a pronounced socioeconomic gradient of chronic cough, bronchitis history, and ear and nose disease. The frequency of chest conditions in children of semiskilled and unskilled workers grew with increasing air pollution.

The cohort of children originally studied by Douglas and Waller[202] was reviewed at the age of 20 by Colley and Reid.[159] They obtained information on respiratory symptoms and illnesses by a self-administered mailed questionnaire. Cigarette smoking was the main determinant of symptoms, but a history of chest illness under 2 years of age was also significant. Neither social factors nor air pollution appeared to influence respiratory disease. However, since Douglas and Waller[202] had originally related air pollution to lower respiratory illnesses during the first 2 years of life, an indirect effect of pollution must clearly have been present.

Many studies of the effects of air pollution on children have been conducted by the Environmental Protection Agency (EPA) as part of their Community Health and Environmental Surveillance System (CHESS).[235-237, 577, 696, 697] This ambitious program was designed to assess the health effects of air pollutants in a large number of communities in many parts of the United States. Acute and chronic diseases (mainly respiratory), variation in lung function, and concentrations of different pollutants in the body are related to estimated pollutant concentrations in the areas of the subjects' residence. Studies have been conducted in communities in the Salt Lake Basin, Utah;[577] the Rocky Mountain area, Idaho/Montana;[236] Chicago/Indiana;[235] New York metropolitan areas;[696] Cincinnati;[697] and other areas. Although elementary schoolchildren less than 12 years old have received the most interest, some studies included high school children, and one study concentrated on nursery children aged 2 to 5 years. Other family members have been included in most of the studies.

Information has been obtained by questionnaires (usually completed by mothers or guardians), by telephone surveillance of illnesses, and, in some studies, by ventilatory lung function testing. Various indices of respiratory disease have been related to present and past measurements of pollution or to estimates of probable pollution based on emissions. The adequacy of the allowances usually made for social circumstances and parents' smoking habits is sometimes debatable.

Attack rates of lower respiratory illnesses, particularly croup, were higher in the most polluted communites of the Salt Lake Basin and Rocky Mountain areas. Asthmatic children were particularly susceptible. The validity of these conclusions is, however, open to doubt. The attack rates in the Salt Lake Basin were strikingly higher only in the area of highest pollution. The excess bronchitis was 100% for children aged 9 to 12, but less than 50% for the younger and older children. Sulfate concentrations varied almost threefold (0.005 to 0.014 mg/m^3); sulfur dioxide concentrations differed even more widely; while suspended particulates were either higher or similar in the low pollution areas. Furthermore, unadjusted 3-year attack rates in the Rocky Mountain areas showed marked variation within the low pollution areas. Rates in the lowest pollution area were often higher than in more polluted areas. Rates differed little in the two high pollution areas despite a doubling of the pollutant concentrations. Adjustment for socioeconomic factors would probably not affect these attack rates very much. Pooling of the high and low pollution areas permitted statistically significant associations to be shown. But the legitimacy of such pooling is open to question.

In Chicago and New York, illnesses were monitored by twice-weekly telephone interviews of mothers. Significantly increased incidence of acute respiratory illnesses, restricted activity, and otitis media were reported in subjects living in high pollution areas of Chicago. In New York, both the incidence and severity of acute lower respiratory illnesses were higher in the more polluted areas. Again, these data show some inconsistencies that lead to questions concerning some of the conclusions.

Ventilatory lung function has been related to pollution in several of the EPA studies. Shy and his colleagues[695] studied second grade schoolchildren in Cincinnati and Chattanooga and kindergarten through sixth-grade children in New York City. The forced expiratory volume ($FEV_{0.75}$) and forced vital capacity (FVC) were measured weekly for a 2- to 3-month period in Cincinnati and Chattanooga and 4 times during the 1970–1971 school year in New York City. In Cincinnati, the FEV was lower among white children living in the more polluted areas, but among black children no differences attributable to pollution could be demonstrated. The FEV tended to be lower in the winter months when pollution was high. But the measurements did not appear to be closely related to the levels of pollution during the lung function testing. Children aged 9 to 13 living in relatively unpolluted New York City suburbs had a significantly higher FEV than children of a similar age living in two more polluted central communites. In contrast, children aged 5 to 8 showed no differences that could be related to variations in pollution. Shy *et al.* suggested that reduced pollution in New York City could have caused these findings.

The authors did not, however, control for the possible effects of cigarette smoking by the older children. Moreover, in a later report they indicated that their measurements of lung function were in doubt because of instrumental malfunctioning.

Lebowitz and his colleagues[453] have recently described studies of children, adolescents, and adults in two Arizona communities. Ventilatory lung function tests were administered after exercise in Tucson and in a smelter mining town in southern Arizona. A reduction in $FEV_{1.0}$ and midmaximum expiratory flow rates (MMEF) after exercise was observed on days of high pollution and high temperature. The extent to which the effects were due to pollution or heat is not clear. The authors indicated that longitudinal studies and possibly experiments are needed to confirm and quantify these results.

In the CHESS studies, the EPA studied pollution effects in adults as well as in children.[134, 233, 234, 258, 282, 283, 320, 341, 367, 487] Although most adults were the parents of the CHESS children, military inductees and panels of patients with cardiorespiratory disease were also studied. Although effects of pollutant concentrations in adults reflected those found in children with regard to prevalence of respiratory symptoms, there are sufficient discrepancies in the attacks of respiratory illnesses or worsening of bronchopulmonary and cardiac conditions to indicate a need for further study. It is particularly difficult to be confident about the levels of individual pollutants that are likely to produce these effects, particularly to exacerbations of asthma with which several of these studies have been concerned. An important finding in several CHESS studies is that particulate sulfates have a more important effect on health than either total suspended particulates or sulfur dioxide. Future studies should include sulfate measurements, including, if possible, different sulfate compounds, to clarify the dose-response relationships involved.

Cohen and his colleagues[156] studied the effects of pollution on 43 asthmatics who lived near a power plant. These subjects had all experienced three or more attacks of respiratory distress with wheezing during the preceding year. Twenty-nine of them were observed for 7 months or more. The investigators correlated reported attacks with daily temperature and with concentrations of such pollutants as total suspended particulates, sulfur dioxide, soiling index, suspended sulfates, and suspended nitrates. Attack rates were highest at low temperatures and high pollution. The effect of pollution on attacks was greater at moderate than at low temperatures. The authors concluded that increased asthma attack rates might occur at pollutant concentrations commonly encountered in polluted cities. The results show an upward trend in the average attack rates as pollution increased. Attack rates rose from 35% in the absence of pollution to 50% or 60%, depending on pollutant and concentration. On the other hand, there is wide variation around this mean regression.

Pollution probably played a relatively minor role in asthma attacks. The findings preclude identification of pollutant concentrations that would be innocuous to such attacks.

EFFECT OF REDUCING POLLUTION

Reductions in pollution appear to affect respiratory disease and lung function. In 1967 Ferris and his colleagues[229] reviewed the Berlin, New Hampshire subjects who were first studied in 1961. In the intervening 6 years, pollution as measured by sulfation rates or total suspended particulate concentrations had declined from 30% to 40%. The prevalence of chronic respiratory disease was less in 1967 than in 1961 after allowances were made for aging and changes in cigarette smoking. Lung function measurements were also slightly better than predicted. Concentrations measured during these studies were:

Sulfation:	0.731 ± 0.241 mg SO_3/cm^2/day in 1961
	0.469 ± 0.111 mg SO_3/cm^2/day in 1967
Particulates:	0.180 ± 0.710 mg/m^3 in 1961
	0.132 ± 0.830 mg/m^3 in 1967

Changes in smoking habits or the use of filter-tipped cigarettes could not explain these differences. These findings apparently supported the conclusion drawn in the Berlin-Chilliwack comparison, i.e., that the 1961 pollutant concentrations in Berlin were associated with decreased ventilatory lung function.

Dose-Response Relationships

Table 8-3 provides estimated concentrations of particulates and sulfur dioxide that may affect health. To reiterate, these two pollutants may not be the most important. They serve only as indices of other, perhaps more important, pollutants. In London, mortality has clearly resulted when 24-hr smoke concentrations have exceeded 1.0 mg/m^3 and sulfur dioxide concentrations have reached 0.750 mg/m^3 (0.288 ppm). These peaks used to occur in London during average annual background concentrations of 0.3 to 0.4 mg/m^3 of smoke and 0.25 to 0.30 mg/m^3 (0.1 to 0.12 ppm) sulfur dioxide. Such concentrations are now fortunately only of historical interest. They should certainly not be tolerated.

In London, 24-hr concentrations of ~0.5 mg/m^3 of smoke and 0.4 mg/m^3 (0.15 ppm) of sulfur dioxide exacerbated bronchitis. With the present lower concentrations, such exacerbations are infrequent. Some correlation still existed when the average annual concentration of smoke was 0.06 mg/m^3 and of sulfur dioxide, 1.70 mg/m^3 (0.654 ppm). Since then, pollution has declined further in London but it is not clear if exacerbations still occur with increases in pollution.

In Britain, sick leave attributed to bronchitis appeared to correlate linearly with winter smoke and sulfur dioxide concentrations over 0.1 mg/m^3 (0.038 ppm). It would be very interesting to know if similar correlations can still be demonstrated at the present lower pollutant concentrations.

In New York City, 24-hr coefficient-of-haze units (COHS) of 5 or more and sulfur dioxide of 2.0 mg/m^3 (0.769 ppm) have resulted in deaths; 3 COHS and 0.7 mg/m^3 (0.269 ppm) sulfur dioxide have caused illness. Studies of daily mortality in relation to pollution suggest that excess deaths may occur when sulfur dioxide is as low as 0.35 mg/m^3 (0.013 ppm). In Chicago, exacerbations of bronchitis were associated with daily sulfur dioxide concentrations of ~0.75 mg/m^3 (0.288 ppm), probably in the presence of high concentrations of particulates.

Table 8-3. Health effects and dose-response relationships for particulates and sulfur dioxide

Averaging time for pollution measurements	Place	Particles (mg/m^3)	SO_2 (mg/m^3)	Effect	Reference
24-hour	London	2.00	1.04	Mortality	102
		0.75	0.71	Mortality	446
		0.50	0.50	Exacerbation of bronchitis	450
	New York City	6 COHS[a]	0.50	Mortality	297
		3 COHS	0.70	Morbidity	296
	Chicago	Not Stated	0.70	Exacerbations of bronchitis	125
	New York City	0.145 (+?)	0.286	Increased prevalence of respiratory symptoms	156a
	Birmingham, Ala.	0.18–0.22	0.026	Increased prevalence of respiratory symptoms	156a
	New York City	2.5 COHS	0.52	Mortality	277
Weekly mean	London	0.20	0.40	Increased prevalence or incidence of respiratory illnesses	39
Six winter months	Britain	0.20	0.20	Bronchitis, sickness, absence from work	547
Annual	Britain	0.07	0.09	Lower respiratory infection in children	202
		0.10	0.10	Bronchitis prevalence	432
		0.10	0.12	Respiratory symptoms and lung function in children	497, 498
	Buffalo	0.08	0.45[b]	Mortality	823, 824
	Berlin, N.H.	0.18	0.73[c]	Decreased lung function	229

[a] Coefficient-of-haze units.
[b] mg $SO_3/cm^2/30$ days.
[c] mg $SO_3/100\ cm^2/day$.

Table 8-4. Threshold estimates for adverse health effects attributable to sulfur dioxide, particulate sulfates, and total suspended particulates (TSP), short-term exposures, using epidemiologic research approach[a]

		Exposure level (μg/m³)				Safety margin (% contained in primary standard [std.][b])		
Adverse effect on human health	Type of estimate	SO_2	Particulate effect	TSP	Duration	SO_2 std. 365 μg/m³	Su pended sulfates (no std.)	TSP std. 260 μg/m³
Mortality harvest	Worst case	30	No data	250	24 hr	0	No data	0
	Least case	500	No data	500	24 hr	37	No data	92
	Best judgment	300–400	No data	250–300	24 hr	0–9	No data	0–15
Aggravation of chronic lung disease	Worst case	119	6	100	24 hr	0	No data	0
	Least case	500	No effect	>250	24 hr	37	No data	<100
	Best judgment	365	10	>250	24 hr	0–37	0	0
Aggravation of asthma	Worst case	23	6	75	24 hr	0	0	0
	Least case	>365	10	>260	24 hr	<100	0	<100
	Best judgment	180–250	8–10	100	24 hr	0	0	0
Aggravation of combined heart and lung disease	Worst case	180	6	61	24 hr	0	0	0
	Least case	>365	10–17	260	24 hr	<100	0	<100
	Best judgment	365–500	8–10	70–100	24 hr	0–37	0	0
Irritation of respiratory tract	Worst case	340	No data	170	2–3 days	0	No data	0
	Least case	340	No data	192	2–3 days	0	No data	0
	Point est	340	No data	170	2–3 days	0	No data	0

[a] From Finklea, 1973.[232a]
[b] Safety margin = Effects threshold − standard/standard × 100.

Table 8-5. Threshold estimates for adverse health effects attributable to sulfur dioxide, particulate sulfates, and total suspended particulates (TSP), long-term exposures, using an epidemiologic research approach[a]

		Exposure level ($\mu g/m^3$)				Safety margin (% contained in primary standard [std.][b])		
Adverse effect on human health	Type of estimate	SO_2	Suspended sulfates	TSP	Duration, yrs	SO_2 std. 80 $\mu g/m^3$	Suspended sulfates (no std.)	TSP std. 75 $\mu g/m^3$
Excess mortality	Worst case	120–198	No data	135	Unknown	50–148	0	80
	Least case	250	No data	175	Unknown	212	0	133
	Best judgment	250	No data	175	Unknown	212	0	133
Increase in prevalence of chronic bronchitis	Worst case	50–90	9	60–100	3	0–12	0	0–33
	Least case	404	20	180	12	405	0	140
	Best judgment	95	14	100	5	19	0	33
Increased frequency or severity of acute respiratory illness in otherwise healthy families	Worst case	91	9	75	1	14	0	0
	Least case	200	23	200	3	150	0	167
	Best judgment	91	9	100	3	14	0	33

Increase in family illness during influenza epidemics	Worst case	106	14	126	1	33	0	68
	Least case	250	18	151	Unknown	212	0	101
	Best judgment	106	14	151	3	33	0	101
Increased lower respiratory tract infections in asthmatics	Worst case	32	8	100	Unknown	0	0	33
	Least case	186	20	No effect	Unknown	133	0	—
	Best judgment	91	8	100	Unknown	14	0	33
Subtle decrease in ventilatory function	Worst case	118-131	9	75-141	1	48-64	0	0-88
	Least case	400-500	28	200	9	400-525	0	167
	Best judgment	200	11	100	8-9	150	0	33

[a] From Finklea, 1973.[232a]

[b] Safety margin = Effects threshold − standard/standard × 100

In Buffalo, mortality from respiratory illness appeared to increase progressively from the lowest to the highest pollutant concentrations. The lowest level of smoke was less than 0.08 mg/m^3 and of sulfation, 0.045 $mg/cm^2/day$. A number of other British studies suggest that average annual concentrations of particulates and sulfur dioxide should both be held to under 0.100 mg/m^3.

Conclusions from the CHESS studies are shown in Tables 8-4 and 8-5. Unfortunately, as the least case/worst case estimates indicate, the results are questionable. As suggested earlier, the findings should be confirmed before expensive decisions are based on them.

PARTICULATES IN RELATION TO RESPIRATORY CANCER

The evidence that cigarette smoking is the main cause of lung cancer is now generally well recognized.[200, 201, 312-314] Certain occupational exposures, notably to radioactivity, chromates, nickel, asbestos, arsenic, and the distillation products of coal are also well established causes of lung cancer.[199] Some synergism, such as that between occupation and smoking among uranium miners and asbestos workers, is definite. But occupational exposures are unlikely to cause more than a small fraction of the total number of lung cancer cases.

Urban-rural differences in respiratory cancer mortality have long been a feature of the vital statistics of many countries. They have been supported by incidence data from certain cancer registries and by special surveys. Differences in smoking habits or occupational exposures between urban and rural areas cannot account for all the excess urban cancer rates, nor can better reporting of lung cancer in towns or the movement of people to towns for diagnosis. The presence in urban air of polycyclic hydrocarbons and other particulates known to cause lung cancer suggests that pollution might explain the urban-rural differences. But despite considerable research, this is still uncertain.

Early studies in Britain suggested that pollution exerted a small or negligible effect on lung cancer after allowances had been made for social factors and population density.[732, 736] Among British researchers, only Stocks considered the possibility that pollution contributed substantially to lung cancer mortality. He compared death rates from lung cancer to pollution and smoking in Liverpool and its surrounding areas. In his first paper on the subject,[734] he claimed that approximately half of the lung cancer deaths in Liverpool were attributed to smoking habits and three-quarters of the remainder were due to a factor that was only slightly present in the rural areas. He suggested that benzpyrene "might be one of the factors involved."

Stocks also related age-standardized mortality ratios for lung cancer to 3:4 benzpyrene, 1:12 benzperylene, pyrene, fluoranthene, sulfur dioxide, and smoke obtained from 17 locations in Liverpool and North Wales. Lung cancer standardized mortality ratios (SMRs) correlated both highly and positively with all these pollution measurements. Cancer of the intestine and rectum, on the other hand, correlated neither with smoke nor with population density.

Anderson[36] reanalyzed Stocks' data by stepwise regression. He concluded that although polycyclic hydrocarbons contributed significantly to the multiple correlation coefficient, the contribution was infinitesimal compared with population density. In another study Stocks[735] collected data at 26 localities. He also used a socioeconomic scale, or "social index," based on the proportion of males aged 15 and over in the lowest income group. Smoke and 3:4 benzpyrene correlated positively with lung cancer mortality. Here, too, Anderson reanalyzed the data and found that lung cancer did not correlate significantly with benzpyrene if smoke and social index remained unchanged. In further studies, Stocks found positive correlations between lung cancer mortality rates and smoke, 3:4 benzpyrene, 1:12 benzperylene, arsenic, beryllium, and molybdenum in eight different localities, and between lung cancer and both cigarette and solid fuel consumption in 20 countries.

Carnow and Meier[126] extended Stocks' analyses to include age-specific lung cancer death rates and cigarette consumption for 19 countries. They also analyzed lung cancer death rates in relation to smoking (based on cigarette sales per person over 14 years old in 1963) and benzpyrene (based on a weighting of urban and rural values from unpublished data of R. I. Larsen and J. B. Clements) in the 48 contiguous states of the United States. They concluded that a 5% increase in lung cancer mortality will be produced by an increase of 0.001 mg/1,000 m^3 benzpyrene. The benz[*a*]pyrene figures for the different states are questionable because of the procedures used to estimate them. Even if they are accepted as valid, their correlation with lung cancer death rates is not impressive and appears to be due mainly to some low values in a few rural states.

The most systematic studies of urban-rural ratios of lung cancer death rates in the United States have been those conducted by Haenszel and his colleagues.[307, 308] In 1958 they studied the smoking habits and residential histories of a representative sample of the population and of a 10% sample of victims of lung cancer deaths in the United States. They found the expected gradients of increased lung cancer mortality with increased cigarette smoking. Differences in death rates (SMR's) for different categories of smokers living in urban and rural areas are shown in Figure 8-2. Differences in death rates on nonsmokers by residence were trivial; but the joint effects of smoking and residence were far greater than would have been expected, assuming that smoking and residence exerted additive effects. Finally, there were higher risks among the more

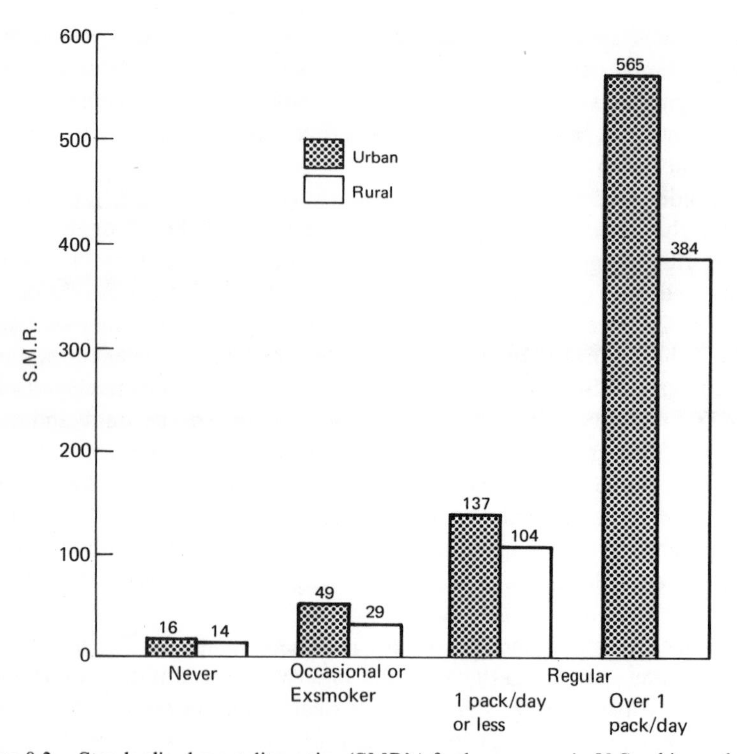

Figure 8-2. Standardized mortality ratios (SMR's) for lung cancer in U.S. white males in 1958 according to smoking and residence. Data from Haenszel *et al.*, 1962.[307] Figure from Higgins, 1976.[336]

mobile groups as compared with lifelong residents. United States farm-born and foreign-born residents of large metropolitan centers were particularly high lung cancer risks.

When allowance is made for mobility and only lifelong residents are compared, the SMR of nonsmokers living in urban areas is appreciably higher than that of nonsmokers living in rural areas. The number of deaths among lifelong residents is, however, small, reflecting the very high mobility of the U.S. population. Mobility, then, should be considered in addition to smoking habits when assessing the effects of pollution on disease.

In 1965 and 1966, Hitosugi[343] studied lung cancer in relation to smoking and air pollution in two Japanese cities near Osaka. The air pollution in these cities was broken down into three categories based on extensive measurements of dust fall, sulfur dioxide, suspended particulates, trace elements, and aromatic hydrocarbons. Hitosugi obtained information on the smoking habits of lung cancer patients as well as a representative sample of healthy persons living in each of the three pollution areas. Death rates for lung cancer rose with increased smoking in each pollution category. The death rates among smokers also increased slightly with increased pollution, but not among nonsmokers. The authors suggested that this might indicate synergism between smoking and air pollution and lung cancer.

Studies of immigrants support the hypothesis that air pollution is a factor in the causation of lung cancer. British immigrants to New Zealand,[209] South Africa,[190] Australia,[191] the United States,[306] or Canada[171] have lower death rates than the British population, but higher rates than natives of their adopted countries. Immigrants who left Britain at age 30 and over have higher rates than younger immigrants. The differences cannot be explained on the basis of differences in smoking habits. Perhaps they reflect environmental exposures before migration. However, it is well established that immigrants are not a representative sample of the country from which they emigrate. Nor has any consideration been given to the types of jobs adopted by immigrants.

One way in which the impact of pollution on lung cancer can be assessed is to see what happens when pollution is reduced. To test the hypothesis that particulate pollution contributes to the development of respiratory cancer, trends in mortality from the disease from 1950 to 1970 in London, where particulate pollution declined greatly, were compared with trends in the rest of England and Wales, where the decline was less. The results are somewhat equivocal. A somewhat greater decline in the age-specific death rates in men under 55 years occurred during this period in London.[335] This did not appear to be entirely attributable to different trends in smoking habits.

SUMMARY

There is good evidence that exceptional episodes of pollution [>1.0 mg/m^3 (0.385 ppm) sulfur dioxide and particulates] caused illness and death. There is also a good deal of evidence that sustained lower levels of pollution [>0.1 mg/m^3 (0.039 ppm) of sulfur dioxide and particulates] for a number of years affect health adversely. Pollution predominantly affects those who are already suffering from disease, particularly of the heart or lungs; however, evidence, especially from studies of children, suggests that pollution can initiate disease as well as exacerbate it. Particulate pollution, especially from sulfur compounds, probably plays a considerable role in the development and progression of bronchitis and emphysema. There have also been suggestions that it plays a role in lung cancer; however, this is much more debatable.

There is insufficient knowledge concerning safe pollutant concentrations. The primary air quality standards for particulates were probably reasonable when they were adopted, bearing in mind the inadequate information then available on the character of the particles. There is now a clear need for much more precise specification of particles and their characteristics, particularly their size distribution. The trend in emphasis on the health effects of sulfates is an indication; however, information on dose-response relationships for sulfates is insufficient to suggest acceptable standards for these compounds.

Much more information is also needed on the effects of interactions between pollutants, or between pollutants and meteorologic, climatic, or personal factors. The possibility of pollutant interaction raises questions about the validity of any national standard for a particular pollutant. The pattern of pollution should be considered when a standard for any particular pollutant is set. Any sulfate standard should take into consideration the type of sulfate.

9

Effects of Particulate Air Pollutants on Vegetation†

Many gaseous air pollutants are readily recognized as the cause of injury to various types of vegetation. In contrast, relatively little is known and only limited studies have been conducted on the effects of particulate air pollutants on vegetation. Published results of experiments are confined principally to settleable dusts emitted from the kilns of cement plants. There are a few reports on effects of fluoride dusts, soot, sulfuric acid mist, lead particles, and particulate matter from various types of metal processing. Most of the research is related to the direct effects of dust on leaves, twigs, and flowers, as opposed to the indirect effects from dust accumulation in the soil.

Gaseous air pollutants have direct access to interior leaf tissues through the stomata, those openings in the epidermis of leaves through which normal gas exchange required for growth takes place. Most crop plants are broad-leaved and the majority of functional stomata are located on the lower leaf surface. Thus, settleable dusts would be deposited in most cases on the upper leaf surface. The mere physical presence of chemically inert particles might alter the quality and intensity of light reaching chlorophyll-bearing cells, to the detriment of photosynthesis. Such dusts could also interfere with biologic control of insects. Those dusts with some chemical activity could induce injury by radical changes in pH or by dissolving or rupturing the cuticle, thus altering water balance or allowing epidermal penetration of toxic chemical constituents. Of course, there is the possibility that some dust may deposit on the lower leaf surface. In so doing it may even cover or clog stomata and interfere with normal gas exchange. But it is unlikely that particles would ever have direct access to inner leaf tissues as do gases.

Because of the dearth of results from experiments, many reports are directed to the question of whether or not particles have deleterious effects on plants rather than to discussions of the extent of plant injury. Because particulate matter is a conglomerate of chemically

† Adapted from Chapter 7, "Particulates," by S. L. Lerman and E. F. Darley, from *Responses of Plants to Air Pollution*, edited by J. Brian Mudd and Theodore T. Kozlowski. Academic Press, 1975.

heterogeneous substances, it is reasonable to anticipate some disagreement concerning its impact on vegetation. The various types of particles and their effects on vegetation are discussed below.

EFFECTS OF SPECIFIC PARTICULATE MATTER ON VEGETATION

Cement-Kiln Dust

This dust is not derived directly from processing of cement. It is contained in waste gases from kilns. Some reports indicate, however, that the composition of wastes from different kilns operating at different efficiencies varies considerably. At times the effluents may contain cementitious materials that belong in the finished product. Reports describing effects of dust deposited on various plants in the field relate to kiln-stack materials, whereas dusts applied in laboratory or field studies are taken from various collectors in the waste-gas system between the kiln and the stack. Differences in results caused by this factor have not been reconciled.

Fallout levels, as well as the physical and chemical properties of kiln dusts, are determined by such factors as the nature of raw material, cement manufacturing processes, and the type of equipment employed for the control of particulate matter emissions. Lerman[460] recorded dust deposits of 1.5 g/m^2/day in the vicinity of a cement plant in California. The rotary kilns in the plant were equipped with a multiple cyclone emission control device. According to reports from Germany, the maximum amount of dust that might be deposited near cement factories varies from 1.5 g/m^2/day[592] to 3.8 g/m^2/day.[76] The diversity in chemical composition and pH reactions of cement-kiln dust samples from various sources are presented in Table 9-1.

Direct effects: field observations and the nature of dust deposition Most reports concerning harmful effects of cement-kiln dust on plants

Table 9-1. Chemical composition of some cement-kiln dusts (%)

Sample	Na	Ca	K	Mg	Al	Fe	pH of water-dust suspension
California (Cal-1)	0.40	31.20	2.16	0.40	0.40	1.15	12.3
Arizona (Ariz-3)	0.13	23.00	8.00	1.00	0.32	0.55	10.8
Germany (G-1)	2.80	17.56	13.00	1.08	0.94	1.10	10.4
Germany (G-2)	7.60	8.00	32.50	0.56	0.48	0.60	7.3

stress the fact that crusts form on leaves, twigs, and flowers. In California, Peirce[605] and Parish[597] noted that settled dust in combination with mist or light rain formed a relatively thick crust on upper leaf surfaces of affected plants. The crust would not wash off and could be removed only with force. The central theme about which Czaja[175-178] builds his case for harmful effects is the crust formation in the presence of free moisture. He states that crust is formed because some portion of the settling dust consists of the calcium silicates that are typical of the clinker (burned limestone) from which cement is made. When this dust is hydrated on the leaf surface, a gelatinous calcium silicate hydrate is formed. This later crystallizes and solidifies to a hard crust. When the crust is removed, a replica of the leaf surface is often found, indicating intimate contact of dust with the leaf. The relatively thick crust formed from continuous deposition is confined to the upper leaf surface of deciduous species but completely encloses needles of conifers. During prolonged dry periods when dust is deposited, the lack of hydration prevents crusts from forming. Dust deposits that are not encrusted are readily removed by wind or hard rain.

Photographs taken in Germany by Darley and Lerman (Figures 9-1 and 9-2) show encrustations of cement-kiln dust on branches of fir trees. Atmospheric concentrations of dust in this area were probably in excess of 1.0 $g/m^2/day$. Encrustations built up on the older twigs (Figure 9-1) and caused needles to fall prematurely. Some of the twigs were dead and encrustations were forming on the newest needles. The net effect was a shortening of each succeeding year's flush of growth. A dead tree had heavy encrustations on the branches (Figure 9-2).

Bohne[76] reported a marked reduction in growth of poplar trees located about 2.2 km from a cement plant in which production had more than doubled. The change in growth rate was determined by the width of annual rings in the wood. Darley[184] observed a reduction of spring growth elongation on conifers in Germany, where the oldest needles were encrusted.

Anderson[38] observed in New York that cherry fruit set was reduced on the side of the tree nearest a cement plant. He demonstrated that the dust on the stigma prevented pollen germination.

Pajenkamp[592] reviewed the unpublished work of several German investigators, some of whom had applied dust artificially to test plants, and stated that he was opposed to the view that dusts are harmful to plants. He concluded that depositions from 0.75 to 1.5 $g/m^2/day$ (the higher amount representing the maximum that might be found near a cement factory) had no harmful effects on plants.

Raymond and Nussbaum[631] concurred that cement dusts have little effect on wild plants. On the other hand, Guderian[300] and Wentzel[793] disagreed with Pajenkamp. They stated that the limited evidence at best

Figure 9-1. Cement-kiln dust on fir branches. Encrustation has built up on the older twigs of a fir tree exposed to cement-kiln dust particles in the vicinity of a cement plant. Needles have fallen prematurely.

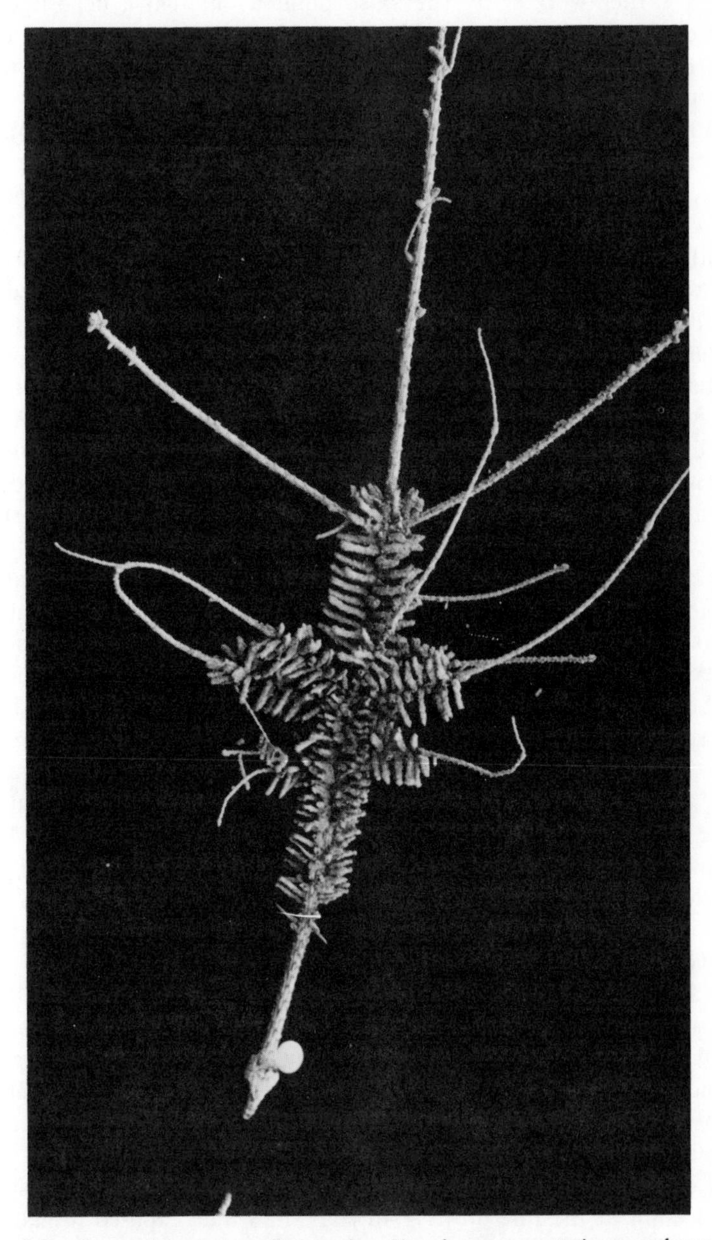

Figure 9-2. Cement-kiln dust on fir branches. Very heavy encrustation on a branch of a dead fir tree exposed to cement-kiln dust particles in the vicinity of a cement plant.

presented a contradictory picture and that Pajenkamp had not cited Czaja's earlier work.[175-178] They also pointed out that a deposit of 1.5 $g/m^2/day$ was not maximum, since other investigators had found up to 2.5 $g/m^2/day$. Bohne[76] has since reported weekly averages of up to 3.8 $g/m^2/day$.

Physical effects Peirce[605] demonstrated that encrustations of cement-kiln dust on citrus leaves not only interfered with the light required for photosynthesis, but also reduced starch formation. This was later confirmed by Czaja[177] and Bohne[76] in a variety of plants. More recently, Steinhübel[722] compared starch reserve changes in undusted common holly leaves with those dusted with foundry dust. He concluded that the critical factor preventing starch formation was the light absorption by the dust layer, and that the influence on transpiration or overheating of leaf tissue was of minor importance. Lecrenier and Piquer[454] attributed the reduced yields from dusted tomato and bean plants to interference with light imposed by the dust layer. Darley[184] demonstrated that dust deposited on bean leaves in the presence of free moisture interfered with the rate of carbon dioxide exchange, but he did not measure starch formation.

Czaja[178] has presented good histologic evidence that stomata of conifers may be plugged by dust, thereby preventing normal gas exchange by the leaf tissue. Uninhibited exchange of carbon dioxide and oxygen by leaf tissue is necessary for normal growth and development.

Lerman[460] demonstrated limited clogging of stomata on bean leaves that were heavily dusted with dry dust. Scanning electron micrographs of upper and lower surfaces of dusted bean leaf are shown in Figures 9-3 and 9-4. Only a few dust particles can be observed on the lower surface of the leaf (Figure 9-3), where most of the stomata are located. Many stomata on the upper surface of the leaf (Figure 9-4) are clear of dust particles in spite of the fact that the leaf was dusted with 6.64 g/m^2, a dust load that would normally cause severe damage to bean plants in the presence of free moisture.

Chemical effects Darley[184] applied kiln dusts of particles $<10\ \mu m$ diam at rates of 0.5 to 3.8 $g/m^2/day$ to leaves for 2 to 3 days in the laboratory. Water mist was applied several times each day. Even though the dust adhered to the leaf in a uniform layer, it was not crustlike, probably because the experiments were of short duration. Reduction in carbon dioxide uptake was reported.

Leaves of bean plants dusted for 2 days with cement-kiln dust 8 to 20 μm diam at the rate of 4.7 $g/m^2/day$, and then exposed to naturally occurring dew, were moderately damaged. The edges (Figure 9-5) of the leaves curled and some interveinal tissues died. Leaves that were dusted but kept dry were not injured.

In 2- to 3-day experiments, Darley[184] dusted the primary leaves of

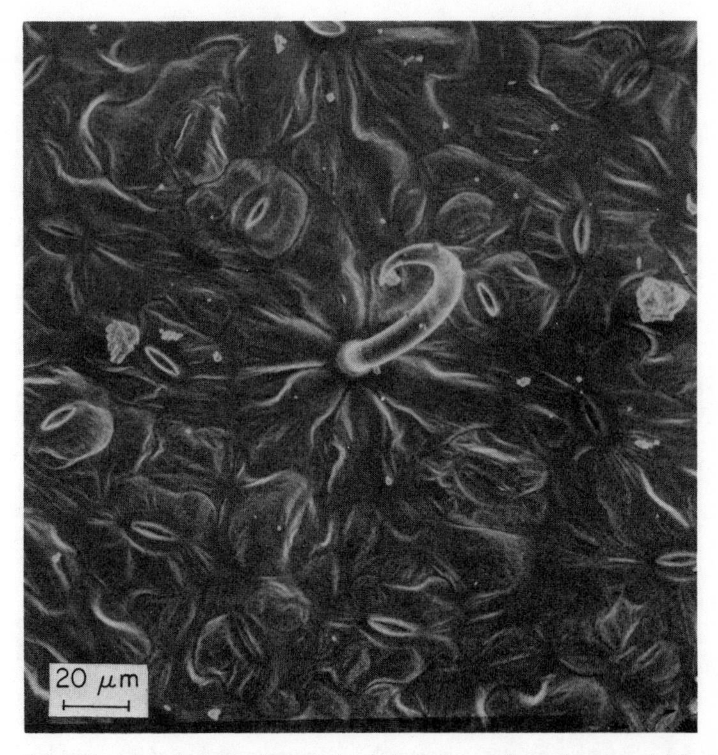

Figure 9-3. Scanning electron micrograph of the lower surface of a cement-kiln-dusted bean leaf. Only a few dust particles can be observed on the lower surface of a leaf that was dusted with a total of 6.64 g/m². (×300)

bean plants with fractionated precipitator dust obtained from Germany. The dust contained relatively high amounts of potassium chloride. When a fine mist was applied to dusted leaves, up to 29% of the leaf tissue was killed. This action was attributed to the potassium chloride. In later experiments other fractions of the same dust containing very little potassium chloride caused an almost equivalent amount of injury, indicating that potassium chloride was apparently not the only factor involved.

Details on the injury to be expected from certain cement-kiln precipitator dusts were given by Czaja.[178] His work is based on comparisons of chemical composition of dusts and resultant injury to leaf cells of a sensitive moss plant, *Mnium punctatum*. A cut leaf was mounted in water on a microscope slide. Dust was placed in contact with the water at the edge of the cover slip. Any effect of the resultant solution on leaf cells could be observed directly. Eighteen of the dusts tested in this way fell into the following categories:

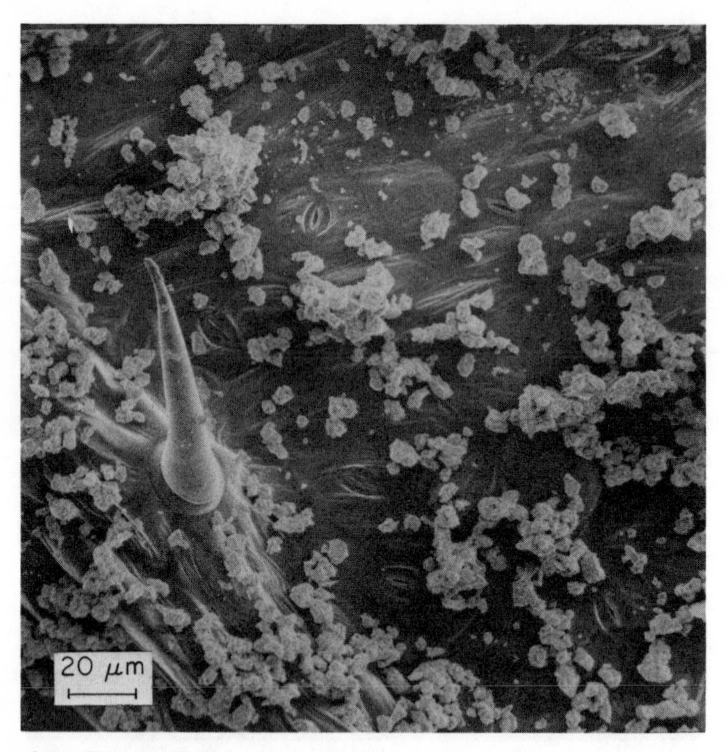

Figure 9-4. Scanning electron micrograph of the upper surface of a cement-kiln-dusted bean leaf. Many stomata of the upper leaf surface are clear of dust particles despite the fact that the leaf was dusted with 6.64 g/m^2. (×300.)

Figure 9-5. The effect of a cement-kiln dust on bean plants. The plants were dusted with particles 8 to 20 μm diam at the rate of 4.7 $g/m^2/day$. (a) Control plant. (b) Dusted but kept dry, no injury symptoms. (c) Dust-dew treated leaf. The injury appeared as a rolling of leaf edges and interveinal necrosis.

No permanent injury to living cells, but some plasmolysis attributed to the solution.
Slight injury to the cells most accessible to the solution, disruption of the cytoplasm, and displacement of chloroplasts.
Severe injury to all cells.

Dusts were described as follows:

Group 1, pH 9.5 to 11.5, relatively high rate of carbonation, an intermediate amount (19% to 29%) of clinker phase (calcium silicates), and a high (36% to 79%) amount of secondary salts.
Group 2, pH approximately 11, a high rate of carbonation, a lower (13% to 16%) clinker phase and a significantly high (81% to 85%) proportion of raw feed.
Group 3, pH 11 to 12, a very slow carbonation rate and a significantly high (17% to 49%) clinker phase.

The greatest injury was related to the largest amounts of clinker phase, which in turn resulted in higher and prolonged alkalinity. But Czaja also pointed out that the composition of dusts within the three groups was not consistent and that, although not yet demonstrated, the constituents of a give dust undoubtedly influence one another.

Czaja[177] stated that the hydration process of crust formation released calcium hydroxide. The hydrated crusts gave solutions of pH 10 to 12. Severe injury of naturally dusted leaves, including killing of palisade and parenchyma cells, was revealed by microscopic examination. The alkaline solutions penetrated stomata on the upper leaf surface, particularly the rows of exposed stomata on needles of conifers, and injured the cells beneath. Czaja[177] stated that on broad-leaved species with stomata only on the lower leaf surface, the alkaline solutions first saponified the protective cuticle on the upper surface, permitting migration of the solution through the epidermis to the palisade and parenchyma tissues. Typical alkaline precipitation reaction with tannins, especially in leaves of rose and strawberry, was evidence that calcium hydroxide penetrated the leaf tissue. Bohne[76] described similar "corrosion" of tissues under the crust formed on oak leaves.

Observations made by Lerman[460] with a scanning electron microscope revealed disorganization of the cuticle on the surface of dust-dew-treated bean plants (Figure 9-6). The breakdown of the layer appeared first in the form of cracks. This later developed into a mass of dust particles surrounded by peeled-off cuticular strips. Figure 9-7 shows the smooth upper surface of a control leaf. Chemical analysis of extracts from the leachate of the treated leaves revealed relatively large quantities of free hexadecanoic acid, and small amounts of free tetradecanoic, dodecanoic, and nonanoic acid. These fatty acids are among the cuticular

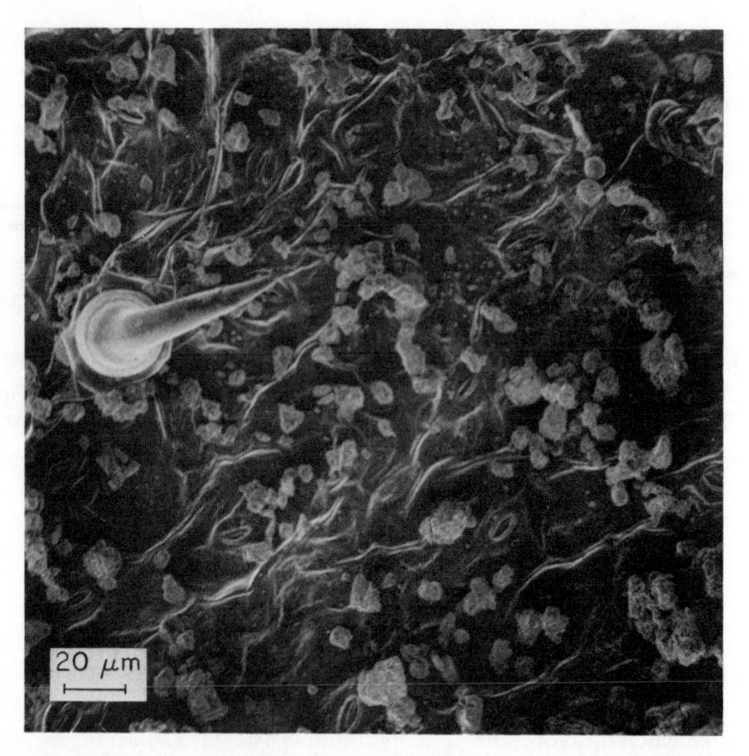

Figure 9-6. Scanning electron micrograph of the damaged upper surface of cement-kiln dust-dew treated leaf. Dust deposit 5.0 g/m^2, an advanced stage in the breakdown of the cuticle, marked by the mass of dust particles surrounded by peeled-off cuticle strips. (×300.)

wax constituents. Extracts from the leachate of control plants did not have any detectable amounts of these fatty acids.

Studies by Lerman and Darley[461] revealed that bean plants respond in various ways to dust-dew treatment using dusts from different sources. Dust from California (Cal-1) caused moderate damage: the pH of the dust-water suspension reached 12.3 (Table 9-1). Arizona (Ariz-3) dust caused severe damage but the pH of the suspension was only 10.8. Dust from one cement plant in Germany (G-1) had a relatively low content of calcium with a pH reaction of 10.4; this dust had no harmful effects on plants. The level of potassium was very high in the dust sample from a second plant (G-2) and the pH reaction was near neutral, yet bean plants dusted with G-2 dust showed damage symptoms and reduced carbon dioxide fixation. This high potassium chloride dust could act on the plant as airborne salinity. Most of the calcium in the dust samples was in the form of calcium oxide, calcium hydroxide, and calcium carbonate. The high pH reactions of the dust-water suspensions were positively correlated with a high content of calcium hydroxide, as was demonstrated by x-ray powder diffractometry. These results indicate that the response of plants to dust-dew treatment is directly related to the chemical composition of the dust. In addition, they indicate that the physical effects of the dust alone are negligible because dust treatment without dew had no detectable harmful effects on plants.

Indirect effects Pajenkamp[592] reported on unpublished work by Scheffer in Germany during two growing seasons. He indicated that even considerable quantities of precipitator dust applied to the soil surface brought about neither harmful effects nor other lasting effects on growth or crop yield of oats, rye grass, red clover, and turnips. The dust content was 29.3% limestone (analyzed as lime, CaO) and 3.1% potassium oxide. The maximum rate of deposit was 1.5 g/m^2/day. Discontinuous dustings, made at 2.5 g/m^2/day, averaged 0.75 g/m^2/day. In 1 year the yield of red clover and the weight of turnips were higher in the dusted plots, although the yield of turnip leaves decreased. Acid manuring of the soil appeared to increase yield but the interaction of dusting and manuring was not understood.

Figure 9-7. Scanning electron micrograph of the upper surface of primary bean leaf. Control undusted leaf. (×300.)

While Scheffer *et al.*[669] found no direct injury to plants, they indicated that there might be indirect effects through changes in soil reaction, which in time might impair yield.

The indirect effects of cement dust on soil fertility and crop growth near a cement factory were reported by Panda *et al.*[596]

In 1956, Stratmann and van Haut studied the effects of cement dust on vegetation in Germany (personal communication). They dusted plants with quantities of dust ranging from 1.0 to 48.0 $g/m^2/day$; dust falling on the soil caused an increase in alkalinity, which was unfavorable to oats but favorable to pasture grass.

Schönbeck[677] treated a field planting of sugar beets biweekly with 2.5 g/m^2 of dust and observed that infection by leaf-spotting fungus, *Cercospora beticola*, was significantly greater than in undusted plots. He postulated that the physiologic balance was altered by dust, increasing susceptibility to infection.

Darley[184] noted that plants were stunted and had few leaves in the heavily dusted portions of an alfalfa field downwind from a cement plant in California. Plants appeared normal in another part of the field where there was no visible dust deposit. The dusted plants were also heavily infested with aphids. Entomologists suggested that the primary effect of the dust was to eliminate the aphid predators, thereby encouraging high aphid populations whose feeding caused poor plant growth. While this effect of cement-kiln dust was not proved, Bartlett[52] had shown that deposits of several kinds of mineral dusts on leaves greatly reduced or eliminated hymenopterous insects that were parasitic to citrus pests. His review of the literature indicates that dusts have a similar effect on a wide range of insect predators.

While assessing the long-term effects of limestone dust on vegetation under natural environmental conditions, Brandt and Rhoades[85] compared dusted and undusted forest communities in the vicinity of limestone quarries and processing plants. At a heavily dusted site, the seedling-shrubs, saplings, and tree strata underwent significant changes in structure and composition. In a comparative study the same investigators[86] demonstrated a reduction of at least 18% in the lateral growth of *Acer rubrum* L. (red maple), *Quercus prinus* L. (chestnut oak), and *Q. rubra* L. (red oak). However, lateral growth of *Liriodendron tulipifera* (yellow poplar) increased by 76%. Changes in soil reaction and nutrient availability were considered as possible factors in both the increase in lateral growth of *L. tulipifera* and in reductions of growth of dominant trees in the dusty site.

Fluorides

Particles containing fluoride appear to be much less injurious to vegetation than gaseous fluorides. Pack *et al.*[589] reported that 15% of the gladiolus leaf area was killed when plants were exposed for 4 weeks to 0.79 $\mu g/m^3$ of fluoride as hydrofluoric acid, but no necrosis developed when plants were exposed to fluoride aerosol averaging 1.9 $\mu g/m^3$ fluoride. Inasmuch as the aerosol was collected from a gas stream that was treated with limestone and hydrated lime, its composition was probably calcium fluoride. Moreover, when the accumulated concentrations of fluoride in leaf tissues were about the same, whether from gas or particulate, injury from the particulates was much less.

McCune *et al.*[524] reported an increase of only 4 mm tipburn on the leaves of gladiolus exposed to cryolite (sodium-aluminum fluoride dust). The washed leaf tissue from this treatment showed an accumulation in the leaf tissue of 29 ppm fluoride on a dry weight basis. A 70-mm increase in tipburn would have been expected if a similar accumulation had occurred from exposure to hydrofluoric acid. Except for this slight tipburn, McCune and his colleagues found that cryolite produced no visible effects on a variety of plants nor did it reduce growth or yield.

It is evident from the work of McCune *et al.*[524] that fluoride accumulates in plant tissue as a result of exposure to cryolite, but that the rate of accumulation is much slower than would be expected from a comparable treatment with hydrofluoric acid. For example, when comparing washed leaf samples, exposure of gladiolus to hydrofluoric acid for 3 days at 1.01 $\mu g/m^3$ fluoride resulted in an accumulation of 26.4 ppm fluoride, whereas only 34 ppm was accumulated from an exposure to cryolite for 40 days at 1.7 $\mu g/m^3$ fluoride. Furthermore, when approximately equivalent amounts of fluoride had been absorbed from dust applications of sodium fluoride, calcium fluoride, and cryolite onto the leaves of tomato, leaf injury occurred only on those leaves treated with sodium fluoride. The authors suggest that the effect of a particulate fluoride is dependent upon its solubility. Very recent experiments by McCune *et al.*[525] show this to be the case. Increasing the dosage of cryolite on mandarin orange leaves from about 120 $\mu g/m^3/hr$ to 900 $\mu g/m^3/hr$ in an atmosphere at 75% RH did not increase leaf absorption over that of controls. At a dosage of 7,800 $\mu g/m^3/hr$ at 85% RH, foliar fluoride was three times that of the controls. The presence of a water mist on the leaves, however, increased foliar concentration by a factor of 5 over the equivalent dose at 85% RH. Similar results were obtained with spinach leaves, but at lower doses of cryolite.

Both of the investigations cited above indicate that much of the particulate matter remains on the surface of the leaf and can be washed off, although that which remains after washing is not necessarily internal fluoride. Reduced phytotoxicity of particulate fluoride is ascribed in part to the inability of the material to penetrate the leaf tissue. In addition, McCune *et al.*[524, 525] suggest that inactivity of particles may be due to their inability to penetrate the leaf in a physiologically active form.

Lead Particles

Published results of several studies[123, 417, 458, 561, 590, 680, 789] show that lead accumulation by vegetation near highways varies with motor vehicle traffic densities and generally declines with distance from heavily traveled roads. Page *et al.*[591] demonstrated that lead accumulations in and on plants next to highways in southern California were caused principally by aerial deposition and to a lesser extent by absorption from lead-contaminated soils. Although lead concentrations near highways have been high, there are no known reports of injury to vegetation.

Soot

Jennings[389] suggested that soot may clog stomata and prevent normal gas exchange but noted that most investigations discount this effect. Microscopic examination failed to show significant clogging of stomata on leaves of broad-leaved species of shade trees. He further stated that interference with light can be more serious but he offers no data from critical experiments to substantiate this theory.

A well-illustrated report by Berge[67] showed plugged stomata on conifers growing near Cologne, West Germany. His report also stated that growth was adversely affected.

Necrotic spotting was observed on leaves of several plants where soot from a nearby smokestack had entered a greenhouse.[544] The necrosis was attributed to acidity of the soot particles. Plants outside the greenhouse were not damaged, possibly because the particles had been removed by rain before severe injury could occur. Ricks and Williams[644] observed oak leaves in the vicinity of a phurnacite smokeless fuel plant in which briquettes composed of a mixture of finely divided coal and pitch are carbonized in ovens at 750°C to 850°C. During the manufacturing process, large quantities of dust and smoke containing gaseous sulfur compounds are emitted. Scanning electron micrographs of the oak leaves revealed aggregation of particles on both leaf surfaces. At high magnification, individual stomata were shown to contain several particles in the pore opening. These leaves had a lower maximal diffusion resistance than that measured in unpolluted leaves.

Magnesium Oxide

The possible indirect effect on vegetation of magnesium oxide falling on agricultural soils was reported by Sievers.[699] He noted poor growth in the vicinity of a magnesite-processing plant in Washington. To prove the role of magnesium as a cause of poor growth, Sievers planted oats in pots containing 1,000 g of soil collected at distances of 0.201, 0.804, and 3.22 km from the calcining plant. Five grams of magnesium carbonate were added to half of the pots in each series; the remaining pots were left untreated. After 10 weeks, plants in the untreated soil collected at 0.201

km exhibited the stunting and discoloring characteristic of affected plants in the field. The addition of the magnesium carbonate caused no further injury. Plants in untreated soil collected at 0.804 km grew somewhat better; however, adding the magnesium carbonate caused reduced growth comparable to the untreated sample at 0.201 km. Plants in the untreated soil at 3.22 km grew normally but the added magnesium carbonate still caused some typical discoloring. After the processing plant ceased operation, injury to crops in the area lessened, indicating that the injurious effect was not permanent.

Iron Oxide

Berge,[68] in Germany, dusted experimental plots with iron oxide at the rate of 0.15 $mg/cm^2/day$ over 1- to 10-day intervals through the growing season for 6 years. The plots were planted with cereal grains or turnips. Effects of treatment on the primary product, on straw, and on leaves were noted. No harmful effect of the dust was detected on either crop. There was a tendency for improvement of yields of grain and turnip roots, but this was not statistically significant.

Foundry Dusts

Changes in starch reserves were compared in common holly leaves that were either untreated or treated with dusts emitted from foundry operations.[722] The critical factor was the amount of light absorbed by the dust layer and not the effects of dust particles on transpiration or temperature of the leaf. These observations agree with some of those reported above concerning the range of effects of cement-kiln dust on vegetation.

Acid Aerosols and Acid Rain

By burning fossil fuels and by various industrial processes, man releases large quantities of sulfur and nitrogen oxides to the atmosphere. These compounds, in addition to naturally occurring gaseous forms of sulfur and nitrogen, are oxidized and hydrolized in the atmosphere to sulfuric and nitric acids at varying rates, depending upon environmental conditions. If these acids are not neutralized by alkaline substances also present in the atmosphere, they will ultimately settle to the ground as mist or precipitation. The impact of acid aerosols and acid rain on vegetation includes direct effects, in the form of injury to above-ground plant parts, or indirect effects, by creating unfavorable conditions for plant growth. This may involve acidification of irrigation water (lakes and river), increases in the leaching rates of soil nutrients, and shifts in the balance of microorganism populations in the soil.

Direct effects Middleton *et al.*[537] observed necrotic spots on the upper leaf surface of vegetation in the Los Angeles area following periods of heavy air pollution accompanied by fog. Thomas *et al.*[753] exposed plants to sulfuric acid mists in concentrations ranging from 108 to 2,160 mg/m³. Necrotic spots developed only on the upper surface of wet leaves.

Gordon[285] exposed several species of western pine to an atomized solution of sulfuric acid having a pH below 4.0. He observed a 50% reduction in needle growth compared to controls. In 1974, growth abnormalities of Christmas trees were attributed by Hindawi and Ratsch[340] to sulfur dioxide and particulate acid aerosol. That same year, Wood and Bormann[829] reported irregular development of leaf tissue and spot necrosis on yellow birch seedling that were exposed to artificial sulfuric acid mist at pH 3.0. In 1975, they also observed[830] an increase in leaching of nutrients from birch foliage by acidification of an artificial mist to pH 4.0. Injury symptoms developed on seedlings of kidney beans and soybeans that were subjected to simulated rain acidified with sulfuric acid to pH 3.2.[693]

Severe reductions in yield and quality occurred on field-grown tomatoes in the Kona district on the island of Hawaii.[427] Tomato plants grown under plastic rain shelters produced normal crops. A definite atmospheric haze that appeared at this time coincided with volcanic eruption. Although the rain water of low pH (4.0) decreased pollen germination and pollen tube growth, the authors did not rule out the possibility that other factors were involved, since appreciable quantities of chlorides and sulfates plus 27 detectable organic compounds in the ppb range were found in the rain water.

Indirect effects In 1971, Norwegian investigators Dahl and Skre[179] showed a clear connection between decreases in productivity of forests and the amount of calcium in the topsoil of forest land. However, liming of the soil did not result in the expected growth increase. Bolin *et al.*[78] tried to explain this contradiction by stating that the relation between calcium and productivity is probably indirect and that the effects of additions or losses only show up after quite a long time. A study in Sweden[394] correlated soil acidification with significant reduction in forest growth between 1945 and 1965. Bolin *et al.*[78] estimated reduction in growth to be 0.3% per year in Sweden. In 1974, Whittaker and his associates,[809] in a study of a New Hampshire hardwood forest, tentatively attributed a striking 18% decrease in volume growth and productivity that took place in 1960 to the effects of acid rain and/or drought.

The effects of acid rain on host-parasite interactions were demonstrated by Shriner.[693] Likens and Bormann[466] suggested numerous indirect effects that could lead to reduction in plant vigor. Direct nutrient leaching and erosion of the cuticle, providing ready access to pathogens, can combine with other impinging stresses to cause an adverse effect. Lerman[460] demonstrated that erosion of the cuticle on the upper leaf surface can also lead to negative water balance in the plant.

10

Effects on Building Materials

Deposition on building materials of such airborne particles as dust and soot can not only greatly reduce the aesthetic appeal of structures, but also, either alone or in concert with other environmental factors, result in direct chemical attack. It is not entirely understood which specific class of particulates, or chemical constituent thereof, under what ambient conditions, actually causes substantial damage or soiling of specific building materials. Therefore, knowledge of the effects of air pollutants on building materials should be reviewed periodically to determine whether additional understanding of these matters has been achieved.

This chapter updates the information on materials contained in the 1969 National Air Pollution Control Administration publication, *Air Quality Criteria for Particulate Matter.*[768] It describes and assesses knowledge concerning the effects of particulate pollutants on building materials. Based on an extensive literature search and discussions with a broad range of building industry professionals and trade association representatives, this chapter focuses on the effects of particulate pollutants on metals, masonry and concrete, paints and finishes, polymeric materials, textiles, porcelain enamels, and asphalts. Available data on the estimated cost of material soiling and degradation due to pollutants are also presented. Conclusions are based on these findings.

The quantity of the literature on building materials does not necessarily correlate with the degree of importance or use of materials in the building industry. For example, although masonry and concrete are used extensively in building construction, there is little information regarding the effects of pollutants on these materials. In comparison, although plastics are used very little, except in building interiors, there is considerable information concerning the effect of air pollutants on them.

Available data are based on exposure of materials to outdoor pollutant concentrations; therefore, the relationship between outdoor and indoor concentrations must be considered if the effects on materials used for building interiors are to be assessed accurately. Anderson[33] showed that over a 7.5-month period the average indoor concentrations of particulates and sulfur dioxide were, respectively, 69% and 51% of simultaneous outdoor values. Worldwide studies of the variance of indoor-outdoor pollutant concentrations also indicate that both particulate and sulfur dioxide concentrations are markedly reduced in indoor atmospheres. The indoor-outdoor ratios vary considerably, presumably because of differences in room sizes, building materials, ventilation rates, external pollutant concentrations, etc.

Particulate matter may react with, influence, or be influenced by other constituents of the atmospheric environment, both natural and manmade. Without consideration of these other pollutants, accurate assessment of the effects of particulate matter on building materials is not possible.

The behavior of particulate matter in relation to other pollutants is particularly important in two respects:

1. Particulates can act as nuclei for absorbed and adsorbed gases, such as sulfur dioxide, hydrogen sulfide, and nitrogen dioxide. This can occur while the particulate is still airborne or after it has settled. Thus, the actual effect of the particulate as a polluting agent can be completely different from that expected of the particulate as emitted. To assess its damage potential, it is necessary to understand both the chemical effects of the particulate and its synergism with pollutants such as absorbed and adsorbed gases whose presence or accumulation on a material surface is due to or increased by the particulate. Because there are very few data on the effects of particulates known to contain adsorbed or absorbed gases, information on the effect of gases alone is included below to provide at least some indication of potential synergism.
2. Gaseous pollutants can convert immediately to liquid acids or salts, which are classified as particulates, on contact with moisture in the air or on the surface of a material. For example, sulfur dioxide (approximately 80% of which results from the oxidation of hydrogen sulfide) reacts with moisture to form liquid sulfur trioxide, sulfuric acids, and sulfate salts. Furthermore, sulfur dioxide in the presence of nitrogen dioxide and hydrocarbons photooxidizes to produce aerosols containing sulfuric acid, which are also classified as particulates. Similarly, nitrogen dioxide combines with moisture to form liquid nitric acid, which in turn can react to form nitrate salts such as ammonium nitrate, both of which are also classified as particulates. There is little information on the effects of liquid particulates. The effect of gases alone are discussed below because most test results probably reflect effects of both gaseous and liquid pollutants.

Gaseous pollutants include sulfur oxides, nitrogen oxides, hydrogen sulfide, ammonia, and ozone. Ozone, although a naturally occurring element of the atmosphere, is considered to be a pollutant because it can be formed by the action of sunlight on oxygen in the presence of pollutants, both particulate and gaseous (e.g., hydrocarbons and nitrogen oxides). In certain areas on days during bad smog, the ozone concentration may reach 0.5 ppm or more for short periods.[304]

Finally, natural environmental factors, such as moisture, temperature, sunlight, and wind can greatly influence the effect of air pollutants on building materials. They are discussed below, where applicable.

METALS

Current application of metals in buildings extends far beyond its traditional uses as roofing sheet for large warehouse-type facilities and as cladding and piping. Some major metals and their most common uses, depending on processing procedure and specific composition, are:

Steel: structural columns and beams, roofing sheets, curtain walls, stacks and storage tanks, cladding, pipe balustrades, stairways, balconies, safety rails, structural elements of bridges, boiler tubes, wire, and bolts, rivets, and other fasteners.

Iron: rods, lath, gas and water pipes, and decorative railing.

Aluminum: curtain walls, siding, roofing sheets, air ducts, window frames, doors, screening, trim, flashing, protective and decorative railings, rainwater leaders, and gutters.

Zinc: flashing and protective coating for iron and steel.

Copper: cladding, flashing, wiring, piping, and decorative cornices and spandrels.

Brass and bronze: decorative doors, window frames and sashes, balustrades, screws and other hardware items, and ornamentation.

Lead: piping and coating for iron and steel.

Moisture and temperature play critical roles in the corrosion of metals, whether or not air pollutants are present.[687, 703] Without moisture in the atmosphere there would be little, if any, electrochemical atmospheric corrosion, even in the most severely polluted environments. Direct chemical attack, of course, could occur in such conditions.

Evidence demonstrates that corrosion of metals can be accelerated by atmospheric pollutants, not only as a result of direct chemical attack by smoke particles, acid aerosols, or mist but also because acids, salts, and other particulates could be contained in or trapped by moisture condensed on a metal surface. Moisture (or electrolyte) thus contaminated is very likely to have greater electrical conductivity and to accelerate corrosion. In addition, moisture can react with sulfur dioxide to form liquid sulfuric acid, which is extremely corrosive to certain metals.

The following paragraphs present evidence, by type of metal, where possible, on the effects of particulates alone, of combined pollutants, and of some gaseous pollutants alone.

Goodwin *et al.*[284] provided the only information on erosion of metals by particulates. Tests of quartz particles on steel and nylon indicated that particles $\lesssim 5$ μm cause relatively little damage. Larger particles cause progressively more damage until saturation is reached. Erosion varied with a simple power of velocity—2.0 for small particles and 2.3 for particles $\gtrsim 100$ μg. Both degree of fragmentation of the impacting particle and properties of the target material influenced erosion behavior. While agreeing to a relationship between impact angle and erosion, the authors stated that observed particle size and velocity effects seem to occur regardless of angle.

Hermance[324] and McKinney and Hermance[528] showed that airborne particulates high in nitrate concentration caused stress-corrosion cracking and ultimate failure of nickel-brass wire spring relays and other electrical equipment in the Los Angeles area. Examinations indicated that particulate matter had collected on insulating surfaces around stress nickel-brass wire in an environment of unregulated humidity. Bulk dust samples from the air of Los Angeles, New York, Philadelphia, Baltimore, and Chicago were analyzed. The highest nitrate concentrations were in the Los Angeles sample, apparently because of the nitrogen dioxide in the air that reacted with moisture to yield nitric acid, much of which eventually associated with fine particulate matter like ammonium and alkali nitrates. The combination of these fine materials in the Los Angeles sample, the absence of soot, and the high proportion of oxidized, polar compounds in the organic fraction probably made the dust deposits from Los Angeles more reactive to moisture than the samples from the eastern areas.

Subsequently, McKinney and Hermance,[528] in their laboratory tests, simulated conditions of electrolysis of hygroscopic salts in the dust deposits. They used ammonium salts of sulfate, nitrate, and chloride. Results showed that, of the major ions in dust deposits, only the nitrate ion was highly active in producing stress corrosion in nickel-brass alloys, depending on relative humidity. Below a critical humidity, which appeared to be between 40% and 50%, stress-corrosion cracking dropped to a negligible rate. They also investigated the influence of applied positive potential and temperature. They found that the rate of stress-corrosion cracking was directly dependent on these factors as well as on nitrate concentrations and humidity. They extrapolated from these results to conclude that noticeable stress-corrosion cracking of the nickel-brass wire could occur in the Los Angeles environment within 2 years. New wire-spring relays made with 80% copper-20% nickel springs are being used because nitrate corrosion does not appear in the absence of zinc. Cracking in existing nickel-brass relays is controlled by high efficiency filters and cooling systems which have been re-designed to maintain the relative humidity (RH) below 50%.

In Japan, Abe *et al.*[1] tested copper specimens exposed for 2 years in marine, industrial, urban, and rural atmospheres. Their analysis of metallic deposits on the surface of exposed specimens revealed basic sulfates, sulfides, and chlorides of copper. This indicated attack by particulate pollution.

When exposed to the atmosphere, copper and its alloys gradually develop a dark color that finally changes to a pale green. Yocom and McCaldin[839] identified this as the familiar green protective patina that can be either basic copper sulfate or, in a marine atmosphere, basic copper chloride—both of which are extremely resistant to further atmospheric attack.

Simpson and Horrobin[703] indicated that the normal time for patina formation is 5 to 10 years, depending on the concentration of sulfur or chlorine pollutants, the amount of moisture in the air, and the temperature. Fink *et al.*[232] reported typical corrosion rates of copper in industrial atmospheres to be approximately 0.05 mils per year (mpy) compared with 0.02 mpy in rural atmospheres.

Even though pollutants accelerate corrosion of copper, the higher rate was not considered sufficient to cause significant damage. The higher rate caused more rapid formation of the patina that retards the attack. Typically, a copper roof or other component will outlast the building or structure of which it is a part whether the air is polluted or not. According to Yocom and McCaldin,[839] 20-year exposure tests with copper specimens conducted by the American Society for Testing and Materials (ASTM) verify this: the specimens showed a loss in thickness of <0.025 mpy and 0.1 mpy in rural and industrial atmospheres, respectively.

In the ASTM studies, corrosion rates of copper in industrial atmospheres were greater than those in marine atmospheres. A number of formulations were tested. With the exception of a high tensile brass, all appeared to be suitable for use in industrial atmospheres. Like the high tensile brass, bronze also may be more susceptible to attack by atmospheric pollutants.

Although aluminum forms an invisible protective film of oxide, it becomes covered with dirt and soot particles in a polluted atmosphere. This tends to mottle and pit the surface. In time the surface appears roughened, but no general thinning occurs. Simpson and Horrobin[703] reported that prolonged exposure of aluminum to industrial atmospheres results in the formation of spots of white crystalline corrosion products on the surface of the aluminum. Depending on the amount of soot in the atmosphere, the metal will acquire a greyish or darkish appearance.

These authors found that corrosion of aluminum is greater in severe industrial atmospheres than in marine or rural atmospheres.

Fink *et al.*,[232] summarizing the work of others, reported that after exposure of many aluminum alloys for 7 years in an atmosphere with sulfur dioxide concentration >0.14 ppm, which implies industrial and therefore high particulate concentration, pits as deep as 13.8 mils formed (see Table 10-1).

In rural clean air, pitting was negligible and loss of thickness was 0.028 mpy compared to rates as high as 0.15 mpy in industrial atmospheres. Most of the pitting occurred during the first 2 years. Afterward, penetration increased only slightly so that the depth after 20 years was similar to that measured after 3 years. Based on these findings, the investigators concluded that corrosion of aluminum and its alloys by air pollution does not greatly affect the life of externally exposed structural elements and components.

Particulate matter is an important factor in the corrosion of metals, especially in the presence of acidic, gaseous pollutants. This was substantiated by field work at the Chemical Research Laboratory in Teddington, England.[839] Investigators there exposed one sample of iron to a moist atmosphere containing traces of sulfur dioxide and particulates, and another sample to the same atmosphere but protected from the particulate matter with a muslin filter. Rusting of the protected sample was negligible compared with the unprotected sample; however, the amount of moisture and sulfur dioxide absorbed by the muslin was not determined.

The influence of particulates on corrosion of metals has also been established from laboratory tests in which various coated and uncoated metallic surfaces were inoculated with fine powders or nuclei of sodium chloride, ammonium sulfate, ammonium chloride, sodium nitrate, and flue dust.[839] The samples were then exposed to atmospheres at various humidities. Corrosion of all surfaces occurred and, except for coated surfaces inoculated with ammonium chloride, corrosion rate increased with increased humidity. The addition of sulfur dioxide traces to the test atmosphere greatly increased corrosion at high humidities. Furthermore, these authors reported that a much more rapid corrosion rate occurs when the air contains both charcoal particles and sulfur dioxide than when it contains only sulfur dioxide. From these observations, they reasoned that the action of the charcoal particles was primarily physical, i.e., they increased the concentration of sulfur dioxide on the surface of the metal by absorbing it, thus creating hot spots at the point of contact between the metal surface and charcoal particle. Tajiri[748] reported this to be the general case for dust particles. Fink *et al.*[232] also stated that when hygroscopic particles settle on a metal surface they usually accelerate the corrosion rate, particularly in the presence of sulfur dioxide, and may cause corrosion at $<70\%$ RH. Dust can disrupt the protective oxide

Table 10-1. Seven-year corrosion results for aluminum alloys exposed to industrial atmospheres[a]

Alloy	Corrosion rate (mpy)				Depth of pitting (mils)		
	State College[b]	Richmond[c]	Chicago[d]	Widnes[e]	Richmond	Chicago	Widnes
1199	—	0.001	0.023	0.048	2.8	7.8	13.8
1135-H14	0.0025	—	—	—	—	—	—
3003-H14	0.0028	0.019	0.044	0.151	3.7	6.9	10.3
3004-H36	—	0.020	0.054	0.091	3.6	4.7	9.7
3004-H34	0.0026	—	—	—	—	—	—
5154-H34	—	0.017	0.055	0.106	8.7	9.3	10.9
5005-H34	0.0025	—	—	—	—	—	—
6061-T6	0.0027	0.015	0.067	0.103	4.1	5.6	10.6
6063-T6		0.010	0.051	0.059	5.1	5.3	10.9

[a] From Fink *et al.*, 1971.[232]
[b] Rural atmosphere (ASTM STP 435).
[c] Mild industrial (0.01 ppm SO_2, avg.).
[d] Industrial (0.14 ppm SO_2, avg., RH, 80%).
[e] Severe industrial, Widnes, England.

films formed on metal surfaces like stainless steel and can result in corrosion. Such general statements attest to the fact that particulates influence corrosion rates, and that airborne particulate matter is a corrosive factor in both industrial and urban regions.

Chandler and Kilcullen[131] investigated the nature of the relationship between the corrosion of two mild steels with different copper contents and atmospheric pollution at 11 sites near Sheffield, England. Their results confirmed that there was an important relationship between the corrosion of steel and the level of atmospheric smoke and sulfur dioxide. However, only about 50% of the variations in corrosion rate at the different sites could be attributed to sulfur dioxide. The influence of smoke intensity on the corrosion rate of steels was similar to that of sulfur dioxide. Because the correlation between sulfur dioxide and smoke was high, their effects on the corrosion rate could not be considered independently. The combined effect of smoke and sulfur dioxide was only slightly greater than their individual effects. The corrosion rate could not be accurately predicted from the quantity of the two pollutants.

Haynie and Upham[322] exposed plain carbon, copper-bearing, and weathering steel samples to both urban and rural atmospheres (see Table 10-2). Twenty-five samples of each steel were mounted at 30° angles facing south.† Replicate samples were examined at 4, 8, 16, 32, and 64 months. Urban exposure sites were selected in or near eight cities in which the pollutant concentrations were being monitored (Table 10-3). Rural areas near the cities were selected as control sites because of their low pollution but similar meteorologic conditions. Pollutant concentrations at these rural sites were not measured during exposures. Results of these exposures, which may or may not have been influenced by particulate concentration, are tabulated in Table 10-4.

These authors reported that the rate of steel corrosion is diffusion-controlled by the type and thickness of the rust film that forms. Some structural carbon steels form a loosely adherent rust that provides very little protection from further corrosion; both copper-bearing and weathering steels form more adherent protective films. Because the residual copper in the carbon steel tested was relatively high, its corrosion behavior was similar to that of copper-bearing steel. In each case a protective film was formed.

Haynie and Upham[322] concluded that more than 90% of the variability in corrosion behavior of steels could be attributed to the variability of the concentrations of both sulfur dioxide, which increases corrosion

---

† Position in space is an important variable in the testing of materials for atmospheric corrosion. Test samples are usually mounted at or near 45° from the horizontal because this more nearly simulates actual use conditions. The undersurfaces often are corroded more rapidly than the upper surfaces because corrosive agents are not washed off as well by rain.

Table 10-2. Chemical analysis of steels used in study[a]

Type of steel	Code	Composition (%)							
		C	P	S	Si	Mn	Cu	Ni	Fe
Plain carbon	A	0.096	0.015	0.022	0.003	0.36	0.17	—	Remainder
Copper-bearing	B	0.036	0.018	0.028	0.003	0.26	0.22	—	Remainder
Weathering	C	0.130	0.058	0.012	0.100	0.82	0.60	0.71	Remainder

[a] From Haynie and Upham, 1971.[322]

Table 10-3. Average[a] pollutant concentrations at urban sites ($\mu g/m^3$)[b]

City	Nominal exposure (months)	SO_2	O_3	NO
Chicago	4	419	39	164
	8	479	43	145
	16	400	47	130
Cincinnati	4	76	65	53
	8	97	53	49
	16	68	61	46
	32	86	58	44
	64	79	59	44
Detroit	4	139	16	119
	8	139	19	106
	16	132	22	96
	32	121	23	92
	64	118	24	80
Los Angeles	4	45	106	65
	8	47	92	119
	16	42	84	93
	32	39	76	97
	64	39	75	100
New Orleans	4	34	39	23
	8	29	34	25
	16	26	37	20
	32	24	35	20
Philadelphia	4	191	45	82
	8	194	45	57
	16	197	53	58
	32	194	56	67
	64	218	57	63
San Francisco	4	10	39	119
	8	13	38	111
	16	29	37	102
	32	34	37	102
	64	34	37	102
Washington	4	222	33	62
	8	207	41	58
	16	141	47	50
	32	136	49	46
	64	126	50	47

[a] Averages for total periods of exposure, calculated from monthly averages weighted by the actual number of days of exposure in each month.

[b] From Haynie and Upham, 1971.[322]

rate, and oxidants, which decrease it. Because these pollutants are counteractive, steel corrosion behavior can vary considerably among cities as the cleanliness of the environment is improved. A national reduction in pollution should result in less steel corrosion; however, in cities where sulfur dioxide concentrations are low and oxidant concentrations are high, lowering the total pollution level will increase steel corrosion.

Average relative humidity, rainfall, and temperature usually correlate well with steel corrosion. Although atmospheric factors were not significant in the results obtained by Haynie and Upham,[322] the authors did not rule out their influence. The range of average relative humidities among cities was within 15%. It is quite likely that the differences in time-of-wetness of the samples was also slight. Another possible explanation was that the rust film at the relative humidities encountered remained moist most of the time.

Oma *et al.*[587] investigated the influences of meteorologic factors and pollutant concentrations on the outdoor corrosion rate of steel near Tokyo. Their statistical analysis showed that, within a 95% confidence limit, a combined relation does exist. Based on this, these investigators developed an equation to calculate and predict corrosion rates in industrial areas. The equation considers: level of humidity; concentration of sulfur dioxide, chlorides, soluble and insoluble substances, sulfates, and tar substances; amount of precipitation; temperature; and wind velocity.

Some investigators have stated that industrial air pollution is the greatest factor in zinc corrosion because acids, gases, and vapors destroy the basic carbonate protective coating that normally forms on zinc.[839] Tajiri[748] and Simpson and Horrobin[703] pointed to sulfurous pollution as the cause of the greatest increase in the corrosion rate of zinc, especially when the material surface is wet. Sulfurous and sulfuric acids reportedly react with the basic zinc carbonate film and convert it to zinc sulfate, which is soluble and washes off with rainfall. The zinc carbonate layer then rapidly renews and remains protective until subjected to further attack. More specifically, McLeod and Rogers[531] reported an average corrosion rate of 24.7 mpy during the first 7 days that zinc is exposed to a moving mixture of sulfurous acid and air. The rate decreased until it became constant at 5 mpy after 500 days. Corrosion products were sulfur, zinc sulfite, and zinc sulfate. The corrosion rate of zinc completely immersed in an unaerated solution of sulfurous acid at 21.11°C to 23.89°C was much higher and increased as the acid concentration was increased. In a solution containing 8 g/liter of sulfur dioxide, the average corrosion rate was 279 mpy; in a solution containing 16 g/liter, the average rate was 496 mpy. As the sulfurous acid content of the unaerated

Table 10-4. Effect of location and time on corrosion of steel[a]

City	Site	Alloy[c]	Average depth of corrosion (μm[b]); exposure (months)				
			4	8	16	32	64
Chicago	Urban	A	19.22	30.48	52.80	[d]	[d]
		B	18.59	30.42	45.46	[d]	[d]
		C	16.89	23.68	39.10	[d]	[d]
	Rural	A	14.80	28.96	41.11	[d]	[d]
		B	14.06	29.10	40.94	[d]	[d]
		C	12.25	21.84	30.80	[d]	[d]
Cincinnati	Urban	A	14.64	18.94	23.18	28.32	37.98
		B	14.05	18.13	22.51	28.10	37.34
		C	11.36	13.53	16.48	19.50	25.44
	Rural	A	14.86	21.45	36.55	51.19	79.79
		B	13.64	20.75	37.04	53.05	82.64
		C	11.01	14.96	22.94	29.14	39.07
Detroit	Urban	A	12.24	25.12	34.19	49.95	74.13
		B	16.12	24.39	33.64	50.15	74.07
		C	13.61	19.16	26.78	35.87	49.14
	Rural	A	10.04	18.61	28.80	40.95	60.58
		B	9.00	18.16	28.41	40.64	62.09
		C	8.16	13.91	21.70	29.42	39.31
Los Angeles	Urban	A	9.72	11.63	15.97	19.57	25.81
		B	8.98	10.42	14.82	18.71	24.83
		C	9.62	10.91	14.05	17.31	23.21
	Rural	A	10.85	14.93	21.40	28.97	40.02
		B	8.96	13.30	20.21	27.87	40.58
		C	9.63	12.34	16.79	21.47	28.93
New Orleans	Urban	A	10.12	16.51	24.52	35.38	[d]
		B	8.56	14.82	23.48	34.85	[d]
		C	8.93	13.05	19.06	25.61	[d]
	Rural	A	4.03	9.90	19.84	32.96	56.98
		B	3.86	8.48	18.31	31.41	55.18
		C	4.24	8.47	16.45	26.04	41.14
Philadelphia	Urban	A	19.50	24.16	32.09	41.37	51.50
		B	18.09	23.16	31.54	40.41	50.95
		C	15.73	17.88	23.20	27.70	33.38
	Rural	A	22.01	28.38	43.84	60.91	79.71
		B	20.80	27.29	43.60	60.59	80.67
		C	17.52	20.93	29.48	36.78[e]	43.17
San Francisco	Urban	A	10.80	17.74	28.61	41.44	57.36
		B	8.57	16.34	27.95	41.18	47.94
		C	8.72	13.44	19.81	27.22	45.35
	Rural	A	10.20	12.06	21.54	[d]	[d]
		B	9.16	10.87	20.80	[d]	[d]
		C	9.30	10.97	17.31	[d]	[d]
Washington	Urban	A	14.58	20.50	25.87	33.94	43.04
		B	13.14	19.34	25.16	33.35	42.59
		C	10.72	12.89	16.45	21.17	26.61
	Rural	A	13.39[e]	20.22[e]	33.49	47.61	66.88[e]
		B	12.02	18.99	32.70	46.96	65.51
		C	10.15[f]	14.45[f]	21.54	27.78	35.77

solutions decreased due to corrosion, hyposulfite, sulfide, thiosulfate, and elemental sulfur were produced. Guttman's[302] mathematical expressions for predicting corrosion rates of zinc consider time-of-wetness and sulfur dioxide content of the atmosphere. Acknowledging a relationship between time-of-wetness and relative humidity, he provided a means of estimating time-of-wetness for relative humidity data.

A striking example of the relationship between zinc corrosion and pollution level is the almost fourfold reduction in the corrosion rate of zinc in Pittsburgh associated with a twofold reduction in dust fall and a threefold reduction in sulfur dioxide (from 0.15 ppm to 0.05 ppm) from 1926 to 1960.[769]

Yocom and McCaldin[839] conducted tests on exposed mild steel plates at 35 sites in metropolitan St. Louis, Missouri. Plates were removed and examined after exposures of 2, 4, 8, and 16 months. Corrosion rates were 30% to 80% greater in urban and industrial locations than in suburban and rural areas. No relationship between dust fall and corrosion was detected. The relationship between corrosion and sulfation rates was strongest during the early stages of exposure. In similar studies made at 20 stations in the Chicago area, downtown corrosion rates were ~70% greater than in the suburbs after a 12-month exposure. The measurements of sulfur dioxide concentrations at seven of the Chicago sites indicated a direct relationship with corrosion rates (Figure 10-1). In similar experiments, Palmer[593] found that iron specimens exposed as long as 12 weeks in Tulsa, Oklahoma averaged corrosion rates 3 times greater than residential areas near Tulsa.

Simpson and Horrobin[703] reported that high chromium and chromium-nickel stainless steels are very resistant to atmospheric corrosion. They attributed this to a thin oxide film that is always present on the surface and is self-repairing when damaged. Superficial deterioration, staining, or soiling in highly polluted, aggressive atmospheres could be prevented by regular washing.

Simpson and Horrobin[703] reported that lead is very resistant to atmospheric corrosion whether pollutants are present or not. In clean atmospheres, the protective layer of lead carbonates and sulfates that forms on the metal surface gives the metal a whitish-grey appearance, which might darken in industrial areas. Weight loss measured over 20

Footnotes for Table 10-4 (opposite):

[a] From Haynie and Upham, 1971.[322]

[b] Depth of corrosion calculated from weight-loss measurements, averages from five specimens.

[c] Code for alloys: A = plain carbon steel, B = copper-bearing steel, and C = "weathering" steel.

[d] Exposure discontinued because of wind damage or vandalism.

[e] Average determined from four specimens.

[f] Average determined from three specimens.

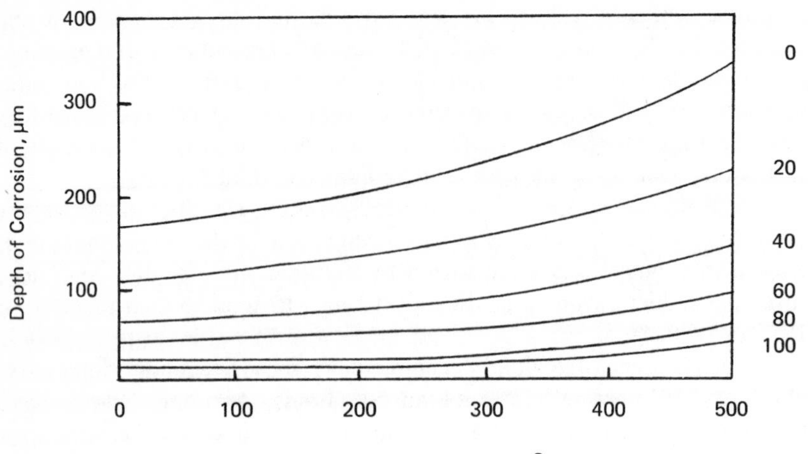

Figure 10-1. Relationship between corrosion of mild steel and corresponding mean sulfur dioxide concentration at seven Chicago sites. From Yocom and McCaldin, 1968.[839]

years at seven U.S. cities indicates decreasing corrosion rates in all but coastal atmospheres. Both basic and antimonial lead were extremely durable; their corrosion rates were ~0.02 mpy. At this rate, a 3/32 inch (~2.34 mm) thick lead sheet would corrode to half its thickness in 2,500 years.

Several investigators exposed various metals to atmospheric pollutants. Gibbons[270] studied the corrosion behavior of major architectural and structural metals exposed to various outdoor Canadian atmospheres for 10 years. He tested three aluminum alloys, two magnesium alloys, three steel alloys, three stainless steels, and rolled zinc. He positioned all specimens at 30° angles to the horizontal and facing south. The eight sites used were classified as rural, semirural, industrial, marine, or marine-industrial. Although most results were reported in terms of sulfur dioxide effects, because they are based on exposure to actual outdoor atmospheres, particulates may or may not have contributed.

The three stainless steels had the best corrosion resistance. Their corrosion was negligible at all but the marine-industrial sites. The aluminum alloys were next in resistance with little difference in performance among the three alloys exposed. Rolled zinc rated third. The three steel alloys and magnesium alloys had the highest corrosion rates. The marine-industrial atmosphere was the most corrosive. One semirural and all rural-marine and industrial atmospheres had similar corrosion rates. They were next in order or corrosivity. Following was an urban site; then a marine, rural, and a second semirural site. The rate of atmospheric corrosion was markedly influenced by the sulfur dioxide content of the atmosphere; the high sulfur dioxide concentration at the marine-industrial site made it the most aggressive atmosphere. The corrosion rate usually decreased after the first year of exposure. With the exception of the rolled zinc and the magnesium alloys, the earthward sides of the specimens were more severely attacked than the skyward sides.

The corrosion rates of all three stainless steels decreased after a 1-year exposure, and the stainless appearance of two types, although somewhat dulled, was retained after 10 years. The sample that was exposed to the marine-industrial atmosphere remained very black even after cleaning; but specimens of the same type exposed at all other sites not only retained their brightness and luster after cleaning, but, except for staining, were quite bright even before cleaning.

After the initial years of exposure, corrosion rates of the steel specimens decreased significantly at all but the rural sites where the corrosion rates were lowest from the start. The residual copper content of the low carbon steel was sufficient to decrease its corrosion rate. The copper-nickel steel alloy formed an adherent protective oxide film that maintained the low corrosion rate after 5 years, particularly at the rural and industrial sites. The corrosion rate of zinc was fairly constant at all sites. Corrosivity was related to sulfur dioxide concentration; therefore, zinc corrosion rates correlated with atmospheric content of sulfur dioxide at the various sites. The combined effect of atmospheric chlorides and/or time-of-wetness on zinc specimens was evident at some sites. The corrosive attack was essentially uniform, although some pitting occurred at the more aggressive sites.

In a second series of experiments, Gibbons[271] tested two lead alloys, a nickel-copper alloy, copper sheet, and a copper-zinc alloy for 10 years. He exposed specimens in the same manner to the same Canadian atmospheric sites used in the previously described experiments. The corrosion rates of all the metals tested were very low at all sites. The atmosphere at the marine-industrial site again proved the most aggressive. But even at that site, only the copper-zinc alloy corroded more than 0.1 mpy at the end of the 10-year exposure. The rural site provided the least corrosive exposure conditions; several materials were only slightly dulled on completion of the test period. Lead was an exception. It exhibited brown-red corrosion products identified as lead peroxide. At the marine-industrial site, unidentified black soot deposits formed on lead. Apart from some minor pitting in areas adjacent to the insulators supporting the samples, no pitting of any consequence was observed at either site. In general, the corrosion rates decreased with years of exposure for all metals with the possible exception of the copper-zinc alloy, for which the corrosion rate was fairly constant throughout the

10-year study. There was little doubt that atmospheric sulfur dioxide played a predominant role in the corrosion behavior of different metals. Additional factors including time-of-wetness, temperature, weather conditions at the time of exposure, and atmospheric chlorides as well as location and orientation of the metal on the building could contribute to corrosion. Gibbons cautioned that all factors should be considered when selecting a metal for service.

In another series of similar experiments, Guttman and Gibbons[303] exposed nine different metal-coated panels to the same Canadian atmospheres and investigated their corrosion behavior over 14 years. Coatings tested were: cadmium electroplated steel, 0.2 to 0.5 mils thick; hot-dipped galvanized steel, 2.2 to 2.4 mils thick; zinc electroplated steel, 1.0 mil thick; continuous galvanized steel, 1.0 mil thick; corrosion-resistant hot-dipped aluminized steel, average thickness 2.0 mils; unsealed zinc metallized mild steel, 5.0 to 7.0 mils thick; zinc metallized mild steel sealed with vinyl copolymer with 10% nonleafing aluminum flake, 5.0 to 7.0 mils thick; unsealed aluminum metallized mild steel, 5.0 to 6.0 mils thick; and aluminum metallized mild steel sealed with one coat of polyvinyl butyral wash primer and one coat of vinyl copolymer with 10% nonleafing aluminum flake, 5.0 to 6.0 mils thick.

The sealed aluminum and zinc metallized coatings afforded heavy-duty protection under very aggressive atmospheric conditions. The wash-primer vinyl sealant was virtually unaffected at all test locations after 11 years of exposure. These results suggest that thinner metallized coatings plus the sealant are capable of offering long, acceptable service at most Canadian locations. The aluminum-zinc metallized coatings, in the thicknesses tested, provided protection in excess of 12 years at the severe marine-industrial site. The authors suggested that sealed coatings should be selected when conditions are similar to those at the marine-industrial sites.

The hot-dipped aluminized coating performed well at all sites. Staining appeared at cut edges early in the testing but did not progress to any serious extent. There were no signs of noteworthy base-metal corrosion at any site, including the marine-industrial site where roughness with an underlying white corrosion product developed. The failure of the electroplated and hot-dipped applied zinc coatings at the aggressive sites was predictable because of the thinness of the coating and site corrosivity. Guttman and Gibbons recommended that the zinc coatings be painted if they are to be used in industrial atmospheres. Elsewhere, even the relatively thin coating can provide a long, useful service life. The thin cadmium coatings had relatively short service lives at all but the non-aggressive rural locations. Similar coatings are applied in practice to fasteners, small stampings, and miscellaneous small articles.

Hukui and Yamamoto[370] described tests of 29 metals, including iron, steel, stainless steel, aluminum, and zinc plating, which were exposed to falling dust, sulfur dioxide, and salt particles at 13 randomly selected locations in Japan. Severity of corrosion was dependent on the level of air pollution. It was related specifically to concentrations of sulfur dioxide in the ambient air and of sulfates and chloride ions in the falling dust. Iron experienced the greatest degree of corrosion, followed by steel, SUS 22 stainless, aluminum, and SUS 27 stainless. Zinc plating exhibited little tendency (about 0.01 that of iron) to corrode. Results of corrosion experiments conducted in the laboratory using sulfur dioxide and periodic exposure to clean water, ultraviolet radiation (arc lamp), and heat did not correlate well with those obtained from the outdoor exposure tests. As a result, Hukui and Yamamoto concluded that particulate matter should be included when conducting laboratory corrosion tests.

The corrosion rate of metals in acids with a normality between 1 and 10,000, such as sulfurous acid, nitric acid, sulfuric acid, and hydrochloric acid, is expressed mathematically by McLeod and Rogers[530] in terms of type of metal and acid concentration. By determining values of constants in the developed equation, corrosion rates of various metals in sulfurous acid can be compared with those of the same metals in nitric, sulfuric, and hydrochloric acids. The corrosion rate of the metals in sulfurous acid of various normalities can then be determined. Results thus obtained showed that nonstainless steel, with or without nickel, is highly susceptible to sulfurous acid corrosion; however, when a substantial proportion of chromium is present in an alloy that contains nickel, steel is not attacked by either sulfurous or nitric acid. Copper and chromium are not appreciably susceptible to sulfurous acid corrosion in solutions lower than $N/1.3$. Tin corrodes more rapidly than other nonferrous metals between $N/100$ and $N/1,000$ but less than cadmium and zinc at $N/10$. Cadmium, lead, and zinc corrode more rapidly in sulfurous acid then in nitric acid. Lead corrodes less rapidly in sulfurous than in nitric acid. Aluminum is rapidly corroded by hydrochloric acid but less so by nitric or sulfurous acid. Hukui and Yamamoto concluded that sulfurous acid solutions causing the greatest damage in urban and industrial areas have normalities between $N/1$ and $N/10,000$.

Ailor[4] conducted 7-year exposure tests of seven aluminum alloys, mild steel, and pure zinc, at Richmond, Virginia; Chicago, Illinois; Phoenix, Arizona; Manila, The Philippines; and Widnes, England. Examination of the test panels at the end of the first, second, and seventh years indicated that corrosion of aluminum in industrial atmospheres generally takes the form of pitting. Pitting and corrosion of a clad material was usually confined to the cladding alloy. No loss of mechanical properties was noted for the clad-alloy 3003-H14 after 7 years. There were property losses for bare aluminum alloys in severe environments of Chicago and Widnes, but the bare aluminum panels were much superior in corrosion resistance to the mild steel in industrial locations. The bare aluminum alloys showed relatively high initial corrosion rates that decreased rapidly over the 7 years. Depth of pitting increased at a much reduced rate after 2 years. The Widnes site was the most severe, followed by Chicago. The Manila and Richmond locations were comparable in corrosivity, and the Phoenix atmosphere was the least corrosive to all metals exposed.

Several investigations of the effect of only sulfur dioxide on metals have been conducted. In one, the influence of the dew-quantity and dew-cycle interval on the corrosion rate was examined. Mild steel panels were exposed to air containing 20 ppm sulfur dioxide at 95% to 100% RH. Results indicated that corrosion was greatest when dewing was heaviest and time-of-wetness to time-of-drying intervals were shortest.[833] Other studies indicated rather direct relationships between metal corrosion rate and sulfur oxide pollution in several ambient atmospheres whose annual average lead peroxide sulfation rates ranged from ~ 0 to 12 mg SO_3/100 cm^2/day.[769] The increase in corrosion per unit increase in sulfation was greater at lower sulfation rates. For example, approximately half of the corrosion of iron at 12 mg SO_3/100 cm^2/day occurred at a sulfation rate of 2 mg SO_3/100 cm^2/day; in the case of zinc, one-fifth had occurred at 2 mg SO_3/100 cm^2/day. A lead peroxide sulfation rate of 2 mg SO_3/100 cm^2/day is equivalent to an average concentration of 0.056 ppm of sulfur dioxide. The investigators reasoned that average annual sulfur dioxide concentrations of <0.05 ppm can result in considerable corrosion, particularly on iron.

Yocom and McCaldin[839] have summarized the results of various researchers. They reported that there appears to be a critical atmospheric humidity that, when exceeded in the presence of sulfur dioxide, produces a sharp rise in the corrosion rate. Critical humidities reported include 80% for aluminum, 60% and 75% for mild steel, 70% for nickel, 63% for copper, 70% for zinc, and 90% for magnesium.

Yocom and McCaldin[839] also reported that aluminum and its alloys are resistant to sulfur dioxide concentrations normally found in polluted atmospheres. Under laboratory conditions in air containing 280 ppm sulfur dioxide at 52% RH, a highly refined grade of aluminum and an aluminum alloy (AA 3003) exhibited the same weight increase and reached a limited value essentially the same as aluminum in air with sulfur dioxide. These authors concluded that, at relatively low humidities, sulfur dioxide does not influence atmospheric corrosion of aluminum nor does it cause any visually discernible changes to the metal surfaces. At higher humidities (72% and 85%), both kinds of aluminum corroded

Table 10-5. Effects of 20- and 60-day exposures to hydrogen sulfide (H_2S)[a]

Base material	Finish	Results of exposures to 100–120 ppm H_2S in air	
		After 20 days	After 60 days
Pure copper	None	Complete blackening Heavy sulfide formation	Heavy black corrosion
Pure copper	Silver-plated (0.0001 inches thick)	Heavy sulfide formation	Heavy black corrosion
Pure copper	Nickel-plated (0.0003 inches thick)	Slight staining	Slight local pitting
Pure copper	Tin-plated (0.0002 inches thick)	Unaffected	Unaffected
Brass (64% Cu)	No finish	General tarnishing	Discoloration and corrosion
Brass (64% Cu)	Nickel-plated (0.0003 inches thick)	Unaffected	Very slight pitting
Brass (64% Cu)	Tin-plated (0.0002 inches thick)	Unaffected	Unaffected
Phosphor bronze	No finish	Complete blackening Heavy sulfide formation	Heavy black corrosion
Phosphor bronze	Nickel-plated (0.0003 inches thick)	Pitted areas of corrosion	Local pitting
Phosphor bronze	Tin-plated (0.0002 inches thick)	Unaffected	Unaffected
Nickel silver (18% Ni)	No finish	Tarnishing and local pitting	Discoloration and corrosion
Nickel silver (18% Ni)	Nickel-plated (0.0003 inches thick)	Unaffected	Slight local pitting
Nickel silver (18% Ni)	Tin-plated (0.0002 inches thick)	Unaffected	Unaffected

[a] From Elliott and Franks, 1969.[213]

much faster, but the highly refined grade lost only approximately half the weight of the AA 3003 alloy. The white powdery deposit formed on the surface of samples under these conditions was aluminum sulfate, an indication that sulfur dioxide played an essential role in the corrosion.

Haynie and Upham[321] exposed zinc panels, positioned at 30° to the horizontal, to the atmospheres of eight cities for 5 years during which continuous air monitoring facilities correlated corrosion rates of metals with atmospheric sulfur dioxide concentrations. Rural sites near each city, assumed to have the same meteorologic conditions but less pollution, were used as control sites, but sulfur dioxide was not measured. Corrosion rates were generally higher at urban sites than at rural sites. In San Francisco, Cincinnati, and New Orleans the corrosion rates were initially higher at the rural sites but eventually converged with the urban rates. In Philadelphia, the corrosion rates at the rural site were exceptionally high, but not as high as at the urban site. Estimates of sulfur dioxide at the rural sites were revised to conform with the amount of corrosion observed. Haynie and Upham concluded that atmospheric sulfur dioxide concentration can be estimated from average zinc corrosion rates and average relative humidities.

There has been little research to determine the influence of other gaseous pollutants, such as nitrogen oxides and hydrogen sulfide, on corrosion. However, Fink *et al.*[232] reported that nitric acid may react with traces of ammonia. It may then be absorbed by hygroscopic particles that may ultimately settle on surfaces, thereby enhancing corrosion rates. Elliott and Franks[213] described the effects of hydrogen sulfide in concentrations of 100 to 120 ppm on various metallic finishes of a variety of base metals (Table 10-5).

MASONRY AND CONCRETE

Masonry and concrete materials used in the building industry include brick, stone, clay tile, mortar, small precast masonry units of concrete, and large poured-in-place or precast concrete units. Brick, both calcareous and noncalcareous (basically aluminum silcate), is applied extensively as a veneer facing for buildings and as walls and chimneys or stacks. Clay tile, made of basically the same materials as brick, is employed as roofing, interior partitions, exterior walls, furring, and flooring. Stones, including limestone, marble, slate, and sandstone, are excellent materials for load-bearing walls, facing, roofing, trim, and ornamentation. Small precast concrete masonry units are used as backing for veneer brick walls, partitions, facing, furring, lintels, trim, and cornices. Concrete, either poured-in-place or precast to form larger building components, consists essentially of a calcareous-based cementitious material, predominantly portland cement, and aggregates of various sizes that have been mixed with water. Concrete components serve as structural elements of buildings (e.g., columns and beams, curtain walls, foundations, and floors), structural elements of bridges, and simple slabs for patios.

Atmospheric pollutants that affect masonry and concrete materials include particulates (such as soot), sulfuric and sulfurous acids (formed when atmospheric sulfur dioxide reacts with water), and several of the acidic gases. Deleterious effects include simple discoloration or staining, erosion and corrosion, and leaching.

Natural environmental factors, such as relative humidity, rain, and wind, can also influence the effect of pollutants. For example, intense rain followed by bright warm sunshine that quickly evaporates moisture from the masonry or concrete surface can provide a cleansing action. However, Wilson[816] reported that high humidity slows the natural evaporative process, especially in sheltered areas. The moisture that is then prevented from leaving the material, if polluted with atmospheric acids, can dissolve the calcareous elements in masonry or concrete. Wind velocity largely determines the force with which both rain and particulate matter strike a building and, thus, the abrasiveness. These factors plus the porosity and texture of the surface determine the amount of particulates that remain on that surface. Yocom and McCaldin[839] reported that under high wind conditions large particulates entrained in the windstream actually result in a slow erosion of surfaces similar to sandblasting.

Effects on building stones depend on their chemical composition. Schaffer[668] reported that soot fills the surface pores of many sandstones, causing them to become uniformly darkened; however, erosion of exposed surfaces of limestones precludes the retention of soot so that only the sheltered sides become blackened. He further stated that soot also contributes to chemical decay of stone because it carries with it acids and soluble salts. Softer building stones, such as marble, limestone, and dolomite, are readily attacked by acids formed when atmospheric smoke and moisture interact.[769] All are calcareous minerals containing carbonates that convert to relatively soluble sulfates by sulfuric acid and can then be leached away by rainwater. The calcium sulfate formed on the surface of such stones is twice as bulky as the carbonate in the stone from which it was formed.

Stones such as granite and some sandstones in which the grains are cemented together with materials containing no carbonate, well-baked bricks, and glazed tiles are relatively unaffected by atmospheric sulfur dioxide.[769, 839] Pressley[619] reported, however, that refractory brick can be influenced by particulate acids and salts at high temperatures. Thermal cycling of magnesite-containing bricks in sulfur trioxide at temperatures between 760°C and 1,040°C considerably increased the strength of the bricks. This was attributed to the formation of magnesium sulfate, which filled the pores in the brick. Under similar test conditions, sodium metavanadate considerably reduced the strength of magnesite brick. It also reduced the strength of magnesite-chrome brick, but less severely.

The effect of atmospheric acids on sandstone, concrete, brick, and mortar is not confined to the surface. Water can transport the aggressive agents to the interior of the materials. Brown[92] wrote that substances taken into solution usually come out of solution and form secondary encrustations elsewhere. With sulfuric acid, the reaction product would be calcium sulfate or gypsum. McBurney[518] indicated that sulfuric acid displaces chlorides and carbonates from alkali (sodium and potassium) and alkaline (calcium and magnesium) components of masonry. He cited the reaction of sulfuric acid with sodium sulfate or sodium bisulfate to form a gas that, in the presence of water, forms hydrochloric acid. When water carrying the reaction products in solution moves to the surface, both the water and hydrochloric acid evaporate, leaving behind efflorescent sodium sulfate to discolor the surface.

Litvin[477] studied the effects of outdoor exposure on concrete samples containing white Portland cement, white silica sand, and grey-white coarse marble aggregate. The samples were exposed 30° from the horizontal at a site in Buffington, Indiana, which is adjacent to two highly industrialized areas. The aggregate became slightly affected, but the white matrix areas had weathered to a light brown color and showed evidence of etching, probably due to acid attack. Litvin compared these results with results obtained using similar samples protected with 14 different commercial polymeric coatings. (These tests are discussed under POLYMERIC MATERIALS, p. 244.) Results indicated that concrete may fare better uncoated than with a coating of unproven service, particularly in nonindustrial areas. Unfortunately, sealing of the surface of porous materials like concrete often leads to spalling, which is usually more deleterious than surface deterioration or discoloration.

Wilson,[816] reporting on variables affecting concrete exposed to the atmosphere, stated that performance is a function of both the uniform quality of the material and the nature of the surface. Aggregates of angular shape and crystalline surface texture, such as crushed rocks of all types, hold dirt more readily than aggregates of a rounded shape and glass texture, such as river or seashore gravels. Although patterned or profiled slabs or curtain walls are subject to general darkening, they collect more particulates on certain facets of the pattern than on others. This often results in a highly interesting and pleasing appearance. Similarly, Adams[3] reported that on smooth facings stains show up more clearly but are easier to wash off. Surfaces like those with exposed variegated stone aggregates seem to dispense dirt readily and their roughness mitigates unsightliness. Wilson[816] concedes that it would be more reasonable to use dark colored surfaces in highly industrial areas, and that dense concrete would reduce the chance for water-carrying atmospheric acids to enter the material.

The effect of sulfur dioxide on oölitic limestone was reported by Spedding.[713] Table 10-6 shows that the uptake of sulfur dioxide is very dependent on humidity and that saturation appears to be reached rapidly—in at least 10 min during the high humidity experiments. From these findings, Spedding reported that a freshly exposed matrix would be rapidly saturated with sulfur dioxide at normal atmospheric humidities and that the erosion process associated with calcium sulfate formation would be virtually continuous.

Few quantitative data are available on the actual chemical effect of pollutants on masonry and concrete structures; however, their potential deleterious effect to works of art or historical monuments has been reported. One investigation[769] indicated that sulfate, resulting from the attack on the stone by atmospheric acids, was found in cracks and fissures as deep as 50 cm beneath the surface of the British Houses of Parliament. An increase from 1.0 to 4.2 in the volume of crystallization of calcium and magnesium salts along the cleavage planes of the stones in these buildings had thrown off great pieces.

Cleopatra's Needle, the large stone obelisk moved from Egypt to New York City's Central Park in the late nineteenth century,[570] suffered more deterioration in the last half century than in the first 3,000 or more years of its existence. In the clean, dry air of Egypt, this monument retained its elaborate hieroglyphic surface carvings. But now, two of its carved faces are completely obliterated. The other two faces, turned from prevailing winds, are less heavily weathered.

Marble structures in Venice, all basically made of calcium carbonate, have also corroded extensively.[396, 570, 771] Virtually all this weathering has occurred in the last 50 years. Photographs of these centuries-old structures taken at the turn of the century reveal almost no serious weathering effects. A recently enacted law provides support for refurbishing approximately 200 pallazzi, churches, and other historical buildings

Table 10-6. Uptake of sulfur dioxide (SO_2) by oölitic limestone[a]

Relative humidity (%)	SO_2 concentration ($\mu g/m^3$)	Time of exposure (min)	$\mu g\ SO_2/cm^2$ of surface
11	360	20	0.069
13	280	40	0.061
79	100	48	0.24
81	370	10	0.28

[a] From Spedding, 1969.[713]

and also provides Venetian homeowners with financial aid to convert oil heating systems to less polluting methane gas systems. Interesting, however, is a recent investigation by Brown[92] that showed no evidence of attack by sulfurous particulates on the Lincoln Memorial in Washington, D.C., but that there were dust and soot particles in open joints.

PAINTS AND FINISHES

Paints (opaque film coatings) and finishes (varnishes and lacquers) are used extensively in the building industry to beautify and protect all types of substrates, from metals serving as a critical structural element in bridge construction to wood used simply as a decorative fence. The paint or finish acts as a protective shield between the construction materials and the environmental factors that can attack and deteriorate them.

Paints basically consist of two components: the film-forming component, which is generally referred to as the vehicle and which may have a legion of components, and pigments, which make the film opaque and provide the desired decorative feature. Generally speaking, the degradation of these two components by environmental factors results in the undesirable appearance of painted surfaces. The principal substances employed as vehicles, either alone or in combination with other additives, include oils (e.g., linseed) and natural, modified natural, and synthetic resins. Of the resin-type paints, the synthetics are most commonly used now, particularly the vinyls and acrylics. Pigments are divided into two major classes, colored and white. Commonly used white pigments include silicates, barium sulfates, white lead, titanium oxides, zinc oxide, basic lead silicate, and zirconium oxide. Organic, mineral, and metal compounds are used as colored pigments.

Natural environmental factors, such as sunlight (particularly ultraviolet), temperature variations, moisture, and fungi, greatly influence the appearance, performance, and service life of paints and finishes. Natural chemical environments, such as those laden with salt water, also can greatly affect the appearance and performance of paints.

In addition to these natural factors, evidence indicates that degradation of paint properties can be caused by many air pollutants, especially particulates, hydrogen sulfide, sulfur oxides, ammonia, and ozone. The effects of air pollutants on paints are loss of gloss, scratch resistance, adhesion, and strength; discoloration; increased drying time; and unattractive, dirty appearance. Significant and prolonged attack without preventive maintenance also can result in exposure and subsequent attack of the substrate being protected. Obviously, exterior paints encounter more destructive influences than those used on interior surfaces.

Particulate matter (dust, soot, hydrocarbons, and other solid matter) affects primarily the aesthetic quality of paints because its accumulation on surfaces (due to thermal, electrostatic, and mechanical forces) gives the paint a dirty appearance. Spence and Haynie[716] reported that painted fences at various U.S. locations became blackened from soot deposits. Efforts to remove it or clean the tarry matter embedded in the paint destroyed the film. Furthermore, the two coats of paint used to cover the soiled paint had a shortened lifespan, and the paints' appearance was marred.

According to Cowling and Roberts,[168] particulates also promote the chemical deterioration of wet or tacky paints in a moist environment by acting as wicks that transfer corrosive agents to the underlying surface or serve as nucleation sites at which other pollutants can concentrate. These authors state that moderately chalky paint removes the particulate matter. Hence, for exterior surfaces in industrial communities, they suggest using paints with the maximum pigment concentration that is consistent with durability requirements.

Severe staining of paint films also can be affected by 0.5 ppm copper and 0.1 ppm iron salts, two metals identified as airborne particulate matter but also found in rainwater in the United States, according to Holbrow[355] and Spence and Haynie.[716] A specific example of particulates reacting with painted surfaces was cited by Yocum and McCaldin.[839] They described the grinding by iron particles on surfaces of cars. Brown stains surrounded each particle. Many cars had to be repainted. The deposit resulted from the formation of ferrous hydroxide in the presence of moisture. In colloidal form this substance reportedly diffused in the paint film. Upon drying and oxidizing, it left the brown stain, ferric oxide. Spence and Haynie reported damage to automobile paints by alkali mortar dust from the demolition of brick buildings. These vehicles also had to be repainted.

Spence and Haynie, summarizing investigations of others, also reported the effect of particulate matter on paints based on frequency of painting and maintenance practices of homeowners. One investigation involved three suburbs of Washington, D.C.—Suitland and Rockville, Maryland, and Fairfax, Virginia—and two cities in the upper Ohio valley—Steubenville and Uniontown (see Table 10-7). The number of years between repaintings decreased as the particulate concentrations increased. Repainting in Steubenville, which has particulate concentrations as high as 234 $\mu g/m^3$, was necessary almost every year. In Fairfax, where particulate concentrations were $\sim 60\ \mu g/m^3$, repainting occurred every 4 years. This maintenance frequency increased as local particulate concentration increased (see Figure 10-2).

Spence and Haynie suggested that there is a significant relationship between the frequency of repainting and particulate concentration. However, extrapolation of the line in Figure 10-2 indicates that repainting does not occur at zero particulate concentration and that repainting

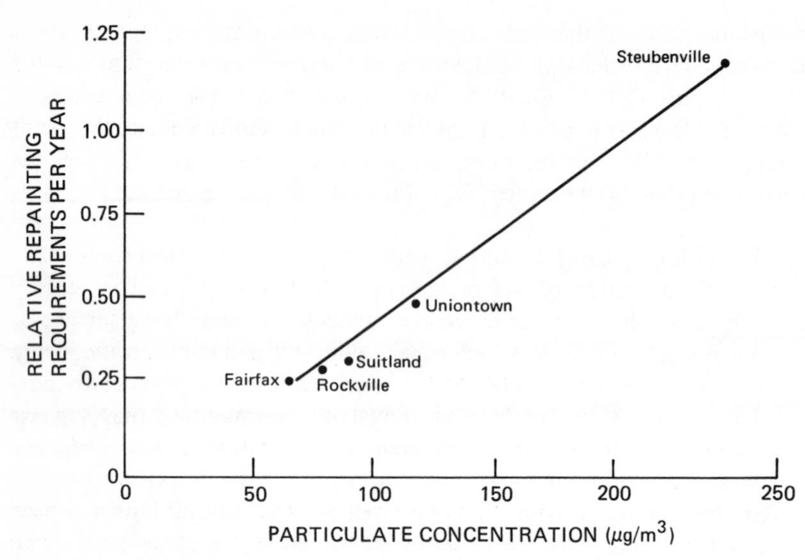

Figure 10-2. Relationship of maintenance frequency for exterior repainting to particulate concentration. From Spence and Haynie, 1972.[716]

frequency continues to increase as particulate concentration increases. A leveling out of the line to form a plateau at the low and high extremes of particulate concentration should be expected. These authors indicated that additional data are needed, particularly for cities with mean annual particulate concentrations > 150 $\mu g/m^3$.

In their study of particulate concentrations in the Philadelphia area (see Table 10-8), Spence and Haynie considered socioeconomic and other factors. They suggested that the inclusion of these factors partially explains why no statistically significant difference in painting frequency as a function of particulate concentration was detected.

Table 10-7. Interval for exterior repainting as a function of particulate concentration in five United States cities[a]

	Steubenville	Uniontown	Suitland	Rockville	Fairfax
Particulate concentration ($\mu g/m^3$)	235.00	115.00	85.00	75.00	60.00
Maintenance interval (yr)	0.88	1.89	2.93	3.62	3.90
Maintenance frequency (number per yr[b])	1.14	0.53	0.34	0.28	0.26

[a] From Spence and Haynie, 1972.[716]
[b] Reciprocal of maintenance interval in years.

Table 10-8. Mean annual frequency for exterior wall painting in Philadelphia area as a function of particulate concentration[a]

Particulate concentration ranges ($\mu g/m^3$)	Exterior wall painting	
	Mean annual frequency	Standard error of mean
<75	0.28	0.016
75–100	0.35	0.053
100–125	0.35	0.041
>125	0.29	0.055

[a] From Spence and Haynie, 1972.[716]

Cowling and Roberts[168] stated that sulfur dioxide is readily absorbed in wet paints and reacts with the vehicle to form soluble compounds. Unspecified concentrations of sulfur dioxide with high humidity during application may cause paints to be tacky. This facilitates additional contamination by particulates whose electrical charge can affect the adhesiveness of the paint. Holbrow[355] also reported that sulfur dioxide concentrations from 1 to 2 ppm increase the drying time of linseed, tung, and bodied dehydrated castor oil paint films from 50% to 100%. At 7 to 10 ppm, drying was delayed by up to 2 or 3 days. Only unbodied castor oil paint was not affected greatly. The drying times of paints containing basic pigments such as white lead or zinc oxide were not greatly affected, but oleoresinous and alkyd paints pigmented with titanium dioxide had both their touch- and hard-dry times increased substantially. Analysis of the dried films showed that sulfur dioxide had reacted with the drying oil, probably modifying the oxidation-polymerization process. Holbrow concluded that sulfur dioxide concentrations in fogs near industrial sites can increase the drying and hardening times of certain paints.

Yocom and McCaldin[839] could find no studies of the effects of sulfur dioxide on drying time of latex-base (synthetic resin) paints. They surmised that when the paint is applied, the sulfur dioxide may interfere with the evaporation of water and, therefore, the coalescence of polymeric-pigment particles. Upon reaction with water in the paint, sulfur dioxide forms sulfurous acid, which could cause instability of the protective colloid around latex particles and, consequently, flocculation. Poor film formation would be possible with resulting loss of aesthetic as well as protective value. Some investigators reported that the effect of sulfur dioxide (presumably between 1 to 2 ppm) is negligible when the paint films are hard and dry, in contrast to the effect on wet paint.

In studies of gloss, Holbrow[355] exposed paint films to a 1.2% sulfur dioxide solution for 15 min after the films were first dried for 24 hr under normal laboratory conditions, then cooled in a refrigerator to effect moisture condensation on the surface. Sulfur dioxide and moisture caused only a small decrease in gloss; however, after further exposure to moisture and warmth in an accelerated weathering chamber, a significant change occurred in the gloss of some paints. A pentaerythritol alkyd was scarcely affected, a linseed stand oil paint experienced the greatest change, and the extent of gloss loss was proportional to the water sensitivity of the films. Very high concentrations of sulfur dioxide rendered the films water-sensitive; during the accelerated weathering, the moisture absorbed by the paint resulted in excessive swelling. Holbrow reported the same effect when films were exposed to hydrogen chloride vapor. Another closely related effect was the blueing of green paint during the early life of the film when it still had a high gloss. Sulfur dioxide and moisture together need not cause a color change, but when followed by exposure to warmth and humidity (accelerated weathering), the change can occur. The blueing was attributed to a bleaching of the lead chromate pigment when the film was exposed to sulfur dioxide shortly after application. The paint loses it sensitivity to this effect after a few hours of exposure to concentrations of ~0.2 ppm sulfur dioxide, but remains sensitive to higher concentrations from 1 to 3 days. The paints containing large quantities of oil were most prone to the effect and remained sensitive longer. The introduction of basic pigments and additives reduced the effect, but large amounts were necessary (e.g., 10% zinc oxide). The presence of copper, cerium, and vanadium was especially effective in preventing blueing. The addition of copper compounds, such as copper phthalate, prevented blueing under laboratory test conditions. Similar quantities of oil-soluble soaps—1% copper, 0.5% cerium, or 0.5% vanadium—as naphthenates in the oil component also were effective.

Although no quantitative data could be found on the effects of sulfuric acid on painted surfaces, Spence and Haynie[716] and Cowling and Roberts[168] reported that the acid may cause damage by reacting with such paint ingredients as pigments, fungicides, and the vehicle itself. The effects of sulfuric acid are implied in the data on the effects of sulfur dioxide, which converts readily to this acid in the presence of moisture.

Sulfur dioxide and moisture in the presence of ammonia cause crystal-line bloom. Holbrow[355] ascribes this to the formation of small crystals (0.5 to 1.0 μm diam) of ammonium sulfate on the surface of varnish and paint films. Formation of the bloom is influenced greatly by moisture and paint composition. Quantitative measurements indicated that a deposit of 3 to 4 $\mu g/cm^2$ of ammonium sulfate can cause a moderate bloom in either a pentaerythritol or glycerol alkyd paint film, and a deposit of 10 $\mu g/cm^2$ can result in a severe bloom. Holbrow calculated that sulfur dioxide concentrations of ~0.1 ppm in 2 m^3 of air would be sufficient to cause a moderate bloom (4 $\mu g/cm^2$) over 100 cm^2. Because only 5 times this volume of air would probably be required to supply the ammonia and because the volume of air that passes over a paint film daily is considerable, blooming can easily occur.

Berger *et al.*[69] reported results of tests conducted on both oil-base (alkyd and linseed) and synthetic resin-base (polyvinyl acetate latex and acrylic latex) paints on aluminum, wood, masonite, and transite after separate exposures to sulfur dioxide and ozone. Specimens were exposed for 750 hr in an accelerated weathering chamber in air under the conditions of 37.78°C, 65% RH, 2 ppm sulfur dioxide, and 2.2 mWhr/hr ultraviolet radiation. Conditions were similar for the ozone tests except that the exposure lasted 1,000 hr and the ozone concentration was 0.5 ppm. Exposed specimens were examined for, among other changes, loss of gloss, scratch resistance, and adhesions. They were then compared with control specimens and specimens exposed to two natural outdoor environments, a heavily polluted urban industrial area high in sulfur dioxide content and a rural area free of sulfur dioxide but high in ozone content. The paint gloss on all substrates decreased with duration of exposure to either pollutant. Ozone seemed to cause the least change and sulfur dioxide the most. The alkyd paint incurred greatest gloss loss on control specimens regardless of substrate. Outdoor exposures to sulfur dioxide caused greater gloss loss than outdoor exposures to ozone. Scratches that resulted from exposures to sulfur dioxide or to outdoor exposures with high sulfur dioxide content were consistently wider than those caused by exposures to ozone or outdoor exposures with high ozone content. Most scratch widths ranged from 0.1 mm to 0.5 mm. Some scratches on paints covering aluminum substrates were much wider, indicating major adhesions loss. No comparisons were made with control specimens. Although some researchers reported that ozone seemed to affect adhesion the least and sulfur dioxide the most, they questioned the scrape-adhesion test method used to evaluate pollutant effects.

Spence and Haynie[716] reported that tests conducted by the Illinois Institute of Technology Research indicated a definite reduction in creep compliance of oil-base paints exposed to ozone. The extent of reduction for second-quality paints was about 3 times greater than that for the first-quality paints tested. The films of the first-quality paints retained more flexibility than the second-quality paints. These paints were exposed to 6% ozone at 35°C and 65°C for up to 505 hr.

Hydrogen sulfide attacks painted surfaces. Spence and Haynie[716] noted that in the southern portion of the San Francisco Bay area it darkened house paints to various tones of grey and even jet black. The discoloration occurred around doors and windows and under eaves. These locations tend to remain moist, indicating that the attack was enhanced by moisture. The most severe discoloration occurred during winter, although the maximum 2-hr average air concentration of hydrogen sulfide was twice as great in the summer. This supported the contention that high humidity increases damage. The actual concentration of hydrogen sulfide at which paint begins to darken is not known; however, Spence and Haynie quote investigators who indicate that, under certain conditions, paint darkening can occur at hydrogen sulfide concentrations between 70 and 140 $\mu g/m^3$. Furthermore, they reported that some laboratory studies have indicated that darkening of paint does not occur unless the films are actually wet with water. Although no studies produced dose-response data on the effects of relative humidity and hydrogen sulfide on paint films, the darkening effect has been attributed to the formation of dark-colored metal sulfide by the chemical reaction of hydrogen sulfide on lead additives and on organometallic driers and preservatives in the paint.

Yocom and McCaldin[839] similarly reported that house paints containing lead compound pigments are rapidly darkened by hydrogen sulfide, which causes the formation of black lead sulfide. They reported that the severity of discoloration is apparently related to the lead content of the paint, amount of hydrogen sulfide in the air, duration of exposure (several hours of exposure to as little as 0.05 ppm was reported to effect a change), and moisture during exposure. They further reported that little damage occurs if both paint surface and air are dry. Holbrow[355] confirms a reaction between hydrogen sulfide and the metal component of the drier or pigment, particularly lead.

POLYMERIC MATERIALS

Plastics and elastomers (including rubber) are considered jointly in this section. Chemically, both are high molecular weight polymeric materials and most such materials are now largely synthetic. The types of polymers and possible variations of each are innumerable. In the course of their manufacture or processing, they can be compounded with varying amounts of a variety of ingredients, including fillers, plasticizers, pigments, dyes, vulcanizers, stabilizers, antioxidants, and a host of other substances, each of which is intended to enhance some special property or properties or to minimize others. Generally, polymeric materials are classified as thermoplastics [e.g., polyvinyl chloride (PVC) or polyethylene] or as thermosets (e.g., epoxy or silicone); however, depending on formulation and processing, certain thermoplastics can be made thermosetting and vice versa. Molecularly, thermoplastics are linear or chain-like polymers capable of experiencing repeated softening by heat and hardening by cooling. Thermosets are highly interconnected and extensively cross-braced, molecularly. As a result, deterioration of such materials is usually a surface phenomenon.[465]

Vinyls constitute approximately 50% of all polymeric materials used in the building industry. They are used extensively as floor tile, wall covering and paneling, siding, piping, vapor barriers, protective coatings, cladding, window frames, and electrical insulation. PVC, which is the most extensively used vinyl, has recently been applied as a membrane material for air structures. Other polymers are polyethylenes, polyesters, polystyrenes, acrylics, polyurethanes, phenolics, polycarbonate, and silicone.

Few investigations have been made to determine the effect of atmospheric pollutants on these materials. In general, particulate matter like dust, soot, and smoke dirty the surfaces, and liquid and gaseous acids appear to cause fading, gloss loss, or disintegration.[195]

Stedman[721] examined the possible effects of atmospheric pollutants on rigid PVC under exposure to ultraviolet radiation. In these tests, 0.75 inch (~18.75 mm) strips of white, rigid PVC were inserted in sealed Vycor test tubes, each of which was filled with one of the following: ammonia, carbon monoxide, hydrogen sulfide, methane, nitrous oxide, nitric oxide, nitrogen dioxide, sulfur dioxide, or atmospheric air. In addition, a PVC strip of a similar compound and one of a different compound, but known to yellow extensively under ultraviolet radiation (fluorescent sunlamp/black lamp), were placed in test tubes opened at both ends. All tubes were exposed outdoors on a 45° angle test deck. Although cloudy weather prevailed for several days at the start of the exposure, after 22 hr the sample in sulfur dioxide had begun to darken. After 144 hr, the strip was almost black and the inside wall of the tube was hazy; the reverse side of the strip facing the deck was uniform, but slightly yellowed.

In a similar test using sulfur dioxide, strips of white, rigid PVC of a different compound, which were exposed to fluorescent sunlamp/black lamp radiation, produced similar results. The only difference was the type of discoloration of the PVC. The primary reactions appeared to be the reduction of the sulfur dioxide to elemental sulfur, recombination of the stabilizers' heavy metal to form colored sulfides, such as tin, barium, and cadmium, and some oxidation of the PVC.

The sample enclosed in the nitrogen dioxide exhibited a different phenomenon. Initial discoloration of the PVC could not be detected visually because of the color of the gas, but after 29 days the strip was yellowish-tan and the gas colorless. When the tube was opened, the brown color of nitrogen dioxide reappeared immediately, indicating a redox reaction whereby the PVC reacted with the oxygen provided by the reduction of nitrogen dioxide to nitric oxide. The yellowness of the strip had increased from 1.1 to about 17.3. Twenty-six days later, during which time the strip was stored in the dark, it had increased to 79.1.

After exposing different rigid PVC compounds in nitrogen dioxide to fluorescent sunlamp/black lamp radiation, differences in discoloration again appeared, the degree depending on the particular compound in the strip.

The yellowness index of PVC strips having the same composition as those exposed to the atmospheric pollutants increased only about 4.8 points after exposure to fluorescent sunlamp/black lamp radiation and 27 days in the dark. This could indicate differences between artificial and natural radiation. The yellowness index of the sample enclosed in ammonia increased from 1.1 to 7.7 in 33 days, while its green tristimulus value changed only from about 75.0 to 73.6. In comparison, after 63 days of exposure to hydrogen sulfide, the yellowness index of the sample increased only about 0.1 points, but its green tristimulus value decreased from about 75.2 to 63.8, indicating general darkening or greying. Nitrous oxide showed the least change in yellowness index, an increase of about 5.5 points, even after 123 days.

During a 63-day period, the control strip in the closed-end test tube showed a slight decrease in yellowness index, while the sample of the same composition in the open-end tube showed an increase. After an additional 25-day exposure, the yellowness index of the open-end control specimen increased about 6.2 points and that of the closed-end specimen about 1.9 points. The open-end specimen with the different compound increased about 2.0 points in yellowness index. After a total of 123 days, the specimen having the different compound had changed very little, but the control strips continued to increase in yellowness, the one in the open-end tube at a higher rate presumably because of the continual supply of oxygen.

Stedman[721] also reported polymeric changes in the samples as a result of the various exposures. For example, cracking of PVC samples as produced in some cases by fluorescent sunlamp/black lamp radiation was duplicated outdoors following exposure to nitrogen dioxide in the Vycor tubes for 29 days and storage in the dark for over 26 additional days. The sample cracked approximately 90% through its thickness when folded 180° on itself, indicating in-depth polymeric degradation resulting from oxidation.

Jellinek[386] studied chain scission of various polymers in the form of thin films exposed to 1 atm of air, near ultraviolet radiation (wavelength greater than 2,800 Å) in the presence and absence of sulfur dioxide and nitrogen dioxide in various concentrations. He concluded that most vinyl polymers are scarcely affected by long exposures to either gas at concentrations of 1 to 5 ppm at or near room temperature. Polyethylene and polypropylene suffered some cross-linking. Chain scission by nitrogen dioxide occurred with nylon. Isotactic polystyrene exposed to the same concentrations of sulfur dioxide in the presence of air and ultraviolet radiation suffered only negligible chain scission after a 1-hr exposure. In other investigations, no chain scission occurred without the small sulfur dioxide pressure (0.85 mm Hg).[387, 388]

Jellinek[386] reported studies in which sulfur dioxide concentrations of 2 ppm in air had no appreciable effect on the tensile strength and elongation of vinyl plastics. The investigators concluded that from 2 to 6 months exposure to 100 ppm sulfur dioxide is required to cause significant changes in the tensile strength of PVC polymers. Other investigations indicated that 18 ppm sulfur dioxide and ultraviolet radiation only slightly affected vinyl polymers—small amounts of carbon dioxide and water evolved at 7.22°C and some discoloration took place. No change was noted in the flexibility of polyethylene, polyethylene-terephthalate, polyester (cross-linked), and PVC after 500 hr exposure to 18 ppm sulfur dioxide; only polystyrene became brittle after this time. Very little effect on the infrared spectra of the polymers was observed. There were some changes in PVC, polyvinyl fluoride, and polystyrene, probably due to the incorporation of sulfur dioxide groups along the polymer backbones and loss of some chlorine and fluorine. Jellinek stated that nitrogen dioxide might also be incorporated along the polymer backbone and that polystyrene becomes quite polar if sufficient nitrogen dioxide groups are thus incorporated, indicating that chain scission is not the only possible kind of deterioration.

Berger *et al.*[69] provided results of tensile tests on plastic films (polycarbonate, polypropylene, polyvinyl fluoride, and PVC acetate) and gloss and adhesion tests on coatings of PVC and polyvinyl fluoride on aluminum siding after exposure to sulfur dioxide and ozone. Specimens were exposed for 750 hr in an accelerated weathering chamber in air under conditions of 37.78°C, 65% RH, 2 ppm sulfur dioxide, and 2.2 mWhr/hr ultraviolet radiation. Conditions were similar for the ozone tests, except that exposure lasted 1,000 hr and ozone concentration was 0.5 ppm. In addition, results were compared with control specimens and specimens exposed to two natural outdoor environments, one a heavily polluted urban industrial area high in sulfur dioxide content and the second free of sulfur dioxide but high in ozone content.

Partial results of these tests indicated that polyvinyl fluoride and PVC films were unaffected by any of the exposure conditions. The polycarbonate film incurred reduced tensile strength from exposures to xenon arc alone and with ozone. The most severe effect was a reduction of approximately 17% in tensile strength after a 250-hr exposure to the xenon arc plus sulfur dioxide. Outdoor natural weathering of the PVC and polypropylene samples was discontinued after 4 and 2 months, respectively, because the samples had become so brittle they could no longer undergo tensile testing. The polycarbonate sample lost little tensile strength after 5-month exposure; however, the sample in the urban outdoor area with high ozone content incurred an approximately 45% loss in the seventh and ninth months of exposure. This reduction was attributed to biodegradation of undetermined source.

After all exposures to air pollutants, PVC siding maintained its gloss better than polyvinyl fluoride, on which sulfur dioxide caused the greatest change. The control PVC samples exhibited the maximum change. Ozone caused the least change in gloss of either coating. Gloss appeared to be affected less at the two outdoor sites than under artificial weathering conditions. The rural exposure did not affect the samples as severely as the heavily polluted industrial site. Ozone affected adhesion of PVC samples more than sulfur dioxide. Sulfur dioxide had essentially no effect on adhesion of the polyvinyl fluoride; however, adhesion decreased with exposure to ozone.

TEXTILES

Textiles are used by the building industry almost exclusively in accessory items such as draperies, carpeting, and furniture. The textiles may be made from both natural fibers (cotton, wool, hemp, jute, linen, silk) and modified natural or synthetic fibers (regenerated animal and vegetable fibers, viscous and acetate rayons, polyamides, acrylonitriles, polyesters, polyvinyls, nylons). Generally, textiles consist of the fiber component and such additives as dyes, water repellants, and finishes.

As in the case of paints and finishes, natural environmental factors, such as sunlight, oxygen, and water vapor, and air pollutants, such as particulate matter, sulfur oxides, hydrogen sulfide, nitrogen oxides, and ozone,[795, 839] can greatly affect the performance and service life of textiles. The degree and rate of effect depend on many variables, notably concentrations and number of different environmental factors to which the fabric is exposed, type of fabric, and types and number of additives. Although apportionment of textile degradation between natural environmental factors and pollutants is difficult, the relative effect of the various pollutants can be gleaned from the existing literature.

In a discussion of textile fibers, Yocom and McCaldin[839] reported that cellulose fibers such as linen, hemp, cotton, and rayon are especially susceptible to damage by acid aerosols, which attack and weaken the cellulose chain in the glucosidic linkage. According to Waller,[780] sulfuric acid reacts with the cellulose fibers to produce a water-soluble product with little tensile strength. Animal fibers such as wool, fur, and hair are more resistant to sulfur dioxide and nitrogen dioxide than the synthetic fibers because they already contain compounds of both nitrogen and sulfur.[839]

Although particulate matter obviously soils fabrics, Yocom and McCaldin[839] noted that it is only damaging when the particles are highly abrasive and the fabric is flexed frequently. Deterioration results mainly from repeated attempts to clean the fabric. Other investigators have concluded that the more tightly woven the cloth, the more resistant it is to soiling. Soiling resulting from thermal precipitation is directly related to the number of degrees below the ambient air temperature to which the surface temperature of the fabric is cooled. Thicker samples of cloth would collect less dust because their surface temperatures would be higher.

Soiling by particulate matter also results from electrostatic attraction. Certain fabrics, such as acetate rayon, become electrostatically charged by friction during their manufacture.[839] Nuessle[583] described laboratory tests in which equal positive and negative charges were applied separately to a cotton fabric. Although soiling increased with the magnitude of both charges, it was greater with the positive charge, perhaps because of the predominance of negatively charged particles in the atmosphere. Nuessle referred to reports that indicate that two-thirds of all airborne particulate matter may be negatively charged and to experiments in which 1,000 volts applied to a cotton fabric tripled the attraction of airborne particles. In these experiments, loose fibers protruding from the surface of the fabric were soiled the most.

Curtains soil greatly in polluted areas. Hanging at open windows, they serve as filters, not only for particulate matter but also for acid droplets. Weakened as a result of such exposure, curtains often split in parallel lines along the folds, on which impinged acidic materials reach the inner fibers.[839]

Zeronian *et al.*[844] investigated the weathering of polymeric fabrics (modacrylic, acrylic, nylon, and polyester) by exposing them to sunlight and polluted air containing 0.2 ppm nitrogen dioxide, 0.2 ppm sulfur dioxide, or ozone. Compared with control specimens exposed to sunlight and unpolluted air, nylon incurred the greatest degree of degradation on exposure to sulfur dioxide. Breaking load, rupture energy, and breaking extension were greatly reduced, the relative viscosity and number of amine groups decreased, and the number of carboxyl end groups increased. In contrast to nylon degraded by acid hydrolysis, there was no yield point in the load extension curve of the nylon degraded by sulfur dioxide. There was some evidence that nitrogen dioxide affected the properties of the acrylic, nylon, and polyester fabrics, and that ozone might affect the properties of acrylic and nylon fabrics. The modacrylic fabric was not affected by any of the pollutants. The polyester was not affected by either sulfur dioxide or ozone.

Cotton broadcloth, cotton broadcloth with a wash-and-wear finish, polyester, polyester and cotton blend with a permanent-press/soil-release finish, and a polyester and cotton blend were exposed to ambient air, air containing nitrogen dioxide, and carbon-filtered clean air (control) for 90 days. Following the exposures, Hosking[365] reported that the cotton with the wash-and-wear finish had the poorest resistance to abrasion and the polyester and cotton blend the best. The polyester fabric had the greatest initial abrasion resistance, but at the end of 90-day test period had deteriorated to a greater extent than either of the polyester blends. Next to the cotton with the wash-and-wear finish, the polyester also incurred the greatest percentage of weight loss following exposure to nitrogen dioxide. Nitrogen dioxide was the most damaging to the fabrics.

Fading of dyed fabrics by both gaseous and particulate pollutants has received considerable attention in the literature. Fujii and Tsuda[263] substantiated the effect of dye formulation on resistance to fading of fabrics exposed to nitrogen dioxide, sulfur dioxide, nitric oxide, carbon monoxide, and particulates. Seven kinds of fabrics with three types of dyes (dispersed, direct, and acidic) were tested. Overall, acidic dye was affected most by the pollutants. The dispersed dye was affected most by nitrogen dioxide and particulates; the direct dye, by sulfur dioxide. The investigators concluded that fading caused by sulfur dioxide and particulates was greater than that caused by nitric oxide, nitrogen dioxide, or carbon monoxide.

Fujii and Hirata[262] conducted tests on viscous rayon, acetate, tetron, and wool fabrics dyed blue. They determined that particulates caused a greater degree of fading than sulfur dioxide. Other investigations indicated the synergistic effect of sulfur dioxide, nitrogen dioxide, and carbon monoxide on the fading of dyed fabrics.[839] Exposure of fabrics to filtered air with 50% RH and 1 ppm sulfur dioxide produced negligible dye fading. The same result occurred when dilute auto exhaust (111 ppm carbon monoxide, 0.2 ppm nitrogen dioxide) was added. However, irradiated auto exhaust (88 ppm carbon monoxide, 0.7 ppm nitrogen dioxide) produced substantial dye fading, and the addition of 1 ppm sulfur dioxide to the irradiated auto exhaust caused the greatest color change.

Ray *et al.*[630] reported similar findings from tests conducted to determine the comparative effects of small quantities of nitrogen oxides and sulfur dioxide on the color and strength of representative rayon fabrics. The oxides of nitrogen caused by far the greatest color changes; sulfur dioxide caused little or no fading. A combination of the two pollutants caused more fading than sulfur dioxide alone, but decidedly less than the nitrogen oxides alone. The sulfur dioxide appeared to reduce fading when used in combination with nitrogen oxides, undoubtedly because of an interaction between the gases under prevailing conditions. The fadings caused by the nitrogen oxides were most similar to those encountered during actual use.

Salvin[661] reported the results of laboratory tests conducted on 28 fabrics to assess the individual and combined effects of nitrogen oxides, ozone, and acid or alkali surface conditions on lightfastness. The most important changes occurred with direct dyes on cottons. Direct Red 75 and Direct Blue 76 dyes were changed appreciably in shade by nitrogen oxides and, along with Direct Blue 78, suffered a substantial reduction in color when the fabric surface was acid. The laboratory ozone tests resulted in slight fading, which was more evident in acid fabrics. In general, there were no significant changes on the dyed nylon and wool fabrics tested.

The investigators concluded that the variations in absorbed acid gases, nitrogen oxides, and ozone influence results of outdoor exposure tests. Such testing is consequently not reproducible in different locations and seasons. Salvin also has stated that nitrogen oxides in the air cause reddening of blue curtains and draperies and that ozone can bleach colored acetate, cotton, and nylon.[189] Beloin[63] confirms that appreciable fading of dyed fabrics results from exposure to nitrogen dioxide, sulfur dioxide, and ozone as well as variations in temperature and humidity.

PORCELAIN ENAMELS

Porcelain or vitreous enamels are protective glossy coatings that are fired on metals. They are composed chiefly of quartz, fieldspar, clay, soda, and borax. Although enameled metals are used primarily as siding and roofing, since World War II their use has expanded rapidly. In modern architectural design, they are used as exterior finishes, especially on buildings employing curtain-wall construction. Their popularity may be due to the wide variety of colors available and to the ease of cleaning.

The increased use of porcelain enamels led to a National Bureau of Standards (NBS) program involving 1-, 3-, 7-, and 15-year outdoor exposure tests to determine the weatherability of post-war enamels. Exposure conditions were the same for each test. Pertinent results of the 7-year test program and some from a 3-year test are discussed below. This 15-year test is not yet complete.

Rushmer and Burdick[655] reported that there were seven exposure sites in the NBS program. The two at Kure Beach, North Carolina, were temperate, rural, seacoast locations. One site was 24 m from the ocean; the second, 240 m. Washington, D.C. and Pittsburgh, Pennsylvania were temperate, commercial locations; New Orleans, Louisiana, a semi-tropical, commercial location; and Dallas, Texas and Los Angeles, California were commercial locations. Except for Kure Beach, samples were exposed on rooftops at 45° to the horizontal* and faced south. At

* Moore and Potter[550] reported that after the 3-year exposure test, enamels of poor weathering resistance exposed vertically at Washington, D.C., exhibited significantly smaller changes in gloss and color than similar specimens exposed at 45° to the horizontal. Enamels with good weather resistance exhibited only minor differences in the degree of attack with exposure angle.

Kure Beach, samples were exposed at ground level and faced the ocean at east-southeast. Weather data and air quality measurements for these sites are given in Table 10-9. Air pollution measurements made during the 3-year tests are given in Table 10-10. These data indicate the content of acid gases in the various atmospheres. Several enamel-coated metals were exposed at each site.

After 7 years, all specimens except those in Pittsburgh and Los Angeles could be cleaned easily; at Los Angeles a gumlike film had been deposited on the specimens and at Pittsburgh the panels were covered with a dirt film that consisted mostly of soot and fly ash. Corrosion of the base metal was noted only on 1 specimen at New Orleans, 10 at Kure Beach (240 m), and 29 at Kure Beach (24 m). In most cases, corrosion occurred on enamels exhibiting pinhole defects or blisters prior to exposure.

Figures 10-3 and 10-4 show the effect of exposure length on gloss and color retention. Neither white porcelain enamels, which normally retain their initial color after even a severe surface attack, nor screening paste enamels, which incur abnormally large color changes, were considered in computations of average color changes. The results from

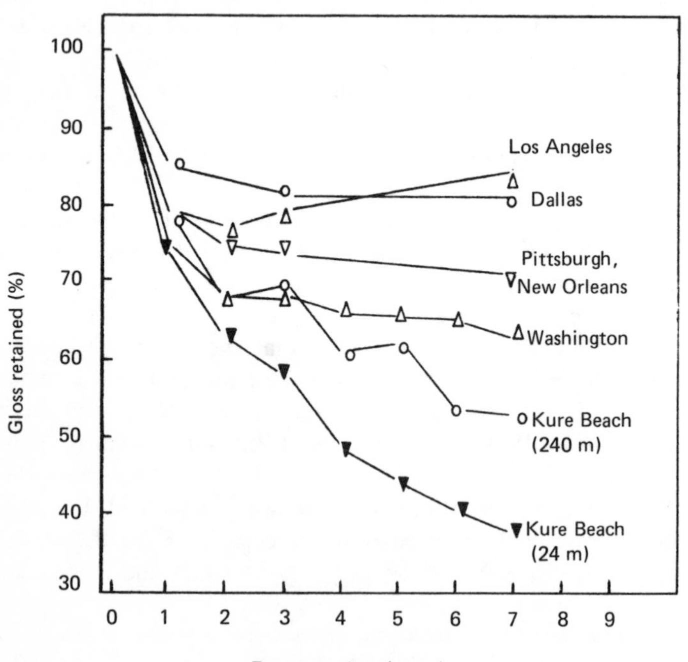

Figure 10-3. Change in average percentage of gloss retained with exposure time. (Points are averaged for all enamels except screening pastes.) From Rushmer and Burdick, 1966.[655]

006

Table 10-9. Weather data and air quality measurements for the 7-year exposure period[a]

Exposure site	pH[b,c]	Average annual relative humidity (%[d])	Average annual temperature (°C[d])	Total annual sunshine (hr[d])	Total annual precipitation (in[d])	Suspended particulate matter ($\mu g/m^3$)[b]
Kure Beach[e]	4.9	77	15.56	2,669	58	31
Washington	6.9	67	13.89	2,576	30	108
Pittsburgh	6.7	68	10.00	2,202	34	160
New Orleans	7.2	74	21.11	2,744	63	89
Dallas	7.3	63	18.89	2,911	40	95
Los Angeles	6.4	62	18.89	3,284	11	169

[a] From Rushmer and Burdick, 1966.[655]

[b] Averages from National Air Sampling Network of the Department of Health, Education, and Welfare.

[c] Measured for solutions prepared by refluxing an 8% aliquot of particulate matter from the atmosphere with 50 ml of distilled water and diluting to 80 ml. Values are averages of measurements made at approximately biweekly intervals.

[d] Averages from the United States Weather Bureau Records.

[e] Data from Cape Hatteras, N.C., rather than Kure Beach, for which no data were available.

Pittsburgh and New Orleans were averaged together because there was no appreciable difference in the degree of attack on the enamels at these locations.

All the enamels lost gloss most rapidly during the first 3 years of exposure; during the remainder of the period, the retained gloss remained nearly constant on all enamels except those at both Kure Beach locations, where it continued to decrease. The slight increase in gloss at Los Angeles after the second year was attributed to incomplete removal of the gumlike film during cleaning. This resulted in a doubly reflecting surface which increased gloss readings. Color change was greatest during the first 2 years, after which it became nearly constant, although different, at all seven sites.

Concentration of acidic air contaminants correlate well with site severity. At the most severe sites the particulates had more acidic pH values, while at the mild sites the pH was nearly neutral. The lack of correlation of enamels exposed at Pittsburgh and Los Angeles might be partially explained by the protective action of the adherent films on the specimens exposed at these locations. A good correlation was also observed between relative humidity and changes in gloss color. Rushmer and Burdick suggested that this correlation could be attributed to the moisture from high humidity, which leaches salts from the deposited particulate matter, producing acidic solutions that vary according to the contaminants.

The glossy, acid-resistant enamels on steel had the best color retention; the screening paste type, the poorest. Some enamels on steel retained their color well even when exposed to the salt air at Kure Beach. The best of the enamels on aluminum incurred little color change at

Table 10-10. Averages of air pollution measurements made during 1958 by the National Air Sampling Network of the United States Department of Health, Education, and Welfare[a]

City	Total suspended particulates[b]	Composition of particulates[b]				
		Organic matter	SO_4	NO_3	Other	pH[c]
Los Angeles	214	30.4	16.1	9.5	158.0	6.0
Pittsburgh	167	13.0	15.2	2.7	136.1	6.4
Washington, D.C.	111	12.9	12.4	3.1	82.6	6.7
New Orleans	92	11.4	9.5	2.2	68.9	7.0
Dallas	113	10.2	7.1	2.3	93.4	7.1

[a] From Moore and Potter, 1962.[550]

[b] $\mu g/m^3$ of air. Values are averages of measurements taken at approximately biweekly intervals.

[c] Measured for solutions prepared by refluxing an 8% aliquot of particulates with 50 ml of distilled water, then diluting to 80 ml. Values are averages of measurements made at approximately biweekly intervals.

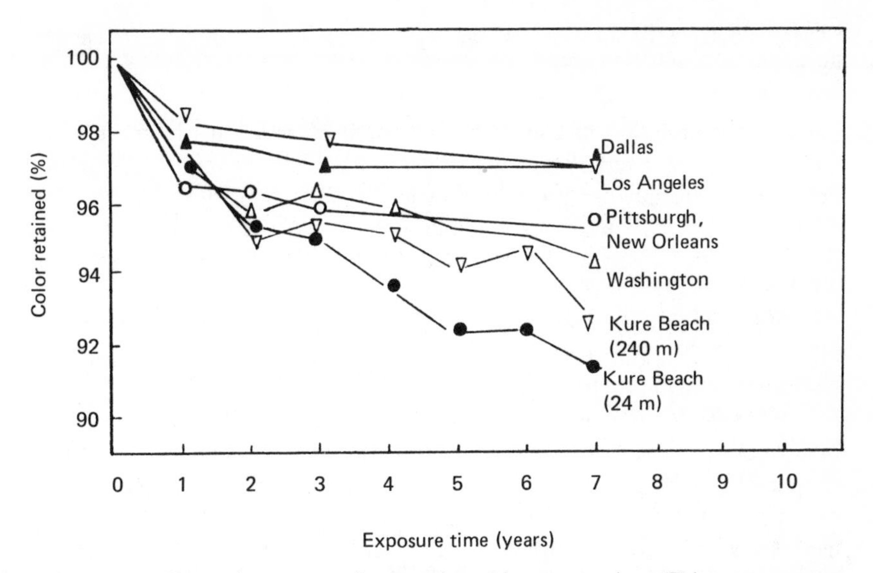

Figure 10-4. Change in average color retention with exposure time. (Points are averages for all nonwhite enamels except screening pastes.) From Rushmer and Burdick, 1966.[655]

Dallas, Los Angeles, and Pittsburgh, but moderate changes in Washington, D.C. and fairly important changes at the two Kure Beach sites and in New Orleans. The severe effects in New Orleans were unexpected. Although the cause was not determined, the authors did not believe that it could be attributed to salt particles carried from either the Gulf of Mexico or Lake Pontchartrain because the site was on a four-story building several kilometers from the nearest salt water. The attack at Kure Beach was tentatively ascribed to chlorides. In most cases, the retained gloss correlated with the color change.

Similar types of enamels exhibited a direct correlation between acid resistance, as measured by either the acid spot test or the boiling acid solubility test, and weather resistance, as measured by changes in gloss and color. The correlation with the citric acid spot test, however, was evident only when averages were considered. The boiling acid solubility test was more reliable in predicting the color retention of the regular enamels on steel than was the citric acid spot test.

ASPHALTS

Asphalt is a dark brown to jet-black material that can be solid or semisolid. The predominating constituents are either natural bitumens or bitumens made from residuals of petroleum refining. Asphalt and its products are used in road construction and as roofing, waterproofing paper, electrical insulation, an adhesive, and as an additive to certain paints.

Asphalt is affected by natural constituents of the atmosphere, particularly sunlight and humidity. In time it becomes brittle and cracks. General property changes that occur include loss of color and gloss; deterioration in uniformity; increases in density, viscosity, hardness, hardening point, softening point, flash point, and other properties; and decreases in spreading rate, shear strength, adhesive power, and volatility.

Hamada *et al.*[311] have provided the only information concerning the effects of air pollutants on asphalt. They conducted tests to determine the resistance of asphalts to outdoor weathering and to acids.

In their tests to determine the resistance of asphalts to acids the investigators used 19 blown, 8 straight, and 1 compound asphalt produced from the same crude oil; 1 blown, 4 straight, and 1 compound asphalt produced from other crude oils; and 7 compounds of Nos. 18-23 asphalts with rubber or other inorganic fillers added. Acids used in the tests were sulfuric acid (80%, 50%, and 12%), hydrochloric acid (10%), and ammonium sulfate (15%). Steel rods approximately 170 mm long were immersed in asphalt. These rods were then dipped to a depth of 150 mm in the acid solutions contained in test tubes, then left standing at room temperature. Changes were observed at different intervals.

In tests of 28 specimens exposed to 80% sulfuric acid, there was no corrosion, but cracks, bulges, and air bubbles formed.[311] Hamada *et al.* attributed this to penetration through pinholes in the asphalt by the acid solution and subsequent reaction with the steel to form sulfate, which caused expansion and consequent cracking and exfoliation.

Compounds of No. 18 and No. 23 asphalt were also exposed to 80% sulfuric acid.[311] All specimens cracked and flaked. This was attributed to the effect of additives which, by decreasing the impermeability of the asphalt, allowed the acid solution to penetrate deeply. Hamada *et al.* provided the test results for asphalts produced from other crude oils immersed in the acids of all concentrations. In each case, there was some degree of transformation. In some cases, change of color in the solution occurred.

The investigators concluded that sulfuric acid, hydrochloric acid, and ammonium sulfate cause some transformation, exfoliation, or crumbling in asphalts. Generally speaking, blown asphalts appear to have good acid resistance, but the coating quality does not appear to be good. Straight asphalts, on the other hand, have a good quality of coating, but, compared with blown asphalt, their acid resistance is not so good. The authors suggest that the application of a regular asphalt coating to blown asphalt would provide an adequate degree of both acid resistance and coating quality.

COST

Cost data often include damage to materials other than those used in buildings (e.g., tires and clothing) and to agricultural goods and human health. In most cases, it is therefore impossible to isolate costs of damage to building materials. Accordingly, such data are not included in this report.

Table 10-11, compiled by Salmon,[660] presents the results of perhaps the most comprehensive effort undertaken to estimate the economic losses due to the effects of air pollution on materials. The values of interaction are estimates of the difference between the rate of material deterioration in polluted atmospheres and in unpolluted atmospheres. This is expressed in dollars lost per year. The in-place values of exposed materials, which include an estimated labor factor, are based on the product of annual production volume in dollars × a weighted average economic material life based on usage × a weighted average factor for the percentage of the material that is exposed to air pollution. Economic losses or damage factors reported in the table represent the product of the interaction values and in-place values. When assessing these economic losses, it is important to remember that the values were derived simply to rank materials in order of their relative importance in air-pollution-induced damage. These values should not be interpreted as actual economic loss, although Barrett and Waddell[50] suggest that they appear reasonable. The reported losses represent only actual damaged materials or impaired serviceability, but not losses associated with surface soiling.

In 1968, Barrett and Waddell estimated the total cost to the nation of air pollution damage to be $16.132 billion. They arrived at this figure after reviewing both published and unpublished estimates. They attributed $9.952 billion to damage on residential properties and materials and the balance, $6.180 billion, to damage to health and vegetation. The investigators considered various pollutants both individually and synergistically. Their estimates were allocated in direct proportion to emission levels (see Table 10-12).

The liability for the costs in Table 10-12 are assigned in Table 10-13 to sources of the total particulates and sulfur oxides emitted nationally. For example, 24%, or $1.248 billion, of the $5.2 billion in residential property losses is assigned to industrial processes because such sources account for 24% of the nation's sulfur oxide and particulate emissions. Table 10-14 presents estimates by the Office of Air Programs, Environmental Protection Agency, of 1968 national emissions of principal pollutants by major source.

The Battelle Memorial Institute estimated that the extra service cost resulting from air pollution corrosion of metals, particularly steel and

Table 10-11. Summary and rankings of damage factors[a]

Rank	Material	Value of interaction ($/year)	In-place value of materials exposed (billion $)	Loss (million $)
1	Paint	0.50×10^{-1}	23.90	1,195.0
2	Zinc	0.29×10^{-1}	26.83	778.0
3	Cement and concrete materials	0.10×10^{-2}	316.21	316.0
4	Nickel	0.25×10^{-1}	10.40	260.0
5	Cotton (fiber)	0.40×10^{-1}	3.80	152.0
6	Tin	0.26×10^{-1}	5.53	144.0
7	Synthetic rubber	0.10×10^{-0}	14.00	140.0
8	Aluminum	0.21×10^{-2}	54.08	114.0
9	Copper	0.20×10^{-2}	54.88	110.0
10	Wool (fiber)	0.40×10^{-1}	2.48	99.2
11	Natural rubber	0.10×10^{-0}	0.54	54.0
12	Carbon steel	0.50×10^{-2}	10.76	53.8
13	Nylon (fiber)	0.40×10^{-1}	0.95	38.0
14	Cellulose ester (fiber)	0.40×10^{-1}	0.82	32.8
15	Building brick	0.10×10^{-2}	24.15	24.2
16	Urea and melamine (plastic)	0.10×10^{-1}	2.27	22.7
17	Paper	0.30×10^{-2}	7.53	22.6
18	Leather	0.40×10^{-2}	5.15	20.6
19	Phenolics (plastic)	0.10×10^{-1}	1.98	19.8
20	Wood	0.10×10^{-2}	17.61	17.6
21	Building stone	0.23×10^{-2}	7.65	17.6
22	Polyvinyl chloride (plastic)	0.10×10^{-1}	1.54	15.4
23	Brass and bronze	0.42×10^{-3}	33.12	13.9
24	Polyesters (plastic)	0.10×10^{-1}	1.37	13.7
25	Rayon (fiber)	0.40×10^{-1}	0.33	13.2
26	Magnesium	0.20×10^{-2}	6.50	13.0
27	Polyethylene (plastic)	0.10×10^{-1}	1.17	11.7
28	Acrylics (plastic)	0.10×10^{-1}	1.00	10.0
29	Alloy steel	0.40×10^{-2}	2.18	8.7
30	Polystyrene (plastic)	0.10×10^{-1}	0.85	8.5
31	Acrylics (fiber)	0.40×10^{-1}	0.19	7.6
32	Acetate (fiber)	0.40×10^{-1}	0.19	7.6
33	Polyesters (fiber)	0.40×10^{-1}	0.16	6.4
34	Polypropylene (plastic)	0.10×10^{-1}	0.64	6.4
35	Acrylonitrile-butadiene-styrene (plastic)	0.10×10^{-1}	0.61	6.1
36	Epoxies (plastic)	0.10×10^{-1}	0.47	4.7
37	Cellulosics (plastic)	0.10×10^{-1}	0.40	4.0
38	Bituminous materials	0.10×10^{-3}	22.45	2.2
39	Grey iron	0.50×10^{-3}	3.86	1.9
40	Nylon (plastic)	0.10×10^{-1}	0.17	1.7
41	Polyolefins (fiber)	0.40×10^{-1}	0.04	1.6

Table 10-11. Continued

Rank	Material	Value of interaction ($/year)	In-place value of materials exposed (billion $)	Loss (million $)
42	Stainless steel	0.85×10^{-4}	18.90	1.5
43	Clay pipe	0.10×10^{-2}	1.44	1.4
44	Acetate (plastic)	0.10×10^{-1}	0.12	1.2
45	Malleable iron	0.16×10^{-2}	0.58	0.9
46	Chromium	0.75×10^{-3}	1.08	0.8
47	Silver	0.12×10^{-2}	0.57	0.7
48	Gold	0.10×10^{-3}	5.80	0.6
49	Flat glass	0.10×10^{-4}	28.59	0.3
50	Lead	0.11×10^{-3}	2.18	0.2
51	Molybdenum	0.25×10^{-3}	0.51	0.1
52	Refractory ceramics	0.10×10^{-4}	1.93	0.02
53	Carbon and graphite	0.10×10^{-5}	0.30	0.00
	TOTAL			3,800.00

[a] From Salmon, 1970.[660]

zinc, totalled approximately $1.45 billion in 1970.[232] This value is based on the nine major classifications of external metal structures for which annual losses have been estimated (see Table 10-15).

Fink *et al.*[232] projected total national cost of corrosion by air pollution from 1970 to 1980. This cost will be influenced greatly by variations in the total metal surface exposed to attack and by changes in the corrosivity of the atmosphere, which, in turn, will be influenced by such factors as increased population and energy production. Table 10-16 lists these estimated changes along with the estimated changes in levels of sulfur oxide pollution in four projected regulatory situations. Although the authors recognize that humidity changes affect corrosivity, the litera-

Table 10-12. National costs of pollution damage, by pollutants, 1968 (billion $)[a]

Pollutant	Loss category		
	Residential property	Materials	Totals
SO_x	2.808	2.202	5.010
Particulates	2.392	0.691	3.083
Oxidants		1.127	1.127
NO_x		0.732	0.732
Totals	5.200	4.752	9.952

[a] From Barrett and Waddell, 1973.[50]

Table 10-13. Costs of pollution damage in United States in 1968, by source and effect (billion $)[a]

Source	Effects: Residential property	Materials	Totals
Stationary source fuel combustion	2.802	1.853	4.655
Transportation	0.156	1.093	1.249
Industrial processes	1.248	0.808	2.056
Solid waste	0.104	0.143	0.247
Miscellaneous	0.884	0.855	1.739
Totals	5.200	4.752	9.952

[a] From Barrett and Waddell, 1973.[50]

ture does not indicate that atmospheric corrosivity has been altered greatly in any recent 10-year period by changes in moisture and related meteorologic factors. Notable changes in corrosion rates have usually been accompanied by changes in pollution levels, often in association with a higher rate of consumption of sulfur-bearing fuels.

Based on the estimate that 80% of the 1970 steel systems would still be in use in 1980, the authors projected an increase in maintenance costs even with improved technology in corrosion control and increased use of labor-saving techniques. However, these costs would probably be offset by the lower maintenance costs of newer structures that have been designed to reduce maintenance expense. Allowing for some increase in both steelwork and population and assuming no change in pollution, they estimated that the per capita annual corrosion cost would be essentially the same in 1980 as it was in 1970. The per capita extra annual maintenance costs were established, then converted to a national basis.

Table 10-14. Estimates of nationwide emissions in 1968 (10^9 kg/yr)[a]

Source	CO	Particulates	SO_x	HC	NO_x
Transportation	63.8	1.2	0.8	16.6	8.1
Fuel combustion in stationary sources	1.9	8.9	24.4	0.7	10.0
Industrial processes	9.7	7.5	7.3	4.6	0.2
Solid waste disposal	7.8	1.1	0.1	1.6	0.6
Miscellaneous	16.9	9.6	0.6	8.5	1.7
Total	100.1	28.3	33.2	32.0	20.6

[a] From Barrett and Waddell, 1973.[50]

Table 10-15. Summation of annual extra losses from corrosion damage by air pollution to external metal structures in 1970[a]

Steel system or structure	Extra costs	Annual loss (thousands $)
Steel storage tanks	Maintenance	$ 46,310
Highway and rail bridges	Maintenance	30,400
Power transformers	Maintenance	7,450
Street lighting fixtures	Maintenance	11,910
Outdoor metal work	Maintenance	914,015
Pole-line hardware	Replacement	161,000
Chain-link fencing	Maintenance and replacement	165,800
Galvanized wire and rope	Replacement	111,800
Transmission towers	Maintenance	1,480
		$1,450,165

[a] From Fink *et al.*, 1971.[232]

The Stanford study also disclosed that more money is spent combating the effects of sulfur dioxide and hydrogen sulfide on low-voltage electrical contacts than is spent combating all other air pollution effects on all electrical devices. Organic gases form "frictional" polymers on sliding contacts. Particulate pollutants present special problems because

Table 10-16. Economic and pollution factors used to assess probable cost of corrosion damage by air pollution, 1970 to 1980[a]

	Change from 1970–1980 (%)
Population Increase	
For nation	11
For metropolitan districts	12
Energy Production	
Increase in power plant capacity	78
Space heating plant capacity	15
Sulfur Oxide Pollution	
No regulation	55
With regulation and improved technology	<10
With strictly enforced regulation and major technological breakthroughs	−40
Complete enforcement of current legislation	−60
External Structures	
Old steel structures (decrease)	−20
Replacement steel	20
New steel	10
New other materials	20
Total change	30
Total increase in steel	10

[a] From Fink *et al.*, 1971.[232]

they are excellent absorbers of water and corrosive agents. The effects of organic gases and particulates are less important, however, than the effects of sulfur dioxide and hydrogen sulfide.

Salvin conservatively estimates the total annual cost of damage to textiles by air pollution to be $2 billion per year.[189] He considered disintegration of natural and man-made fibers before the end of a normal wear cycle (specifically $300 to $400 million for cotton and nylon); extra laundering and dry cleaning due to soiling by oily dust particles ($800 million); fading of colors of nylon carpet, acetate fabrics, cottons, and permanent press fabrics ($350 million); and discoloration of white fabrics due to the effect of hydrogen sulfide and various acids on the fibers and finishes (approximately $540 to $550 million). Barrett and Waddell[50] reported that Salvin is planning a comprehensive investigation to determine more accurately the annual estimated cost of air pollution damage to dyed textiles. The estimate will include not only the cost resulting from damage but also the cost of the steps taken to mitigate the damage, such as substitution of more costly materials, additional protection, modified production techniques, implementation of better quality control techniques, research and development, and environmental testing.

CONCLUSION

The literature dealing with the effects of pollutants on building materials is easily misleading and should be interpreted with caution. Although the literature generally implies a detrimental effect, however small, it contains no guidelines by which one can assess impact on intended function and normal service life of the affected materials. This applies even to documentation of catastrophic damage, particularly in the case of historical buildings and irreplaceable sculptures.

Particular care must be taken when assessing estimates of costs resulting from damage or soiling of materials by pollutants. In many instances all costs associated with services or materials are attributed to air pollutants, although there is inadequate justification for including even portions of such costs. These estimates are often used by others as a base for additional, supposedly more inclusive, estimates, thereby compounding the problem. Furthermore, except for tests involving extended, natural outdoor exposures, principally of metals, most investigators use pollutant concentrations greatly in excess of those found in even the most polluted atmospheres. Results from accelerated weathering tests like these are difficult, if not impossible, to correlate with actual natural weathering conditions.

To place effects of air pollutants on building materials in true perspective, several additional factors must be considered. Notably, sulfur oxides and the liquid acids formed from them are the most detrimental pollutants because of their chemical attack on a material. Although particulate matter, such as soot and fly ash, can also result in chemical attack of materials, particularly in concert with the sulfur oxides and their acids, its primary effect on materials is surface dirtying. In many cases this can be mitigated by proper cleaning procedures. However, atmospheric concentrations of both sulfur oxides and particulates are presently being reduced with introduction and use of low-sulfur fuels and emission control equipment. Based on the realistic assumption that newer fuels and control technologies will emerge, pollutant concentrations should continue to decrease. In addition, new paints and materials more resistant to specific pollutants or whose surfaces are more readily cleanable are being developed.

Another factor to be considered is the synergistic effect pollutants have on materials. Considerably more testing with particulates of known composition acting in concert with all other pollutants is essential if the true cause of the detrimental effect is to be understood adequately. In planning such studies, it should be recognized that results from accelerated tests involving higher than normal concentrations of only one pollutant and normal concentrations of others are not likely to correlate well with results from actual outdoor exposures.

The structure itself and the constraints placed upon its designer also must be taken into account. The initial design parameters (e.g., function, intended service life, initial cost, and aesthetic value) greatly influence choice of materials used in construction and, thus, the structure's performance under either polluted or nonpolluted conditions. For example, in building warehouses, initial cost is of paramount concern. Decisions may be made during the planning stage to use materials more susceptible to pollutants than others—if the intended service life can still be assured—simply because aesthetic qualities are of little concern.

The facts that today's buildings generally are not considered to be as durable as those built in the past and that future buildings will probably not be more durable reflect the constraints placed on designers who try to meet society's demand for low cost and fast construction. Consumers simply can no longer afford buildings designed and constructed with expensive safety factors and excessive materials. To remain competitive, during the past 20 years or longer, designers have employed newly developed materials simply because they were less expensive or because they facilitated the construction process, regardless of their susceptibility to air pollutants. Designers have only recently begun to select materials for different building elements because of their resistance to particular air pollutants. Similarly, only recently have design criteria for mitigating the effects of air pollutants begun to emerge.

Proper selection of materials, use of emerging design criteria, and application of proper maintenance procedures should appreciably reduce the effect of air pollutants on building materials. Consequently, the effect of pollutants on construction materials used today cannot be assessed simply in terms of past performance.

11

Summary, Conclusions, and Recommendations

SUMMARY AND CONCLUSIONS

During the past 5 years there has been a revolution in our knowledge and our understanding of atmospheric aerosols as a result of both routine monitoring and special aerosol characterization studies. These studies have emphasized the importance of size distribution and chemical composition in understanding the sources, effects, and sinks of atmospheric aerosol.

Aerosols: Characteristics, Behavior, and Measurement

Recent studies have elucidated several important features of the aerosol size distribution. This distribution consists of three separate modes. The first mode, named the nuclei or Aitken mode, includes particles that range in diameter from approximately 0.01 to 0.1 μm. This mode is formed by a dynamic equilibrium between new nuclei formed in the atmosphere plus nuclei from combustion and coagulation into the next larger mode, the accumulation mode. The first mode numerically accounts for most of the particles in the atmosphere except for very well-aged background aerosols, most of which are found in the accumulation mode. Most of the submicrometer mass tends to accumulate in this second mode because of the twin mechanisms of coagulation and heterogeneous nucleation (condensation of one material on another). The particles in this accumulation mode range in diameter from about 0.1 to 2.0 μm and comprise most of the total surface area of the atmospheric aerosol. The nuclei and accumulation modes are collectively referred to as "fine particles."

Little mass is transferred from the accumulation mode into the coarse particle mode. This third mode consists of particles >2 μm diam and mass that, on a long-term average, is about equal to that in the accumulation mode. The almost complete independence of the origin, behavior, and removal processes of fine and coarse particles result in different effects on health, visibility, and meteorology. There are variations over several orders of magnitude in the amounts of fine and coarse particles in clean and polluted atmospheres. These factors suggest that fine and coarse particles should be controlled separately.

Both differential and integral measurements of atmospheric aerosol should not respect only these size-dependent differences. It is also important to recognize the pitfalls that are generally associated with sample collection and subsequent laboratory analysis versus *in situ* sampling. The trend in aerosol monitoring is towards the replacement of collection methods for both physical and chemical analyses with *in situ*, automatic methods.

Aerosol Cycles

When studying the life histories of aerosol particles, one must distinguish coarse (≥ 2.0 μm diam) from fine (≤ 2.0 μm diam) particles. Coarse particles arise primarily from mechanical processes, both naturally (windblown dust, sea spray) and anthropogenically (fly ash, comminutive processes). Fine particles are formed in the nuclei mode by condensation of hot, low vapor pressure materials, such as metallic vapor, and of cold, low vapor pressure materials, such as partially oxygenated aromatics.

Aerosol transformation processes are also distinctly different for coarse and fine particles. Coarse particles are not significantly transformed in the atmosphere, whereas fine particles are. These fine particle transformation processes include both gas-particle and particle-particle interactions. Gas-particle interactions usually involve gas-liquid reactions because fine particles have sufficient water associated with them under most atmospheric conditions to present a liquid surface to the gas phase. Particle-particle interactions (e.g., coagulation) are governed by Brownian diffusion. Because aerosol size ditributions are rarely monodisperse, coagulation usually occurs between unlike size particles. Coagulation of particles from the nuclei mode to the accumulation mode is important in the atmosphere, whereas coagulation from the accumulation mode into the coarse mode is not.

Most coarse particles are emitted close to the earth's surface, where they are removed by gravitational settling and washout before being transported very far. Fine particles, especially in the accumulation mode, are not strongly affected by these processes and are transported much longer distances ($\sim 10^3$ km).

Urban aerosol is usually dominated by human activity. The size distribution and composition of the average global aerosol is less certain but there are indications that the well-aged aerosol throughout the bulk of the atmosphere is dominated by a single mode, the accumulation mode.

Measurement of the Size Distribution and Concentration

There is an inevitable trade-off between intensity and duration of measurement programs. Results are either continuous aerosol measurements that provide only limited information on size distribution and chemical composition or detailed aerosol data that were collected over only a short time.

Background measurements over the oceans indicate that the aerosol is dominated by coarse particle sea salts mechanically produced at the sea surface and by aged coarse and fine particles above the mixing layer that have been transported from land. Background aerosol over land varies with location. Over remoted snow and ice surfaces, aerosol concentrations are as low as or lower than over the oceans. Background aerosol over dry, desert regions is dominated by coarse particles, usually windblown dust. In areas of average precipitation and vegetative cover, the coarse particle concentrations are on the average much lower than over dry areas, and there is a small but persistent background concentration of accumulation mode aerosol.

In urban areas coarse particle concentrations are highly variable. Although urban accumulation mode aerosol concentrations sometimes exceed several hundred micrograms per cubic meter, concentrations average between 10 and 30 $\mu g/m^3$. Significant concentrations by mass of nuclei mode aerosols are found only near highways. Accumulation mode aerosols produced in urban areas can be transported several hundred kilometers. Increasing evidence indicates that accumulation mode aerosols, a large fraction of which is sulfate, may be transported several thousand kilometers.

Chemical and Trend Data

The chemical composition and molecular character of atmospheric aerosol as a function of particle size should be extensively examined. At moderate or low relative humidities, toxicity, water solubility, hygroscopicity, deliquescence, refractive index, particle shape, and physical state depend on molecular nature rather than on the presence of a single atomic or ionic species, such as sulfate. However, in the case of a highly hydrated droplet aerosol, composed of a large number of anions and cations, the ionic composition clearly dominates the system behavior. The definitive molecular composition measurements as a function of particle size, time, and location have yet to be made.

Recent data suggest that a chemically well-mixed aerosol, one that has a similar composition at all sizes and in all three modes, is probably rare. Rather, certain classes of substances appear to dominate the coarse particle mode while other classes dominate the fine mode. The fine particles consist mainly of sulfates from sulfur dioxide oxidation, nitrates from NO_x reactions, ammonium ions from ammonia reactions, condensed organic matter (partially oxygenated), and primary emitted substances, such as lead, arsenic, antimony, and carbon. The coarse particles consist largely of mechanically produced substances, such as soil or rock dust (e.g., silicon, calcium, iron, and aluminum), road and tire dust, fly ash, and sea salt. The frequent particle-particle interactions within the fine mode suggest a homogeneously mixed submicrometer aerosol; however, the less frequent interactions of the coarse mode suggest discrete coarse particles consisting of relatively pure substances.

Depending on their chemistry, fine particles can grow at high humidity to 1.5 to 2.0 times their initial dry diameter in a fraction of a second. This growth can influence such aerosol effects as deposition in the respiratory tract, visibility in the atmosphere, and condensation processes in clouds. With the exception of sea salt, coarse particles usually do not exhibit this growth. Fine particles, however, almost always do, resulting in a dramatic increase in the amount of material (usually water) in the fine particle mode.

Although total suspended particulate matter mass concentrations in selected urban areas throughout the United States have decreased from 1960 to 1971, nonurban values remained the same. The lack of size resolution and chemical composition precludes a detailed understanding of these trends.

Effects on Atmospheric Processes

Aerosols can influence atmospheric processes by interacting with both incoming solar radiation and cloud processes. The extinction of solar radiation by atmospheric aerosol can be generally classified into scattering and absorption components. The scattering component is theoretically a complex function of the fine particle mass concentration, the total mass concentration, the aerosol size distribution, the relative humidity, the wavelength of light, and the particle's refractive index and shape. However, a strong correlation has been found in the atmosphere between fine particle mass concentration and scattering coefficient. Increases in relative humidity are usually accompanied by increases in particle light-scattering because aerosol growth resulting from water vapor condensation mainly occurs in the fine particle mode, which dominates light-scattering.

Besides degradation of visibility, aerosol extinction also decreases the intensity of direct sunlight reaching the ground. Although most scattered light reaches the surface along with the direct beam, some is absorbed and some is scattered back into space by the aerosol. The turbid air mass centered over the eastern United States has twice the extinction coefficient of the air over the rural western United States. However, disagreement exists on whether there will be a decrease or an increase in temperature as a result of changes in aerosol concentration, both regionally or globally. More measurements are needed on the absorptive component as a function of particle size, the hemispheric backscatter coefficient, and their spatial dependencies. The role of relative humidity on these parameters should also be examined. The relatively few measurements of absorption indicate that the fraction of extinction due to absorption can be as high as 30%. Direct measurements of this absorption component are preferable because size dependence of the aerosol chemical compositon, and thus the refractive index, must be considered when estimating the absorption component from chemical data.

Aerosol interaction with clouds occurs because the transformation of water vapor, at approximately 1% supersaturation, into water droplets and ice crystals is a nucleation phenomenon requiring the presence of a specific size of nuclei. These "weather active" nuclei, which comprise $<1\%$ of the total aerosol number concentration, influence cloud droplet and ice crystal concentrations. These, in turn, influence cloud dynamics and the probability and amount of rainfall. The precipitation probability is a complex function of the concentration, size distribution, and molecular form of these nuclei. In urban areas, the effect is generally an increase in precipitation downwind because of the emission of the nuclei.

Fogs are also regulated by atmospheric aerosol. Due to condensation and coalescence, fog stability is increased by the formation of small (1 to 3 μm diam), unactivated drops, and decreased by large (15 μm diam), activated ones. The factors that determine activation in both fogs and clouds deserve further study.

Aerosol can also affect the chemistry of precipitation through both below-cloud and in-cloud scavenging. Fine particles are scavenged mainly by in-cloud processes, whereas coarse particles are scavenged by both in-cloud and below-cloud processes. This chemical modification of the rainwater is complicated by the fact that such gases as sulfur dioxide can also contribute to the same species in raindrop solutions as the aerosol, e.g., sulfate. The relative importance of rainout and washout mechanisms should be studied further.

Effects of Inhaled Particles on Man and Animals

Particle deposition efficiencies and patterns within the respiratory tract are highly variable. They are determined by dimensions and configurations of the air path in the exposed individual, the pattern and depth of the respirations, and the characteristics of the airborne particles. The size of bronchial airways varies considerably among normal nonsmoking adults and more so among smokers without clinical disease symptoms. However, the greatest variations occur among individuals with clinical evidence of bronchitis.

The distribution of particle deposition sites depends greatly on the aerodynamic diameters of the particles. In normal, healthy humans, inhaled nonhygroscopic particles that impact in the head and ciliated airways of the lungs are concentrated onto a small fraction of the surface. Cigarette smoking and bronchitis produce a proximal shift in the deposition pattern. For nonhygroscopic aerosols with aerodynamic diameters $\lesssim 3$ μm, an increasing fraction remains airborne as particle size decreases and is exhaled. Total respiratory tract deposition in normal subjects reaches a minimum of ~10% to 20% for particles between ~0.2 and 1.0 μm, and increases for particles $\lesssim 0.2$ μm. The major factor determining the probability of deposition of the smaller particles is their transfer from tidal to reserve air. They may remain airborne within the reserve and residual air for a number of breaths before they are actually deposited.

The dominant deposition mechanisms are impaction (for particles $\gtrsim 1.5$ μm), sedimentation (~0.5 to 1.5 μm), and diffusion ($\lesssim 0.5$ μm). Both sedimentation and diffusion produce relatively uniform surface deposits in small bronchioles, alevolar ducts, and alveolar sacs. Hygroscopic particles grow rapidly within the warm, moist airways, becoming dilute aqueous droplets ~3 times larger than the inhaled particles.

For particles that are soluble in respiratory tract fluid, systemic uptake may be relatively complete for all deposition patterns, and local toxic and/or irritant effects may result. On the other hand, slowly soluble particles depositing in the head beyond the anterior nares, or on ciliated tracheobronchial airways, are transported by the surface flow of respiratory tract fluid to the glottis, then swallowed within 24 hr.

Mucociliary transport rates are highly variable, both along the ciliated airways in a given individual and between individuals, depending on the thickness and character of the secretions and the number and beat rate of the cilia. Effective fluid movement depends upon the coupling of the ciliary motion within the sol layer with the overlying mucus layer. A moderate increase in secretions, such as those produced by a few cigarettes, or therapeutic dosages of some adrenergic drugs, can accelerate mucus transport. Larger dosages or long-term exposures, which can cause an increase in the number and/or size of the secretory cells and glands, can produce mucus layers that are too thick to be effectively propelled by the cilia, and clearance stasis and periodic retrograde flow can result.

Mucocilary transport rates decrease distally within the bronchial tree. Measurements of normal mean tracheal transport rates in humans and animals have mostly fallen within the range of 2.5 to 25.0 mm/min. The total duration of bronchial clearance in healthy humans varies from approximately 2.5 to 20.0 hr. The changes in clearance rates produced by drugs, cigarette smoke, and various environmental pollutants can greatly increase or decrease these rates. However, the importance of alterations of mucociliary transport in the pathogenesis of chronic lung disease is not yet clear.

The relatively small particles deposited in nonciliated airways have large surface-to-volume ratios. Therefore, significant clearance by dissolution can occur for materials generally considered insoluble. They can also be cleared as free particles either by passive transport along surface liquids or, after phagocytosis, by transport within alveolar macrophages. If the particles penetrate the epithelium, either bare or within macrophages, they can be sequestered within cells or enter the lymphatic circulation, in which they are transported to pleural, hilar, and more distant lymph nodes. In most cases the quantitative aspects of these clearance pathways vary with the composition of the particles and are poorly described. Nontoxic insoluble particles are cleared from the alveolar region in a series of temporal phases. The earliest, lasting several weeks, appears to involve the clearance of phagocytized particles via the bronchial tree. The terminal phases appear to be related to solubility at interstitial sites. The effects of infectious diseases, cigarette smoking, and various environmental factors on kinetics of alveolar clearance are not known.

The mechanisms and dynamics of particle deposition and clearance have been reasonably well determined, but many quantitative aspects and relative contributions of competing and/or interacting processes have not been established. There is very little information on the relationships among normal physiologic clearance rates, reserve capacity to cope with environmental exposures, and secretion or transport rates that compensate for environmental exposures in the development and progression of temporary or permanent dysfunction. Bronchitis produces marked changes in regional particle deposition and sputum production. However, there are very few data on the quantitative aspects of regional deposition in well characterized bronchitics, or on sputum composition, transport, or volume in healthy subjects. For alveolar deposition, the clearance pathways and rates depend on the physical and chemical characteristics of the particles. There are few well established rate constants adequate for estimating toxic dose levels.

The major uncertainties about the effective toxic doses resulting from inhalation exposures can and should be resolved through further *in vivo* investigations on humans and laboratory animals. Studies of humans would also help to identify susceptible segments of the population that would benefit from personal protective measures during exposures to high concentration. Most investigations can be performed with available techniques. Consequently, an intensified research effort could produce useful data within a few years after its initiation.

Effects of Sulfur Dioxide and Aerosols, Alone and Combined, on Lung Function

The structural, biochemical, functional, and genetic effects of air pollutants may be assessed by experiments on *in vitro* preparations, animals, and, in some instances, human volunteers. Apart from the inherent biologic differences among species, there are well recognized difficulties in extrapolating results from animals to human populations. Only simple approximations of ambient pollution are possible in the laboratory, and the simulation of human cardiorespiratory disease in animals for the purpose of determining the response of vulnerable members of the population is difficult to achieve. Ethical and legal restrictions preclude many studies on human subjects, especially those involving carcinogens and chronic exposure. In Chapter 7 the effects of inhaling sulfur dioxide or aerosols alone are considered before proceeding to the consequences of gas-aerosol interactions on the mechanical and ventilatory functions of the lung.

During quiet breathing, sulfur dioxide is removed almost entirely in the upper airways. The mouth is less efficient than the nose as a scrubber of soluble gases, especially at the high flow rates occurring during exercise. In healthy subjects, inhalation of 2.6 mg/m^3 of sulfur dioxide for 1 to 3 hr at rest increases nasal and pulmonary flow resistance and impairs forced expiratory flow rate. Inhalation of 13.0 mg/m^3 of sulfur dioxide for several hours causes greater effects on these functions and also reduces nasal mucus flow rate. These concentrations of sulfur dioxide are greatly in excess of ambient concentrations.

There is considerable variability in response within and among animal species, and among individuals. It is not known why the effect of sulfur dioxide may vary in an individual from one exposure to the next. Adaptation, measured as a remission of bronchoconstriction, begins in humans after an approximately 10-min exposure to sulfur dioxide, and sooner in cats and dogs. In guinea pigs flow resistance may remain elevated or may increase for several hours. The response in people and cats via the vagal nerve is reflex in origin. It can be abolished by atropine. The possibility that more remote mechanisms may operate either through nasobronchial reflexes or the excretion of sulfur dioxide into the lung from the bloodstream has been suggested. Sulfur dioxide has been administered to guinea pigs in concentrations up to 13.0 mg/m^3 and in concentrations up to 3.4 mg/m^3 to monkeys for 1 to 1.5 yrs without apparent structural or functional consequences.

Chemically inert and chemically active aerosols can change lung mechanics. Aerosols tend to deposit at or near bifurcations, where epithelial nerve endings are concentrated. Nerve endings in central airways are considered especially responsive to mechanical stimuli;

peripheral receptors appear to be more responsive to chemical stimuli. The functional response varies with species, age, level of consciousness, type of anesthesia, activity, nutrition, and state of health. The mass concentration, aerodynamic size, molecular composition, pH, and solubility of the aerosol are additional important factors.

Evidence suggests that many potentially toxic chemicals that are present in airborne particles, such as lead, cadmium, antimony, arsenic, nickel, zinc, and benzo[*a*]pyrene, as well as sulfate and nitrate ions, reside chiefly in the accumulation mode. Studies in guinea pigs have shown that sulfuric acid and other molecular forms of sulfate may increase air flow resistance, but the extent to which this response is due to reflex bronchoconstriction, release of smooth muscle constrictors, excessive secretion of mucus, or submucosal edema is not known. Pulmonary compliance may also be reduced. Such mechanical defects increase the work of breathing and, insofar as they may be unevenly distributed throughout the lung, affect the distribution of inspired air. Functional and structural defects are reportedly greater with <1 μm diam particles than with larger particles, presumably reflecting differences in the sites of deposition based on size and, possibly, differences in the surface area-to-volume ratio of the particles.

Sulfuric acid is more effective than zinc sulfate in causing an increase in flow resistance at equivalent particle sizes and concentration. Over limited ranges, the size of the droplet appears to be more critical than mass concentration in determining the response. Following inhalation, hygroscopic aerosols may undergo changes in aerodynamic and chemical properties that have implications for the biologic response. In one study of healthy subjects, functional changes resulting from 0.5 to 0.7 mg/m^3 of sulfuric acid have been reported; most studies have resorted to higher concentrations to elicit responses.

Animals exposed 18 months to fly ash and sulfuric acid have shown little structural or functional changes to concentrations of 0.16 and 0.46 mg/m^3. In monkeys, higher concentrations of sulfuric acid have caused maldistribution of ventilation, increased breathing rates, and histologic changes, such as epithelial hyperplasia and thickening of bronchial walls. The carbon monoxide diffusing capacity has been reduced by administration of 0.9 mg/m^3 of sulfuric acid ($90\% > 0.5$ μm) for 620 days.

Synergism between gas-aerosol mixtures has been demonstrated in guinea pigs. Several mechanisms are possible: adsorption of the gas onto the surface of the particle (dry), absorption of the gas into the particle (droplet), or chemical reaction of the gas within the particle with the formation of a more toxic compound. It is more likely that the interaction between the gas and particle occurs in ambient air before inhalation rather than following inhalation. An important exception is the possible reaction between inhaled ammonia and inhaled acid sulfate aerosols, which may mitigate the effects of the inhaled acid sulfate aerosols by themselves. Relative humidity may be an important determinant of the reaction. Synergism is reported to occur more rapidly in the presence of certain catalytic salts. Reports in Japan of synergism between sulfur dioxide or nitrogen dioxide and sodium chloride aerosols in humans have not been confirmed in this country.

There is a growing conviction that sulfur dioxide alone is not especially hazardous to health. Aerosols resulting from the oxidation of sulfur dioxide in the atmosphere may be more hazardous. If this supposition is borne out experimentally, it may become necessary to control the emission of sulfur dioxide to keep these oxidative products at acceptable levels. There is virtually no information about the experimental effects of nitrate aerosols.

Epidemiology

Severe episodes of pollution first drew attention to the serious effects of pollution on health. The lethal effect of smog was particularly evident in Europe where fogs occurred mainly in cold weather. Deaths due to pollution were therefore fewer in New York City than in London. Differentiation of the effect of cold from that of pollution proved difficult. Persons with chronic respiratory and cardiovascular diseases appeared to be predominantly affected, though the very young and the aged may also have been vulnerable. Support for an increased susceptibility to fog among those with chronic disease came from a 10-year follow-up of the population of Donora, Pennsylvania.

The interest stimulated by these exceptional episodes of pollution initiated a great deal of research on the effects of air pollution on health. During the 1950s and early 1960s a variety of studies were designed mainly to determine whether or not the more usual pollutant concentrations in most industrial cities adversely affected human health. Temporal and spatial variations in mortality, morbidity, respiratory disease symptom prevalence, lung function levels, and sickness were related to various measures of air pollution in different populations and in different segments of the population. Studies were conducted in more or less representative communities, occupational groups, infants, young children, and patients with respiratory and other diseases.

During the late 1960s, stimulated by the determination to formulate air quality standards and to prepare preliminary air quality criteria documents, investigators focused on quantifying pollution in relation to its effects on health. Exposure-effects relationships or dose-response curves relating various indicators of pollution to different indicators of health impairment were urgently sought. Much of the evidence on which the Federal Air Quality Standards were based came from studies that had been conducted in Europe, mainly Britain, whose relevance to the

American scene was debatable. At this time, information on exposure-effects relationships in the United States was urgently needed. A major effort to provide this evidence came from the Environmental Protection Agency's Community Health and Environmental Surveillance System (CHESS) studies. Perhaps inevitably, in an attempt to provide a great deal of information in a short time, these studies, though well designed, had a number of deficiencies in their execution that have made their interpretation uncertain. Replication is essential if the margin of uncertainty is to be reduced. During the past 5 to 10 years there has also been an increasing awareness that more appropriate measurements of environmental pollutants must be made if firmer conclusions on permissible concentrations of pollutants are to be reached.

Tentatively, however, the exposure concentrations likely to cause mortality or morbidity or reduce lung function are summarized in Table 8-3, while conclusions that have been reached from the CHESS studies, expressed in terms of least case, worst case, and best judgment estimates, are indicated in Tables 8-4 and 8-5.

Effects on Vegetation

There has been little research on the effects of settleable particulate matter on vegetation. Most experiments have dealt with given dusts near specific stationary sources rather than the conglomerate mixture normally encountered in the atmosphere.

The significance of dusts as phytotoxicants is not yet entirely clear, but there is considerable evidence that certain fractions of cement-kiln dusts adversely affect plants when such dusts are naturally deposited on moist leaf surfaces. Dry cement-kiln dusts appear to have little deleterious effect, but when deposited on plants in the field over a long period in the presence of dew, the dust solidifies into a hard adherent crust that can damage leaf tissue and inhibit growth. Dust deposited in excess of 1.0 $g/m^2/day$ solidifies into a hard crust that results in damage to leaf tissue and inhibits growth. The calcium hydroxide in alkaline solutions accompanying crust formation was found within parenchyma tissues of affected leaves. At levels in excess of 1.0 $g/m^2/day$, encrustations, premature needle drop, and shortening of each succeeding year's flush of growth have been observed on branches of fir trees. A marked reduction in the growth of poplar trees 2.2 km from a cement plant was observed after cement production was more than doubled.

Although crusts were not formed, moderate damage has been observed on the leaves of bean plants dusted in the laboratory at the rate of about 4.7 $g/m^2/day$ for 2 days, then followed by exposure to naturally occurring dew. In similar treatments the calcium oxide and calcium hydroxide in the dust-dew solution caused breakdown of the cuticle and release of the fatty acids that are among the cuticular constituents. Thus, while the mechanism by which injury occurs is not entirely understood, screening of light, partial clogging of stomata, and direct injury to the plant tissue by the chemical reaction of the dust on the leaf surface have been demonstrated.

The harmful effect of cement-kiln dusts on vegetation is not fully substantiated and has been questioned by some investigators. This is not surprising in view of the limited research to date and because not all studies have been conducted under identical conditions or with dusts of the same composition. The problem is further complicated by the effects of cement-kiln dust deposits on the soil. Some investigators report no harmful effect at deposit rates from 1.5 to 7.5 $g/m^2/day$; others report that deposits from 1.0 to 48.0 $g/m^2/day$ cause shifts in the soil alkalinity that may favor one crop but harm another.

The great disparity between results of experiments and the conclusions drawn by many investigators exists because the pollutant called "cement-kiln dust" is actually a heterogeneous substance whose constituents and amounts vary with time and location. No general conclusions can be drawn about the effects of cement-kiln dust until each dust source is classified and studied separately.

Fluorides in particulate form are less damaging to vegetation than gaseous fluorides. Although fluoride may be absorbed from depositions of soluble fluoride on leaf surfaces, the amount absorbed is small compared to that entering the plant in gaseous form. The fluoride from particulates apparently has great difficulty penetrating the leaf tissue.

The research evidence suggests that there are few, if any, effects on vegetation attributable to fluoride particulate concentrations below 2 $\mu g/m^3$. Concentrations of this magnitude can be found near sources of fluoride particulates, but rarely in urban atmospheres. Fluorides absorbed by or deposited on plants may be detrimental to animal health. Fluorosis in animals has resulted from ingestion of vegetation covered with particulates containing fluorides.

Although lead concentrations and accumulations by plants in the vicinity of highways have been high, there are no known reports of injury to vegetation.

Soot may clog leaf stomata and may also produce necrotic spotting if it carries with it a soluble toxicant, such as one with excess acidity. Particles from a phurnacite factory aggregating around leaf stomata resulted in a lower maximal diffusion resistance than that measured on clean leaves. Deposits of magnesium oxide on soils reduce plant growth, while deposits of iron oxide (Fe_2O_3) have no harmful effects and may be beneficial.

Emissions of sulfur and nitrogen oxides into the atmosphere may be oxidized and hydrolized to form sulfuric and nitric acid at varying rates, depending on atmospheric conditions. Such acids may ultimately settle to the ground as mist or precipitation. The impact of acid aerosols and acid rain on vegetation include direct effects in the form of leaf necrosis, irregular development of leaf tissue, and leaching of nutrients from leaves and the indirect effects of excessive soil acidification creating an unfavorable environment for plant growth.

Effects on Materials

Studies of effects of particulate matter on materials have been broadened to include other pollutants because under natural conditions exposures consist of a combination of constituents. Airborne particles can act as a nuclei of absorbed or adsorbed gases like sulfur dioxide, hydrogen sulfide, and nitrogen dioxide. Additionally, these gases can form aerosols in the atmosphere.

The normal corrosion of metals can be accelerated by atmospheric pollutants. A notable example is the stress corrosion cracking and ultimate failure of nickel-brass wire spring relays and other electrical equipment in the Los Angeles area caused by nitrate accumulation from the atmosphere. Chapter 10 summarizes the corrosion rates for various metals under different exposure conditions.

Masonry and concrete are affected by particulates, sulfuric acid, sulfurous acid, and several other acidic gases. Deleterious effects include simple discoloration or staining, erosion or corrosion, and leaching. Environmental factors that influence the effects of pollutants (positively or negatively) include relative humidity, rain, and wind.

Paints and finishes can be degraded by particulates, hydrogen sulfied, sulfur oxides, ammonia, and ozone. The effects are generally manifested as loss of gloss, scratch resistance, adhesion, and strength; discoloration; increased drying time; and unattractive, dirty appearance. Appreciable and prolonged attack can result in exposure and deterioration of the substrate being protected.

Plastics and elastomers may be affected by pollution. Among the types of deterioration found are discoloration, cracking, and loss of tensile strength.

Textile fibers can be damaged by atmospheric pollutants. Cellulose fibers such as linen, hemp, cotton, and rayon are especially susceptible to damage by acid aerosols. Particulate matter soils fabrics but does not damage them unless the particles are highly abrasive and the fabric is flexed frequently. Deterioration is caused mainly by repeated attempts to clean the fabric.

Studies of the exposure of porcelain enamels on test panels designed for exterior use at several locations from 1 to 7 years indicated that the loss of gloss correlated well with acidity of the particulate matter. In general the color retention was good.

Test results indicated that the presence of sulfuric acid, hydrochloric acid, and ammonium sulfate caused asphalts to exhibit some transformation, exfoliation, or crumbling. The application of a regular asphalt coating to blown asphalt would provide an adequate degree of acid resistance and coating quality.

Attempts to place a monetary value on the damage to buildings and materials by air pollution have produced estimates that range as high as \$9.9 billion/yr. One study attributes \$3 billion of this damage to particulate matter.

RECOMMENDATIONS

Measurement of Aerosols

It is important to understand the composition of airborne particles as they exist in their suspended state and as a function of size to understand their effects and their sources. This composition is difficult to measure. Current and past methods all have inherent limitations and ambiguities that must be considered in interpreting data.

The chemical composition of particles as a function of size should be measured. The fine particle modes are of particular interest due to their relation to respirable particles, their role in determining the optical properties of the atmosphere, their composition, which includes toxic substances, and the fact that they accumulate the products of all secondary aerosol formation in the atmosphere.

When possible, the molecular nature of the particulate matter should be determined, in view of its role in determining effects.

Difficulties will certainly be incurred in future attempts to implement such broad recommendations. The necessary technology is only partly developed, and the cost of monitoring molecular composition as a function of particle size in many locations could be excessive. Until the techniques are better established, it will be difficult to make comprehensive recommendations. As a minimum, it is recommended that the fine particle and coarse particle fractions be sampled and analyzed for composition separately. However, it is possible to suggest a strategy that has proved effective as a basis for understanding the basic physical/chemical nature of particles in such studies as the Aerosol Characterization Experiment (ACHEX) and the Regional Air Pollution Study (RAPS). Specifically, in these studies, routine monitoring efforts were supplemented by intensive, short-term studies by teams of scientists. The data acquired for such combined efforts have been of much more use than monitoring alone.

Health Effects

Characterization of the health impact of ambient aerosols should be based on sampling techniques that permit the aerodynamic classification of the particles.

Aerodynamic particle classifications should either permit characterization of the overall size-mass distribution of the total aerosol and/or its potentially toxic constituents, or its separation into two size fractions based on the sampler acceptance criteria of the American Conference of Governmental Industrial Hygienists (ACGIH).

Future *in vivo* inhalation studies should determine:

The effects of age, sex, respiratory disease, cigarette smoking, and exposure to other airborne contaminants on regional particle deposition, bronchial clearance, and alveolar clearance.

The effects of particle shape and hygroscopicity on regional particle deposition.

Quantitation of the pathways and rates of alveolar clearance of inert insoluble particles, toxic insoluble particles, such as quartz, asbestos, transuranic oxides, etc., and lipid and water-soluble particles.

The effects of cigarette smoke, sulfur oxides, nitrogen oxides, oxidants, and other common pollutants on the mucociliary clearance of inert particles on mucus production and on morphologic changes in all types.

Agencies sponsoring research on the biologic effects of gases and aerosols should insist that the investigators are competent in both the physical and biologic disciplines needed to produce useful as well as fundamental data.

More emphasis should be given to studies of realistic sizes and concentrations of aerosols shown by atmospheric chemists to be present in urban atmospheres. It is important to assess the effect of chronic as well as acute exposure to these aerosols.

Investigators should be encouraged to study pollutants in combination with other forms of stress to which our population is exposed, including variations in temperature, relative humidity, nutritional state, exercise, and coexistent pulmonary and circulatory disease.

Epidemiology

More precision in measurement of airborne particles and in the assessment of dosage is needed in epidemiologic studies. The time is long past when estimates of particulate pollution can be based on total suspended matter as measured by a high-volume sampler. Particles should be characterized physically and chemically into at least the fine and coarse fractions. Particular attention should be given to their surface absorptive characteristics, hygroscopicity, and size distribution. Not only should particles in the respiratory size range be differentiated from the total suspended particulate matter, but if the significance of submicrometer particles to health is to be adequately assessed, separation of the aerosol at 1 μm radius is desirable. Although the emphasis should probably be on particles within the respiratory size range, larger particles should not be ignored. Pertinent chemical characteristics include pH, oxidation state, and reactive attributes.

Although particular interest is now focused on sulfuric acid and acid sulfates, improved methods for measuring these substances are badly needed. Experiments with animals suggest that there are considerable differences in the irritancy of different acid sulfates. At present, such differentiation in epidemiologic studies is nonexistent. In addition to the sulfates, some attention should also be paid to particulate nitrates.

The problems involved in assessing dosage, even over short periods, from current knowledge of pollution concentrations at different places are considerable. Accurate assessments over a number of years, or life, are especially difficult. Various assumptions can be adopted in deriving estimates of daily exposure based on pollution concentrations both at home and in the work place and the proportion of the 24 hr spent in each. These estimates should be validated by concurrent measures using personal monitors. Biologic estimation of exposure, such as measurement of chemicals in hair samples, although useful for certain particles, appears to offer little for the particles discussed in this document.

More information is needed on the health effects of exposure to low concentrations of pollution and on the changes in these effects that may follow changes in exposure. There has been a surprising lack of interest in any quantification of the benefits of reduction in pollution as shown by reduced mortality or morbidity. Studies of this kind are, of course, liable to be confounded by other changes that may have occurred over the same period. But changes in pollution have varied from place to place and reduction in particulate pollution has sometimes affected some urban areas earlier than others. Consequently, it should be possible to make appropriate allowance for these other changes. In no city has reduction in particulate pollution been more dramatic than in London. The reduction there preceded similar reductions in most British cities. Yet, comparative analyses of mortality or morbidity have seldom been made. Since national morbidity records are available in Britain, the effect of pollution reduction on illness as well as on death could be studied there. The differential effects of pollution from particulates rather than sulfur dioxide can be evaluated in Britain more accurately than elsewhere.

Day-to-day correlations of pollution concentrations with mortality should continue, provided better estimates of pollution over the whole metropolitan area under study can be ensured. More attention also needs to be given to age, sex, and cause of excess deaths in the populations under study. Studies of the kind being conducted in New York City would be useful in other U.S. cities and elsewhere, especially in London, Tokyo, and Osaka. In London, where day-to-day correlations were first shown to be useful, updating of the earlier studies would be particularly interesting.

Mortality, although an extremely valuable index of the effect of pollution on health, should not be overemphasized. It is after all a terminal event in a long disease process, much of which has usually been determined by other factors. Pollution needs to be related to the initiation and exacerbation of illness and to any functional changes, which can be readily measured. Morbidity records, as has been indicated, are sometimes available and suitable for this purpose. But more often the degree of precision now needed will only be possible by special surveys. These surveys should include not only the general community but also groups that are especially vulnerable to the effects of pollution, such as the elderly and patients with chronic respiratory disease. In the general community, representative samples should be selected on the basis of concentrations and types of pollutant exposures and on the expectation of change. Longitudinal observations should be made on respiratory symptoms, concentrations, and changes in lung function and mortality in relation to concentrations of and changes in pollution. Of particular value are studies of children. In contrast to adults, the effects of pollution in children can be more readily separated from the effects of occupational exposures and smoking.

Surveillance of patients with cardiorespiratory disease is the most practicable way of relating short-term concentrations of pollution to deterioration of health. The measurement of lung function is a most valuable, objective addition to such surveillance. It is possible to obtain repeated daily lung function tests in small groups of patients and occupational groups. Daily measurements of this kind, combined with home measurements of pollution and, possibly, personal monitoring, might provide much greater precision at the lower end of the dose-response curves. In these special surveys, obtaining information on the interaction of particles with other pollutants, as well as with meteorologic, climatic, or biologic factors, should be considered. Conversely, situations should be sought in which there may be dissociation between pollutants. In these cases it might be possible to assess the effects of relatively pure pollutants.

Effects on Vegetation

The direct effects of particulates on vegetation affect a much smaller segment of the total population than do effects on climate and visibility, on humans and animals, and on materials, The demand for research into effects on vegetation has not been great and may not increase much except in very localized situations with specific settleable dusts. With the present competition for research dollars, plant biologists may find it difficult to mount any large-scale projects just for the sake of basic research. Research might well be predicated upon local industry/agriculture-related problems that demand an answer. Of course, there may be exceptions. Two that come to mind are: (1) situations in which particles or some of their chemical constituents are known to be harmful to people or animals and would get into food chains via plant foodstuff or forage; and (2) the relatively recent and expanding interest in the possible effects of acid rain in those regions where sulfur compounds are important air pollutants.

It is difficult to make recommendations for research in this area that have as broad an application as those concerning the effects of particulates on humans. With these reservations in mind, recommendations are made as follows.

Now that scanning electron microscopy is generally available, surveys of a number of plant species could be made in areas where settleable particles are a problem in order to determine the prevalence of the clogging of functional stomata. These data should be related to particle deposit rates.

Similarly, the effect of the particles on the cuticle of upper leaf surfaces could be examined by the same scanning electron micrograph (SEM) techniques. Leaf examination should be timed to encompass pertinent weather factors, such as alternate wetting and drying of leaf surfaces. Whether or not particulate deposits tend to form crusts should also be noted.

If particles are to be applied to leaves in laboratory experiments, a special effort should be made to collect particle samples from the atmosphere near the source in an attempt to preserve particle-size integrity and chemical composition. Special equipment would be needed to collect sufficiently large samples in a relatively short time. This approach would be particularly important with chemically complex materials like cement-kiln dusts but less so with more inert materials like iron oxides.

In long-term studies, experiments could be conducted to elucidate further the real importance of particles in preventing essential light energy from reaching leaf surfaces. Such experiments may best be conducted in the laboratory where deposit rates can be monitored and adequate control plants maintained. Field experiments might be developed if appropriate particle-free leaves could be maintained without interfering with normal light.

Although particles containing fluorine and lead appear to have little direct effect on plants, we should remain alert to the possibility of some compounds of these elements getting into the food chain via vegetation.

Because of the new and expanding interest in acid rains, and because of the regional nature of their effects, continued research on the effects on vegetation should be encouraged. Where prepared acid mists are being applied directly to plant leaves, great care should be taken to ensure that the mist is chemically comparable, and perhaps even identical, to the acid rain that is being studied.

Most concern for the effects of particulates on plants is related to the direct effect of the particles settling on plant leaves. Such particles also fall on soils in which the plants are growing. This possible indirect effect, whether harmful or beneficial, should not be ignored. The experiments require careful planning and might require several years to produce results of statistical significance.

References

1. Abe, H., Y. Ishii, and H. Kato. Evaluation of atmospheric factors by analyses of corrosion products and surface deposits on copper plates. Rail. Tech. Res. Inst. 12(3):170–174, 1971.
2. Acheson, E. D., R. H. Cowdell, and E. Rang. Adenocarcinoma of the nasal cavity and sinuses in England and Wales. Brit. J. Ind. Med. 29:21–30, 1972.
3. Adams, A. A Study of Exposed Reinforced Concrete Facings. Division of Building Research Technical Translation 933. (Translated by D. A. Sinclair from Ann. de l'Inst. Techniq. du Batiment et des Travaux Publics (159–160):377–388, 1961) Ottawa: National Research Council of Canada, 1961.
4. Ailor, W. H., Jr. Aluminum corrosion at urban and industrial locations. J. Structural Div. Proc. Amer. Soc. Civil Eng. 95:2141–2160, 1969.
5. Aitken, J. Collected Scientific Papers of John Aitken, L.L.D., F.R.S. Edited by C. G. Knott. London: Cambridge University Press, 1923. 591 pp.
6. Alarie, Y., W. M. Busey, A. A. Krumm, and C. E. Ulrich. Long-term continuous exposure to sulfuric acid mist in cynomolgus monkeys and guinea pigs. Arch. Environ. Health 27:16–24, 1973.
7. Alarie, Y., C. E. Ulrich, W. M. Busey, A. A. Krumm, and H. N. MacFarland. Long-term continuous exposure to sulfur dioxide in cynomolgus monkeys. Arch. Environ. Health 24:115–128, 1972.
8. Alarie, Y., C. E. Ulrich, W. M. Busey, H. E. Swann, Jr., and H. N. MacFarland. Long-term continuous exposure of guinea pigs to sulfur dioxide. Arch. Environ. Health 21:769–777, 1970.
9. Alarie, Y., I. Wakisaka, and S. Oka. Sensory irritation by sulfite aerosols. Environ. Physiol. Biochem. 3:182–184, 1973.
10. Albert, R. E., D. Alessandro, M. Lippmann, and J. Berger. Long-term smoking in the donkey: Effect on tracheobronchial particle clearance. Arch. Environ. Health 22:12–19, 1971.
11. Albert, R. E., and L. C. Arnett. Clearance of radioactive dust from the lung. Arch. Ind. Health 12:99–106, 1955.
12. Albert, R. E., J. Berger, K. Sanborn, and M. Lippmann. Effects of cigarette smoke components on bronchial clearance in the donkey. Arch. Environ. Health 29:96–101, 1974.
13. Albert, R. E., M. Lippmann, and W. Briscoe. The characteristics of bronchial clearance in humans and the effects of cigarette smoking. Arch. Environ. Health 18:738–755, 1969.
14. Albert, R. E., M. Lippmann, H. T. Peterson, Jr., J. Berger, K. Sanborn, and D. Bohning. Bronchial deposition and clearance of aerosols. Arch. Intern. Med. 131:115–127, 1973.
15. Albert, R. E., H. T. Peterson, Jr., D. E. Bohning, and M. Lippmann. Short-term effects of cigarette smoking on bronchial clearance in humans. Arch. Environ. Health 30:361–367, 1975.

16. Allison, D. J., T. P. Clay, J. M. B. Hughes, H. A. Jones, and A. Shevis. Effects of nasal stimulation on total respiratory resistance in the rabbit. J. Physiol. 239:23P–24P, 1974. (Abstr.)
17. Altshuler, B., E. D. Palmes, and N. Nelson. Regional aerosol deposition in the human respiratory tract, pp. 323–335. In C. N. Davies, Ed. Inhaled Particles and Vapours II. Proceedings of an International Symposium organized by the British Occupational Hygiene Society, Cambridge, 28 September–1 October 1965. Oxford: Pergamon Press, 1967.
18. Altshuler, B., L. Yarmus, E. D. Palmes, and N. Nelson. Aerosol deposition in the human respiratory tract. I. Experimental procedures and total deposition. A.M.A. Arch. Ind. Health 15:293–303, 1957.
19. Altshuller, A. P. Atmospheric sulfur dioxide and sulfate. Distribution of concentration at urban and nonurban sites in United States. Environ. Sci. Technol. 7:709–712, 1973.
20. Altshuller, A. P. Characteristics of the chemical composition of the fine particulate fraction in the atmosphere. May 1976. (Unpublished paper)
21. Altshuller, A. P., and J. J. Bufalini. Photochemical aspects of air pollution: A review. Environ. Sci. Technol. 5:39–64, 1971.
22. Altshuller, A. P., D. K. Klosterman, P. W. Leach, I. J. Hindawi, and J. E. Sigsby, Jr. Products and biological effects from irradiation of nitrogen oxides with hydrocarbons or aldehydes under dynamic conditions. Air Water Pollut. Int. J. 10:81–98, 1966.
23. Amdur, M. O. The physiological response of guinea pigs to atmospheric pollutants. Int. J. Air Pollut. 1:170–183, 1959.
24. Amdur, M. O. The effect of aerosols on the response to irritant gases, pp. 281–292. In C. N. Davies, Ed. Inhaled Particles and Vapours. Proceedings of an International Symposium organized by the British Occupational Hygiene Society, Oxford, 29 March–1 April 1960. Oxford: Pergamon Press, 1961.
25. Amdur, M. O. Toxicologic appraisal of particulate matter, oxides of sulfur, and sulfuric acid. J. Air Pollut. Control Assoc. 19:638–644, 1969.
26. Amdur, M. O. Animal studies, pp. 175–205. In Assembly of Life Sciences-National Academy of Sciences-National Research Council. Proceedings of the Conference on Health Effects of Air Pollutants, Washington, D.C., October 3–5, 1973. Washington, D.C.: U.S. Government Printing Office, 1973.
27. Amdur, M. O. The long road from Donora. 1974 Cummings Memorial Lecture. Amer. Ind. Hyg. Assoc. J. 35:589–597, 1974.
28. Amdur, M. O., and M. Corn. The irritant potency of zinc ammonium sulfate of different particle sizes. Amer. Ind. Hyg. Assoc. J. 24:326–333, 1963.
29. Amdur, M. O., and J. Mead. A method for studying the mechanical properties of the lungs of unanesthetized animals: Application to the study of respiratory irritants, pp. 150–159. In Proceedings of the Third National Air Pollution Symposium, Pasadena, California, April 18–20, 1955.
30. Amdur, M. O., W. W. Melvin, and P. Drinker. Effects of inhalation of sulphur dioxide by man. Lancet 2:758–759, 1953.
31. Amdur, M. O., L. Silverman, and P. Drinker. Inhalation of sulfuric acid mist by human subjects. A.M.A. Arch. Ind. Hyg. Occup. Med. 6:305–313, 1952.
32. Amdur, M. O., and D. Underhill. The effect of various aerosols on the responses of guinea pigs to sulfur dioxide. Arch. Environ. Health 16:460–468, 1968.

33. Andersen, I. Relationships between outdoor and indoor air pollution. Atmos. Environ. 6:275–278, 1972. (Technical note)
34. Andersen, Ib. Mucociliary Function in Trachea Exposed to Ionized and Non-ionized Air. Arhus, Denmark: Akademisk Boghandel, 1971. 178 pp.
35. Andersen, Ib., G. R. Lundqvist, P. L. Jensen, and D. F. Proctor. Human response to controlled levels of sulfur dioxide. Arch. Environ. Health 28:31–39, 1974.
36. Anderson, D. O. The effects of air contamination on health: A review. Part II. Canad. Med. Assoc. J. 97:585–593, 1967.
37. Anderson, D. O., B. G. Ferris, Jr., and R. Zickmantel. The Chilliwack respiratory survey, 1963: Part III. The prevalence of respiratory disease in a rural Canadian town. Can. Med. Assoc. J. 92:1007–1016, 1965.
38. Anderson, P. J. The effect of dust from cement mills on the setting of fruit. Plant World 17:57–68, 1914.
39. Angel, J. H., C. M. Fletcher, I. D. Hill, and C. M. Tinker. Respiratory illness in factory and office workers. A study of minor respiratory illnesses in relation to changes in ventilatory capacity, sputum characteristics, and atmospheric pollution. Brit. J. Dis. Chest 59:66–80, 1965.
40. Antweiler, H. Über die Funktion des Flimmerepithels der Luftwege, insbesondere unter Staubbelastung, pp. 509–535. In K. Thomas, Ed. Grundfragen aus der Silikoseforschung, II. Bericht über die von der Bergbau-Berufsgenossenschaft am 30. November und 1. Dezember 1956 in Göttingen veranstaltete Tagung der Arbeitsgemeinschaft für Silikose-Grundfragenforschung. Bochum: Hauptverwaltung der Bergbau-Berufsgenossenschaft, 1956.
41. Asmundsson, T., and K. H. Kilburn. Mucociliary clearance rates at various levels in dog lungs. Amer. Rev. Resp. Dis. 102:388–397, 1970.
42. Asmundsson, T., and K. H. Kilburn. Mechanisms of respiratory tract clearance, pp. 107–180. In M. J. Dulfano, Ed. Sputum. Fundamentals and Clinical Pathology. Springfield: Charles C Thomas, Publisher, 1973.
43. Atwater, M. A. Planetary albedo changes due to aerosols. Science 170:64–66, 1970.
44. Azarniouch, M. K., A. J. Bobkowicz, N. E. Cooke, and E. J. Farkas. Growth of a sulfuric acid droplet exposed to water vapor. Can. J. Chem. Eng. 51:590–595, 1973.
45. Bair, W. J., and J. V. Dilley. Pulmonary clearance of $^{59}Fe_2O_3$ and $^{51}CR_2O_3$ in rats and dogs exposed to cigarette smoke, pp. 251–268. In C. N. Davies, Ed. Inhaled Particles and Vapours II. Proceedings of an International Symposium organized by the British Occupational Hygiene Society, Cambridge, 28 September–1 October 1965. Oxford: Pergamon Press, 1967.
46. Baker, A. P., and J. R. Munro. Multiglycosyltransferase system of canine respiratory tissue. J. Biol. Chem. 246:4358–4362, 1971.
47. Ballenger, J. J. A study of ciliary activity in the respiratory tract of animals. Ann. Otol. Rhinol. Laryngol. 58:351–369, 1949.
48. Barclay, A. E., and K. J. Franklin. The rate of excretion of Indian ink injected into the lungs. J. Physiol. 90:482–484, 1937.
49. Barnett, B., and C. E. Miller. Flow induced by biological mucociliary systems. Ann. N. Y. Acad. Sci. 130:891–901, 1966.
50. Barrett, L. B., and T. E. Waddell. Cost of Air Pollution Damage: A Status Report. Publ. AP-85. Research Triangle Park, N.C.: National Environmental Research Center, Environmental Protection Agency, 1973. 73 pp.
51. Barrie, L. A., and H. W. Georgii. An experimental investigation of the absorption of sulphur dioxide by water drops containing heavy metal ions. Atmos. Environ. 10:743–749, 1976.

52. Bartlett, B. R. The action of certain "inert" dust materials on parasitic Hymenoptera. J. Econ. Entomol. 44:891–896, 1951.
53. Basch, F. P., P. Holinger, and H. G. Poncher. Physical and chemical properties of sputum. II. Influence of drugs, steam, carbon dioxide and oxygen. Amer. J. Dis. Child. 62:1149–1171, 1941.
54. Bates, D. V., G. M. Bell, C. D. Burnham, M. Hazucha, J. Mantha, L. D. Pengelly, and F. Silverman. Short-term effects of ozone on the lung. J. Appl. Physiol. 32:176–181, 1972.
55. Bates, D. V., C. A. Gordon, G. I. Paul, R. E. G. Place, D. P. Snidal, C. R. Woolf, M. Katz, R. G. Fraser, and B. B. Hale. Chronic bronchitis: Report on the third and fourth stages of the co-ordinated study of chronic bronchitis in the Department of Veterans Affairs, Canada. Med. Serv. J. Can. 22:5–61, 1966.
56. Bates, D. V., and M. Hazucha. The short-term effects of ozone on the human lung, pp. 507–540. In Assembly of Life Sciences-National Academy of Sciences-National Research Council. Proceedings of the Conference on Health Effects of Air Pollutants, Washington, D.C., October 3–5, 1973. Washington, D.C.: U.S. Government Printing Office, 1973.
57. Bates, D. V., C. R. Woolf, and G. I. Paul. Chronic bronchitis. A report on the first two stages of the co-ordinated study of chronic bronchitis in the Department of Veterans Affairs, Canada. Med. Serv. J. Can. 18:211–305, 1962.
58. Becker, W. H., F. J. Schilling, and M. P. Verma. The effect on health of the 1966 eastern seaboard air pollution episode. Arch. Environ. Health 16:414–419, 1968.
59. Bedrossian, C. W. M., S. D. Greenberg, and B. S. Duran. Bronchial gland measurements: A continuing search for a "yardstick." Exper. Molec. Pathol. 18:219–224, 1973.
60. Beeckmans, J. M. The deposition of aerosols in the respiratory tract. I. Mathematical analysis and comparison with experimental data. Can. J. Physiol. Pharmacol. 43:157–172, 1965.
61. Bell, K. A., and S. K. Friedlander. Aerosol deposition in models of a human lung bifurcation. Staub-Reinhalt. Luft 33(4):178–182, 1973.
62. Bell, K. A., W. S. Linn, M. Hazucha, J. D. Hackney, and D. V. Bates. Respiratory effects of exposure to ozone plus sulfur dioxide in Southern Californians and Eastern Canadians. Amer. Ind. Hyg. Assoc. J. 38:696–706, 1977.
63. Beloin, N. J. A chamber study. Fading of dyed fabrics exposed to air pollutants. Text. Chem. Color. 5:128/29–133/34, 1973.
64. Bensch, K. G., and E. A. M. Dominquez. Studies on the pulmonary air-tissue barrier. Part IV. Cytochemical tracing of macromolecules during absorption. Yale J. Biol. Med. 43:236–241, 1971.
65. Bensch, K. G., G. B. Gordon, and L. R. Miller. Studies on the bronchial counterpart of the Kultschitzky (argentaffin) cell and innervation of bronchial glands. J. Ultrastruct. Res. 12:668–686, 1965.
66. Berfenstam, R., T. Edlund, and L. Zettergren. Hyaline membrane disease. The influence of high oxygen concentration on ciliary activity in the respiratory tract. An experimental study on rabbits. Acta Paediatr. 47:527–533, 1958.
67. Berge, H. Luftverunreinigungen im Raume Köln. Allgem. Forstz. 20(51/52):834–838, 1965.

68. Berge, H. Emissionsbedingte Eisenstäube und ihre Auswirkungen auf Wachstum und Ertrag landwirtschaftlicher Kulturen. Luftverunreinigung 2:1–7, 1966.
69. Berger, H. W., M. Tryon, E. J. Clark, and L. F. Skoda. Final Project Report on the Effects of Atmospheric Contaminants on the Durability of Building Materials. NBS Report 10 504. For the National Air Pollution Control Administration. Washington, D.C.: National Bureau of Standards, 1970. 36 pp.
70. Bernstein, D., M. T. Kleinman, T. J. Kneip, T. L. Chan, and M. Lippmann. A high-volume sampler for the determination of particle size distributions in ambient air. J. Air Pollut. Control Assoc. 26:1069–1072, 1976.
71. Bielke, S., and H. W. Georgii. Investigation on the incorporation of sulfurdioxide into fog- and rain-droplets. Tellus 20: 435–442, 1968.
72. Bigg, E. K., S. C. Mossop, R. T. Meade, and N. S. C. Thorndike. The measurement of ice nucleus concentrations by means of Millipore filters. J. Appl. Meterorol. 2:266–269, 1963.
73. Blake, J. On the movement of mucus in the lung. J. Biomechanics 8:179–190, 1975.
74. Bodhaine, B. A., and B. G. Mendonca. Preliminary four wavelength nephelometer measurements at Mauna Loa Observatory. Geophys. Res. Letters 1:119–122, 1974.
75. Böhlau, V., Ed. Aerosole in Physik, Medizin und Technik. Proceedings of a conference held in Bad Soden, October 17–18, 1973. Bad Soden, West Germany: Gesellschaft fur Aerosolforschung, 1973. 165 pp.
76. Bohne, H. Schädlichkeit von Staub aus Zementwerken für Waldbestände. Allgem. Forstz. 18(7):108–111, 1963.
77. Bohning, D. E., R. E. Albert, M. Lippmann, and W. M. Foster. Tracheobronchial particle deposition and clearance. A study of the effects of cigarette smoking in monozygotic twins. Arch. Environ. Health 30:457–462, 1975.
78. Bolin, B., L. Granat, L. Ingelstam, M. Johannesson, E. Mattsson, S. Odén, H. Rodhe, and C. O. Tamm. Air Pollution Across National Boundaries. The Impact on the Environment of Sulfur in Air and Precipitation. Royal Ministry for Foreign Affairs. Royal Ministry of Agriculture. Stockholm: P. A. Norstedt & Söner, 1971. 96 pp.
79. Bolin, B., G. Witt, and R. J. Charlson. Stockholm tropospheric aerosol seminar: Measurement of regional to global scale pollution by airborne particles. Bull. Amer. Meteorol. Soc. 55:228–231, 1974.
80. Boyd, E. M., J. W. Clark, and W. F. Perry. Estrogen and their effect on ciliated mucosa. Arch. Otolaryng. 33:909–915, 1941.
81. Boyd, J. T. Climate, air pollution and mortality. Brit. J. Prev. Soc. Med. 14:123–135, 1960.
82. Bracewell, J. M., and D. Gall. The catalytic oxidation of sulphur dioxide in solution at concentrations occurring in fog droplets, pp. 17–26. In Air Pollution. Proceedings of the Symposium on the Physico-chemical Transformation of Sulphur Compounds in the Atmosphere and the Formation of Acid Smogs, 8–9 June, 1967, Mainz, Germany. Paris: Organisation for Economic Co-operation and Development, 1967.
83. Bradley, W. H., W. P. D. Logan, and A. E. Martin. The London fog of December 2nd–5th, 1957. Monthly Bull. Min. Health 17:156–166, 1958.

84. Brain, J. D. Free cells in the lungs. Some aspects of their role, quantitation, and regulation. Arch. Intern. Med. 126:477–487, 1970.
85. Brandt, C. J., and R. W. Rhoades. Effects of limestone dust accumulation on composition of a forest community. Environ. Pollut. 3:217–225, 1972.
86. Brandt, C. J., and R. W. Rhoades. Effects of limestone dust accumulation on lateral growth of forest trees. Environ. Pollut. 4:207–213, 1973.
87. Brantigan, O. C., E. Mueller, and M. B. Kress. A surgical approach to pulmonary emphysema. Amer. Rev. Resp. Dis. 80:194–206, 1959.
88. Bricard, J., G. J. Madelaine, and D. Vigla. The formation and evolution of aerosols generated through the action of ultraviolet and solar radiation, pp. 173–176. In Proceedings of the 7th International Conferences on Condensation and Ice Nuclei, September 18–24, 1969, Prague and Vienna. Prague: Academia, 1969.
89. Brock, J. R. Condensational growth of atmospheric aerosols. J. Coll. I. Sci. 39:32–36, 1972.
90. Brosset, C., K. Andreasson, and M. Ferm. The nature and possible origin of acid particles observed at the Swedish west coast. Atmos. Environ. 9:631–642, 1975.
91. Brown, J. H., K. M. Cook, F. G. Ney, and T. Hatch. Influence of particle size upon the retention of particulate matter in the human lung. Amer. J. Public Health 40:450–458, 480, 1950.
92. Brown, L. S. Evaluation of the Condition of the Lincoln Memorial Stonework. Alexandria, Va.: Value Engineering Laboratory, 1971. 5 pp. (Study conducted for the U.S. Department of the Interior, National Park Service)
93. Brundelet, P. J. Experimental study of the dust-clearance mechanism of the lung. I. Histological study in rats of the intra-pulmonary bronchial route of elimination. Acta Pathol. Microbiol. Scand. Suppl. 175:1–141, 1965.
94. Bryson, R. Climatic modification by air pollution, pp. 134–174. In N. Polunin, Ed. Proceedings of the first International Conference on Environmental Future, held in Finland from 27 June to 3 July 1971. London: The Macmillan Press, 1972.
95. Bryson, R. A. "All other factors being constant . . ." A reconciliation of several theories of climatic change. Weatherwise 21:56–61, 1968.
96. Bryson, R. A. A perspective on climate change. Climate responds rapidly and significantly to small changes of the independent variables. Science 184:753–760, 1974.
97. Bryson, R. A., and D. A. Baerreis. Possibilities of major climatic modification and their implications: Northwest India, a case for study. Bull. Amer. Meteor. Soc. 48:136–142, 1967.
98. Buck, S. F., and D. A. Brown. Mortality from Lung Cancer and Bronchitis in Relation to Smoke and Sulfur Dioxide Concentration, Population Density and Social Index. Tobacco Research Council Research Paper No. 7. London: Waterlow and Sons, Ltd., 1964. 27 pp.
99. Budyko, M. I. The effect of solar radiation variations on the climate of the earth. Tellus 21:611–619, 1969.
100. Buechley, R. W., W. B. Riggan, V. Hasselblad, and J. B. VanBruggen. SO_2 levels and perturbations in mortality. A study in the New York-New Jersey Metropolis. Arch. Environ. Health 27:134–137, 1973.
101. Bufalini, M. Oxidation of sulfur dioxide in polluted atmospheres—A review. Environ. Sci. Technol. 5:685–700, 1971.
102. Burgess, S. G., and C. W. Shaddick. Bronchitis and air pollution. Roy. Soc. Health J. 79:10–24, 1959.

103. Burn, J. H. Functions of Autonomic Transmitters. Baltimore: The Williams and Wilkins Company, 1956. 227 pp.
104. Burn, J. L., and J. Pemberton. Air pollution bronchitis and lung cancer in Salford. Int. J. Air Water Pollut. 7:5–16, 1963.
105. Burrows, B., A. L. Kellogg, and J. Buskey. Relationship of symptoms of chronic bronchitis and emphysema to weather and air pollution. Arch. Environ. Health 16:406–413, 1968.
106. Burton, G. G., M. Corn, J. B. L. Gee, C. Vasallo, and A. P. Thomas. Response of healthy men to inhaled low concentrations of gas-aerosol mixtures. Arch. Environ. Health 18:681–692, 1969.
107. Bushtueva, K. A. The determination of the limit of allowable concentration of sulfuric acid in atmospheric air, pp. 20–36. In V. A. Ryazanov, Ed. Limits of Allowable Concentrations of Atmospheric Pollutants—Book 3. Translated by B. S. Levine. U.S. Department of Commerce, Office of Technical Services, Washington, D.C., 1957.
108. Bushtueva, K. A. The toxicity of H_2SO_4 aerosol. U.S.S.R. Literature Air Pollut. Related Occup. Dis. 1:63–66b, 1960.
109. Bushtueva, K. A. Experimental studies on the effect of low oxides of sulfur concentrations on the animal organism, pp. 92–102. In V. A. Ryazanov, Ed. Limits of Allowable Concentrations of Atmospheric Pollutants—Book 5. Translated by B. S. Levine. U.S. Department of Commerce, Office of Technical Services, Washington, D.C., 1962.
110. Cadle, R. D. Particles in the Atmosphere and Space. New York: Reinhold Publishing Corporation, 1966. 226 pp.
111. Cadle, R. D. The Measurement of Airborne Particles. New York: John Wiley and Sons, Inc., 1975. 342 pp.
112. Cadle, R. D., and P. L. Magill. Chemistry of contaminated atmospheres, pp. 3-1 to 3-27. In P. L. Magill, F. R. Holden, and C. Ackley, Eds. Air Pollution Handbook. New York: McGraw-Hill Book Company, Inc., 1956.
113. Calvert, J. G. Interactions of air pollutants, pp. 19–101. In Assembly of Life Sciences-National Academy of Sciences-National Research Council. Proceedings of the Conference on Health Effects of Air Pollutants, Washington, D.C., October 3–5, 1973. Washington, D.C.: U.S. Government Printing Office, 1973.
114. Camner, P., P. Helström, and K. Philipson. Carbon dust and mucociliary transport. Arch. Environ. Health 26:294–296, 1973.
115. Camner, P., C. Jarstrand, and K. Philipson. Tracheobronchial clearance in patients infected with microplasma pneumoniae, pp. 236–238. In J. F. Ph. Hers and K. C. Winkler, Eds. Airbone Transmission and Airborne Infection: Concepts and Methods. Presented at the VIth International Symposium on Aerobiology. Held at the Technical University at Enschede, The Netherlands, 1972. New York: John Wiley and Sons, 1973.
116. Camner, P., C. Jarstrand, and K. Philipson. Tracheobronchial clearance in patients with influenza. Amer. Rev. Resp. Dis. 108:131–135, 1973.
117. Camner, P., B. Mossberg, and K. Philipson. Tracheobronchial clearance and chronic obstructive lung disease. Scand. J. Resp. Dis. 54:272–281, 1973.
118. Camner, P., and K. Philipson. Tracheobronchial clearance in smoking-discordant twins. Arch. Environ. Health 25:60–63, 1972.
119. Camner, P., K. Philipson, and T. Arvidsson. Cigarette smoking in man. Short-term effect on mucociliary transport. Arch. Environ. Health 23:421–426, 1971.

120. Camner, P., K. Philipson, and T. Arvidsson. Withdrawal of cigarette smoking. A study of tracheobronchial clearance. Arch. Environ. Health 16:90–92, 1973.
121. Camner, P., K. Strandberg, and K. Philipson. Increased mucociliary transport by cholinergic stimulation. Arch. Environ. Health 29:220–224, 1974.
122. Camner, P., K. Strandberg, and K. Philipson. Increased mucociliary transport by adrenergic stimulation. Arch. Environ. Health 31:79–82, 1976.
123. Cannon, H. L., and J. M. Bowles. Contamination of vegetation by tetraethyl lead. Science 137:765–766, 1962.
124. Carlson, T. N., and B. M. Prospero. The large-scale movement of Saharan air outbreaks over the northern equatorial Atlantic. J. Appl. Meteor. 11:283–297, 1972.
125. Carnow, B. W., M. H. Lepper, R. B. Shekelle, and J. Stamler. Chicago air pollution study. SO_2 levels and acute illness in patients with chronic bronchopulmonary disease. Arch. Environ. Health 18:768–776, 1969.
126. Carnow, B. W., and P. Meier. Air pollution and pulmonary cancer. Arch. Environ. Health 27:207–218, 1973.
127. Carson, S., R. Goldhamer, and R. Carpenter. Mucus transport in the respiratory tract. Amer. Rev. Resp. Dis. 93 (Suppl):86–92, 1966.
128. Casarett, L. J. The vital sacs: Alveolar clearance mechanisms in inhalation toxicology, pp. 2–36. In W. J. Hayes, Jr., Ed. Essays in Toxicology, Volume 3. New York: Academic Press, 1972.
129. Cassatt, W. A., and R. S. Maddock, Eds. Aerosol Measurements. The Proceedings of a Seminar on Aerosol Measurements, May 7, 1974 sponsored by The National Bureau of Standards and The Food and Drug Administration. National Bureau of Standards Special Publication 412. Washington, D.C.: U.S. Government Printing Office, 1974. 182 pp.
130. Chakrin, L. W., A. P. Baker, S. S. Spicer, J. R. Wardell, Jr., N. DeSanctis, and C. Dries. Synthesis and secretion of macromolecules by canine trachea. Amer. Rev. Resp. Dis. 105:368–381, 1972.
131. Chandler, K. A., and M. B. Kilcullen. Survey of corrosion and atmospheric pollution in and around Sheffield. Brit. Corros. J. 3:80–84, 1968.
132. Changnon, S. A., Jr. The La Porte weather anomaly—fact or fiction? Bull. Amer. Meteorol. Soc. 49:4–11, 1968.
133. Changnon, S. A., Jr., F. A. Huff, and R. G. Semonin. Metromex: An investigation of inadvertent weather modification. Bull. Amer. Meterol. Soc. 52:958–967, 1971.
134. Chapman, R. S., C. M. Shy, J. F. Finklea, D. E. House, H. E. Goldberg, and C. G. Hayes. Chronic respiratory disease. In military inductees and parents of schoolchildren. Arch. Environ. Health 27:138–142, 1973.
135. Charlson, R. J. Atmospheric visibility related to aerosol mass concentration. A review. Environ. Sci. Technol. 3:913–918, 1969.
136. Charlson, R. J., N. C. Ahlquist, and H. Horvath. On the generality of correlation of atmospheric aerosol mass concentration and light scatter. Atmos. Environ. 2:455–464, 1968.
137. Charlson, R. J., N. C. Ahlquist, H. Selvidge, and P. B. MacCready, Jr. Monitoring of atmospheric aerosol parameters with the integrating nephelometer. J. Air Pollut. Control Assoc. 19:937–942, 1969.
138. Charlson, R. J., D. S. Covert, Y. Tokiwa, and P. K. Mueller. Multiwave-

length nephelometer measurements in Los Angeles smog aerosol. III. Comparison to light extinction by NO_2, pp. 333–338. In G. M. Hidy, Ed. Aerosols and Atmospheric Chemistry. Proceedings of the American Chemical Society. Los Angeles, California, March 28–April 2, 1971. New York: Academic Press, 1972.

139. Charlson, R. J., H. Harrison, and G. Witt. Aerosol concentrations: Effect on planetary temperatures. Science 175:95–96, 1972.
140. Charlson, R. J., and M. J. Pilat. Climate: The influence of aerosols. J. Appl. Meteorol. 8:1001–1002, 1969.
141. Charlson, R. J., and M. J. Pilat. Reply [to S. H. Schneider on climate]. J. Appl. Meteorol. 10:841–842, 1971.
142. Charlson, R. J., W. M. Porch, A. P. Waggoner, and N. C. Ahlquist. Background aerosol light scattering characteristics: Nephelometric observations at Mauna Loa Observatory compared with results at other remote locations. Tellus 26:345–360, 1974.
143. Charlson, R. J., A. H. Vanderpol, D. S. Covert, A. P. Waggoner, and N. C. Ahlquist. $H_2SO_4/(NH_4)_2SO_4$ background aerosol: Optical detection in St. Louis region. Atmos. Environ. 8:1257–1267, 1974.
144. Charlson, R. J., A. H. Vanderpol, D. S. Covert, A. P. Waggoner, and N. C. Ahlquist. Sulfuric acid-ammonium sulfate aerosol: Optical detection in the St. Louis region. Science 184:156–158, 1974.
145. Charman, J., and L. Reid. Sputum viscosity in chronic bronchitis, bronchiectasis, asthma and cystic fibrosis. Biorheology 9:185–199, 1972.
146. Cheng, R. T., M. Corn, and J. O. Frohliger. Contribution to the reaction kinetics of water soluble aerosols and SO_2 in air at ppm concentrations. Atmos. Environ. 5:987–1008, 1971.
147. Chinard, F. P. The permeability characteristics of the pulmonary blood-gas barrier, pp. 106–147. In C. G. Caro, Ed. Advances in Respiratory Physiology. London: Edward Arnold Ltd., 1966.
148. Christensen, O. W., and C. H. Wood. Bronchitis mortality rates in England and Wales and Denmark. Brit. Med. J. 1:620–622, 1958.
149. Chýlek, P., and J. A. Coakley, Jr. Aerosols and climate. Science 183:75–77, 1974.
150. Ciocco, A., and D. J. Thompson. A follow-up of Donora ten years after: Methodolgy and findings. Amer. J. Public Health 51:155–164, 1961.
151. Clara, M. Zur Histobiologie des Bronchalepithels. Z. Mikrosk. Anat. Forsch. 41:321–347, 1937.
152. Clark, W. E. Measurements of Aerosols Produced by the Photochemical Oxidation of Sulfur-Dioxide in Air. Ph.D. thesis, University of Minnesota, Minneapolis, Minn., 1972. 152 pp.
153. Clark, W. E., and K. T. Whitby. Concentration and size distribution measurements of atmospheric aerosols and a test of the theory of self-preserving size distributions. J. Atmos. Sci. 24:677–687, 1967.
154. Clark, W. E., and K. T. Whitby. Measurements of aerosols produced by the photochemical oxidation of SO_2 in air. J. Coll. I. Sci. 51:477–490, 1975.
155. Clements, J. A., J. T. Sharp, R. P. Johnson, and J. O. Elam. Estimation of pulmonary resistance of repetitive interruption of airflow. J. Clin. Invest. 38:1262–1270, 1959.
156. Cohen, A. A., S. Bromberg, R. W. Buechley, M. A. Heiderscheit, and C. M. Shy. Asthma and air pollution from a coal-fueled power plant. Amer. J. Public Health 62:1181–1188, 1972.

156a. Cohen, A. A., C. J. Nelson, S. M. Bromberg, M. Pravda, E. F. Ferrand, and G. Leone. Symptom reporting during recent publicized and unpublicized air pollution episodes. Amer. J. Public Health 64:442–449, 1974.
157. College of General Practitioners. Chronic bronchitis in Great Britain. A national survey carried out by the respiratory disease study group of the College of General Practitioners. Brit. Med. J. 2:973–979, 1961.
158. Colley, J. R. T., and W. W. Holland. Social and environmental factors in respiratory disease. Arch. Environ. Health 14:157–161, 1967.
159. Colley, J. R. T., and D. D. Reid. Urban and social origins of childhood bronchitis in England and Wales. Brit. Med. J. 2:213–217, 1970.
160. Comstock, G. W., R. W. Stone, Y. Sakai, T. Matsuya, and J. A. Tonascia. Respiratory findings and urban living. Arch. Environ. Health 27:143–150, 1973.
161. Constantine, H., L. Dautrebande, N. Kaltreider, F. W. Lovejoy, Jr., P. Morrow, and P. Perkins. Influence of carbachol and of fine dust aerosols upon the breathing mechanics and the lung volumes of normal subjects and of patients with chronic respiratory disease before and after administering sympathomimetic aerosols. Arch. Int. Pharmacodyn. Ther. 123:239–252, 1959.
162. Cooke, D. D., and M. Kerker. Response calculations for light-scattering aerosol particle counters. Appl. Opt. 14:734–739, 1975.
163. Cooper, D. W., R. L. Byers, and J. W. Davis. Measurements of laser light backscattering vs. humidity for salt aerosols. Environ. Sci. Technol. 7:142–146, 1973.
164. Corn, M., N. Kotsko, D. Stanton, W. Bell. and A. P. Thomas. Response of cats to inhaled mixtures of SO_2 and SO_2-NaCl aerosol in air. Arch. Environ. Health 24:248–256, 1972.
165. Cornwall, C. J., and P. A. B. Raffle. Bronchitis—sickness absence in London transport. Brit. J. Ind. Med. 18:24–32, 1961.
166. Covert, D. A. The Relationship between Chemical Composition, Relative Humidity and Light Scattering by Aerosols. M.S. thesis, University of Washington, 1971. (On file at University of Washington Library)
167. Covert, D. S. A Study of the Relationship of Chemical Composition and Humidity to Light Scattering by Aerosols. Ph.D. thesis, University of Washington, 1974. (University Microfilms, Ann Arbor, MI) 167 pp.
168. Cowling, J. E., and M. E. Roberts. Paints, varnishes, enamels and lacquers, pp. 596–645. In G. A. Greathouse and C. J. Wessel, Eds. Deterioration of Materials. Causes and Preventive Techniques. New York: Reinhold Publishing Corporation, 1954.
169. Cox, R. A. Quantum yields for the photooxidation of sulfur dioxide in the first allowed absorption region. J. Phys. Chem. 76:814–820, 1972.
170. Cox, R. A., and S. A. Penkett. Aerosol formation from sulphur dioxide in the presence of ozone and olefinic hydrocarbons. J. Chem. Soc. 68:1735–1753, 1972.
171. Coy, P., S. Grzybowski, and J. F. Rowe. Lung cancer mortality according to birthplace. Can. Med. Assoc. J. 99:476–483, 1968.
172. Crocker, T. T. Nitrogen oxides—animal effects, pp. 345–361. In Assembly of Life Sciences-National Academy of Sciences-National Research Council. Proceedings of the Conference on Health Effects of Air Pollutants, Washington, D.C., October 3–5, 1973. Washington, D.C.: U.S. Government Printing Office, 1973.

173. Cuddihy, R. G., and B. B. Boecker. Controlled administration of respiratory tract burdens of inhaled radioactive aerosols in beagle dogs. Toxicol. Appl. Pharmacol. 25:597–605 (Aug.), 1973.
174. Cutz, E., and P. E. Conen. Ultrastructure and cytochemistry of Clara cells. Amer. J. Path. 62:127–142, 1971.
175. Czaja, A. T. Die Wirkung von verstäubtem Kalk und Zement auf Pflanzen. Qual. Plant. Mater. Veg. 7:184–212, 1960.
176. Czaja, A. T. Zementstaubwirkungen auf Pflanzen: die Entstehung der Zementkrusten. Qual. Plant Mater. Veg. 8:201–238, 1961.
177. Czaja, A. T. Über das Problem der Zementstaubwirkungen auf Pflanzen. Staub 22:228–232, 1962.
178. Czaja, A. T. Über die Einwirkung von Stäuben, speziell von Zementofenstaub auf Pflanzen. Angew. Botan. 40:106–120, 1966.
179. Dahl, E., and O. Skre. An investigation of the effects of acid precipitation on land productivity, pp. 27–40. In Konferens om avsvavling, Stockholm, 11 November 1969. Helsinki: Nordforsk Miljovardssekretariatet, 1971. (in Norwegian)
180. Dalhamn, T. Mucous flow and ciliary activity in the trachea of healthy cats and rats exposed to respiratory irritant gases (SO_2, H_3N, HCNO): A functional and morphologic (light microscopic and electron microscopic) study, with special reference to technique. Acta Physiol. Scand. 36: Suppl. 123, 9–161, 1956.
181. Daly, C. Air pollution and bronchitis. Brit. Med. J. 2:687–688, 1954.
182. Daly, C. Air pollution and causes of death. Brit. J. Prev. Soc. Med. 13:14–27, 1959.
183. Dana, M., J. M. Hales, C. E. Hane, and J. M. Thorp. Some references pertinent to the field of precipitation chemistry, pp. 121–136. In Appendix B of Precipitation Scavenging of Inorganic Pollutants from Metropolitan Sources. Report # EPA-650/3-74-005. Richland, WA: Battelle-Pacific Northwest Laboratories, 1974. (Prepared for U.S. Environmental Protection Agency. Distributed by National Technical Information Service, Springfield, VA)
184. Darley, E. F. Studies on the effect of cement-kiln dust on vegetation. J. Air Pollut. Control Assoc. 16:145–150, 1966.
185. Davies, C. N. Deposition and retention of dust in the human respiratory tract. Ann. Occup. Hyg. 7:169–183, 1964.
186. Davies, C. N. The deposition of aerosol in the human lung, pp. 90–99. In V. Bohlau, Ed. Aerosole in Physik, Medizin und Technik. Proceedings of a conference held in Bad Soden, October 17–18, 1973. Bad Soden, West Germany: Gesellschaft für Aerosolforschung, 1973.
187. Davies, C. N., J. Heyder, and M. C. Subba Ramu. Breathing of half-micron aerosols. I. Experimental. J. Appl. Physiol. 32:591–600, 1972.
188. Davis, D. D., W. A. Payne, and L. J. Stief. The hydroperoxyl radical in atmospheric chemical dynamics: Reaction with carbon monoxide. Science 179:280–282, 1973.
189. Davis, W. Textile Pollution Loss is in Billions. Page 10, Section 4, March 29, 1970. Raleigh, N.C.: The News and Observer, 1970.
190. Dean, G. Lung cancer among white South Africans. Brit. Med. J. 2:852–857, 1959.
191. Dean, G. Lung cancer in South Africans and British immigrants. Proc. Royal Soc. Med. 57(2):984–987, 1964.

192. Dean, G. Lung cancer and bronchitis in northern Ireland, 1960–2. Brit. Med. J. 1:1506–1514, 1966.
193. Deane M., J. R. Goldsmith, and D. Tuma. Respiratory conditions in outside workers. Report on outside plant telephone workers in San Francisco and Los Angeles. Arch. Environ. Health 10:323–331, 1965.
194. Denton, R. Bronchial obstruction in cystic fibrosis: rheological factors. Pediatrics 25:611–620, 1960.
195. Dietz, A. G. H. Plastics for Architects and Builders. Cambridge: The MIT Press, 1969. 129 pp.
196. Dohan, F. C. Air pollutants and incidence of respiratory disease. Arch. Environ. Health 3:387–395, 1961.
197. Dohan, F. C., and E. W. Taylor. Air pollution and respiratory disease. A preliminary report. Amer. J. Med. Sci. 240:337–339, 1960.
198. Dolan, D. F., D. B. Kittelson, and K. T. Whitby. Measurement of Diesel Exhaust Particle Size Distributions. ASME Preprint #75-WA/APC-5. Presented at the Winter Annual Meeting of the American Society of Mechanical Engineers, November 30–December 3, 1975, Houston, Texas.
199. Doll, R., Ed. Methods of Geographical Pathology. Report of the Study Group Convened by the Council for International Organizations of Medical Sciences Established Under the Joint Auspices of UNESCO and WHO. Oxford: Blackwell Scientific Publications, Ltd., 1959. 72 pp.
200. Doll, R., and A. B. Hill. Mortality in relation to smoking: Ten years' observations of British doctors. Brit. Med. J. 1:1399–1410, 1964.
201. Dorn, H. F. Tobacco consumption and mortality from cancer and other diseases. Public Health Rep. 74:581–593, 1959.
202. Douglas, J. W. B., and R. E. Waller. Air pollution and respiratory infection in children. Brit. J. Prev. Soc. Med. 20:1–8, 1966.
203. DuBois, A. B., and L. Dautrebande. Acute effects of breathing inert dust particles and of carbachol aerosol on the mechanical characteristics of the lungs in man. Changes in response after inhaling sympathomimetic aerosols. J. Clin. Invest. 37:1746–1755, 1958.
204. Durham, J. L., W. E. Wilson, T. G. Ellestad, K. Willeke, and K. T. Whitby. Comparison of volume and mass distributions for Denver aerosols. Atmos. Environ. 9:717–722, 1975.
205. Dzubay, T. G., and R. K. Stevens. Ambient air analysis with dichotomous sampler and x-ray fluorescence spectrometer. Environ. Sci. Technol. 9:663–668, 1975.
206. Dzubay, T. G., and R. K. Stevens. The characterization of atmospheric aerosol by physical and chemical methods. American Chemical Society, Division of Environmental Chemistry. Paper No. 41 in Preprints of Papers Presented at the 171st National Meeting, New York, N.Y., April 4–9, 1976. 16(1):136–139, 1976.
207. Dzubay, T. G., R. K. Stevens, and C. M. Peterson. Application of the dichotomous sampler to the characterization of ambient aerosols, pp. 95–106. In T. G. Dzubay, Ed. X-Ray Fluorescence Analysis of Environmental Samples. Ann Arbor: Ann Arbor Science Publishers, Inc., 1977.
208. Eagan, R. C., P. V. Hobbs, and L. F. Radke. Particle emissions from a large Kraft paper mill and their effects on the microstructure of warm clouds. J. Appl. Meteorol. 13:535–552, 1974.
209. Eastcott, D. F. The epidemiology of lung cancer in New Zealand. Lancet 1:37–39, 1956.
210. Ebert, R. V., and M. J. Terracio. Observation of the secretion on the sur-

face of the bronchioles with the scanning electron microscope. Amer. Rev. Resp. Dis. 112:491–496, 1975.

211. Eggert, J. Physical Chemistry. (Translated by S. J. Gregg) New York: D. Van Nostrand Co., Inc. 1933. 632 pp.

212. Eisenbud, M., and T. J. Kneip. Trace Metals in Urban Aerosols. Research Project 117. Final Report. Prepared by New York University for Electric Power Research Institue (EPRI). 1975.

213. Elliott, J. F. E., and A. G. Franks. Design of telephone exchanges for corrosive atmospheres. Systems Tech. No. 6:39–48, 1969.

214. Ellis, H. T., and R. F. Pueschel. Solar radiation: Absence of air pollution trends at Mauna Loa. Science 172:845–846, 1971.

215. Elterman, L., R. B. Toolin, and J. D. Essex. Stratospheric aerosol measurements with implications for global climate. Appl. Opt. 12: 330–337, 1973.

216. Emerson, P. A. Air pollution, atmospheric conditions and chronic airways obstruction. J. Occup. Med. 15:635–638, 1973.

217. Engelmann, R. J., and W. G. N. Slinn. Precipitation Scavenging. Proceedings of a symposium held at Richland, Washington, June 2–4, 1970. Oak Ridge, Tenn.: USAEC Division of Technical Information Extension, 1970. 499 pp.

218. Ensor, D. S., R. J. Charlson, N. C. Ahlquist, K. T. Whitby, R. B. Husar, and B. Y. H. Liu. Multiwavelength nephelometer measurements in Los Angeles smog aerosol. I. Comparison of calculated and measured light scattering, pp. 315–324. In G. M. Hidy, Ed. Aerosols and Atmospheric Chemistry. Proceedings of the American Chemical Society, Los Angeles, California, March 28–April 2, 1971. New York: Academic Press, 1972.

219. Ensor, D. S., and M. J. Pilat. Calculation of smoke plume opacity from particulate air pollutant properties. J. Air Pollut. Control Assoc. 21:496–501, 1971.

220. Ensor, D. S., W. M. Porch, M. J. Pilat, and R. J. Charlson. Influence of the atmospheric aerosol on albedo. J. Appl. Meteorol. 10:1303–1306, 1971.

221. Evans, G. R., and A. F. Roddy. The condensation nuclei formed in polluted air by ultraviolet radiation whose wavelengths are longer than 2900 A, pp. 369–373. In Proceedings of the 7th International Conference on Condensation and Ice Nuclei, September 18–24, 1969, Prague and Vienna. Prague: Academia, 1969.

222. Fairbairn, A. S., and D. D. Reid. Air pollution and other local factors in respiratory disease. Brit. J. Prev. Soc. Med. 12:94–103, 1958.

223. Fairchild, G. A., P. Kane, B. Adams, and D. Coffin. Sulfuric acid and streptococci clearance from respiratory tracts of mice. Arch. Environ. Health 30:538–545, 1975.

224. Feather, E. A., and G. Russell. Sputum viscosity in cystic fibrosis of the pancreas and other pulmonary diseases. Brit. J. Dis. Chest 64:192–200, 1970.

225. Ferin, J. Lung clearance of particles, pp. 64–78. In E. F. Aharonson, A. Ben-David, and M. A. Klingberg, Eds. Air Pollution and the Lung. Proceedings of the Twentieth Annual "OHOLO" Biological Conference, Máalot, Israel, March 16–19, 1975. New York: John Wiley & Sons, Inc., 1976.

226. Ferris, B. G. Effects of air pollution on school absences and differences in lung function in first and second grades in Berlin, New Hampshire, January 1966 to June 1967. Amer. Rev. Resp. Dis. 102:591–606, 1970.

227. Ferris, B. G., Jr., and D. O. Anderson. The prevalence of chronic respiratory disease in a New Hampshire town. Amer. Rev. Resp. Dis. 86:165–177, 1962.
228. Ferris, B. G., Jr., and D. O. Anderson. Epidemiological studies related to air pollution: A comparison of Berlin, New Hampshire, and Chilliwack, British Columbia. Proc. Roy. Soc. Med. 57:979–987, 1964.
229. Ferris, B. G., Jr., I. T. T. Higgins, M. W. Higgins, and J. M. Peters. Chronic nonspecific respiratory disease in Berlin, New Hampshire, 1961 to 1967. A follow-up study. Amer. Rev. Resp. Dis. 107:110–122, 1973.
230. Findeisen, W. Über das Absetzen Kleiner, in der Luft suspendierter Teilchen in der menschlichen Lunge bei der Atmung. Arch. Ges. Physiol. 236:367–379, 1935.
231. Findeisen, W. Zur Frage der Regentropfenfildung in reinen Wasserwolken. Meteorol. Z. 56:365–368, 1939.
232. Fink, F. W., F. H. Buttner, and W. K. Boyd. Technical-Economic Evaluation of Air-Pollution Corrosion Costs on Metals in the United States. Final Report. Prepared for the Air Pollution Control Office, Environmental Protection Agency. APTD-0654. Columbus, Ohio: Battelle Memorial Institute, 1971. [160 pp.]
232a. Finklea, J. F. Conceptual basis for establishing standards, pp. 619–668. In Assembly of Life Science-National Academy of Sciences-National Research Council. Proceedings of the Conference on Health Effects of Air Pollutants, Washington, D.C., October 3–5, 1973. Washington, D.C.: U.S. Government Printing Office, 1973.
233. Finklea, J. F., D. C. Calafiore, C. J. Nelson, W. B. Riggan, and C. G. Hayes. Aggravation of asthma by air pollutants: 1971 Salt Lake Basin studies, pp. 2-75 to 2-91. In U.S. Environmental Protection Agency, Office of Research and Development. Health Consequences of Sulfur Oxides: A Report from CHESS, 1970–1971. EPA-650/1-74-004. Washington, D.C.: U.S. Government Printing Office, 1974.
234. Finklea, J. F., J. H. Farmer, G. J. Love, D. C. Calafiore, and G. W. Sovocool. Aggravation of asthma by air pollutants: 1970–1971 New York studies, pp. 5-71 to 5-84. In U.S. Environmental Protection Agency, Office of Research and Development. Health Consequences of Sulfur Oxides: A Report from CHESS, 1970–1971. EPA-650/1-74-004. Washington, D.C.: U.S. Government Printing Office, 1974.
235. Finklea, J. F., J. G. French, G. R. Lowrimore, J. Goldberg, C. M. Shy, and W. C. Nelson. Prospective surveys of acute respiratory disease in volunteer families: Chicago nursery school study, 1969–1970, pp. 4-37 to 4-55. In U.S. Environmental Protection Agency, Office of Research and Development. Health Consequences of Sulfur Oxides: A Report from CHESS, 1970–1971. EPA-650/1-74-004. Washington, D.C.: U.S. Government Printing Office, 1974.
236. Finklea, J. F., D. I. Hammer, D. E. House, C. R. Sharp, W. C. Nelson, and G. R. Lowrimore. Frequency of acute lower respiratory disease in children: Retrospective survey of five Rocky Mountain communities, 1967–1970, pp. 3-35 to 3-56. In U.S. Environmental Protection Agency, Office of Research and Development. Health Consequences of Sulfur Oxides: A Report from CHESS, 1970–1971. EPA-650/1-74-004. Washington, D.C.: U.S. Government Printing Office, 1974.
237. Finklea, J. F., C. M. Shy, G. J. Love, C. G. Hayes, W. C. Nelson, R. S. Chapman, and D. E. House. Health consequences of sulfur oxides: Sum-

mary and conclusions based upon CHESS studies of 1970–1971, pp. 7-3 to 7-24. In U.S. Environmental Protection Agency, Office of Research and Development. Health Consequences of Sulfur Oxides: A Report from CHESS, 1970–1971. EPA-650/1-74-004. Washington, D.C.: U.S. Government Printing Office, 1974.

238. Firket, J. Sur les causes des accidents survenus dans la vallée de la Meuse, lors des brouillards de décembre 1930. Bull. Acad. Roy. Med. Belg. 11:683–741, 1931.

239. Fischer, K. Bestimmung der Absorption von sichtbarer Strahlung durch Aerosolpartikeln. Beitr. Phys. Atmos. 43:244–254, 1970.

240. Fischer, K. Massenabsorptionskoeffizient natürlicher Aerosolteilchen im Wellenlängenbereich zwischen 0.4 und 2.4 μm. Beitr. Phys. Atmos. 46:89–100, 1973.

241. Fish, B. R. Inhalation of uranium aerosols by mouse, rat, dog and man, pp. 151–165. In C. N. Davies, Ed. Inhaled Particles and Vapours. Proceedings of an International Symposium organized by the British Occupational Hygiene Society, Oxford, 29 March–1 April 1960. Oxford: Pergamon Press, 1961.

242. Fisher, M. V., P. E. Morrow, and C. L. Yuile. Effect of Freund's complete adjuvant upon clearance of iron-59 oxide from rat lungs. J. Reticuloendothel. Soc. 13:536–556, 1973.

243. Fletcher, C. M., C. M. Tinker, I. D. Hill, and F. E. Speizer. A five-year prospective field study of early obstructive airway disease, pp. 249–251. In U.S. Department of Health, Education, and Welfare, Public Health Service. Current Research in Chronic Respiratory Disease, Proceedings of the Eleventh Aspen Conference. Washington, D.C.: U.S. Government Printing Office, 1968.

244. Fletcher, N. H. The Physics of Rainclouds. Cambridge, Eng.: University Press, 1962. 386 pp.

245. Flocchini, R. G., T. A. Cahill, D. J. Shadoan, S. J. Lange, R. A. Eldred, P. J. Feeney, G. W. Wolfe, D. C. Simmeroth, and J. K. Suder. Monitoring California's aerosols by size and elemental composition. Environ. Sci. Technol. 10:76–82, 1976.

246. Florey, H., H. M. Carleton, and A. Q. Wells. Mucus secretion in the trachea. Brit. J. Exp. Path. 13:269–284, 1932.

247. Flowers, E. C., R. A. McCormick, and K. R. Kurfis. Atmospheric turbidity over the United States. J. Appl. Meteorol. 8:955–962, 1969.

248. Flueck, J. A. Statistical Analysis of the Ground Level Precipitation Data. Part V of the Final Report on Project Whitetop, A Convective Cloud Randomized Seeding Project. Chicago: University of Chicago, 1971. 294 pp.

249. Flyger, H., K. Hansen, W. J. Megaw, and L. C. Cox. The background level of the summer tropospheric aerosol over Greenland and the North Atlantic Ocean. J. Appl. Meteor. 12:161–174, 1973.

250. Flyger, H., N. Z. Heidam, K. Hansen, W. J. Megaw, E. G. Walther, and A. W. Hogan. The background level of the summer tropospheric aerosol, sulfur dioxide and ozone over Greenland and the North Atlantic Ocean. J. Aerosol Sci. 7:103–140, 1976.

251. Foster, W. M., E. H. Bergofsky, D. E. Bohning, M. Lippmann, and R. E. Albert. Effect of adrenergic agents and their mode of action on mucociliary clearance in man. J. Appl. Physiol. 41:146–152, 1976.

252. Frank, N. R., M. O. Amdur, and J. L. Whittenberger. A comparison of the acute effects of SO_2 administered alone or in combination with NaCl

particles on the respiratory mechanics of healthy adults. Int. J. Air Wat. Pollut. 8:125–133, 1964.

253. Frank, N. R., M. O. Amdur, J. Worcester, and J. L. Whittenberger. Effects of acute controlled exposure to SO_2 on respiratory mechanics in healthy male adults. J. Appl. Physiol. 17:252–258, 1962.
254. Frank, N. R., and F. E. Speizer. SO_2 effects on the respiratory system in dogs: Changes in mechanical behavior at different levels of the respiratory system during acute exposure to the gas. Arch. Environ. Health 11:624–634, 1965.
255. Frank, N. R., R. E. Yoder, J. D. Brain, and E. Yokoyama. SO_2 (^{35}S-labeled) absorption by the nose and mouth under conditions of varying concentration and flow. Arch. Environ. Health 18:315–322, 1969.
256. Frank, N. R., R. E. Yoder, E. Yokoyama, and F. E. Speizer. The diffusion of $^{35}SO_2$ from tissue fluids into the lungs following exposure of dogs to $^{35}SO_2$. Health Phys. 13:31–36, 1967.
257. Frank, R. The effects of inhaled pollutants on nasal and pulmonary flow-resistance. Ann. Otol. Rhinol. Laryngol. 79:540–546, 1970.
258. French, J. G., G. Lowrimore, W. C. Nelson, J. F. Finklea, T. English, and M. Hertz. The effect of sulfur dioxide and suspended sulfates on acute respiratory disease. Arch. Environ. Health 27:129–133, 1973.
259. Friedlander, S. K. Theoretical considerations for the particle size spectrum of the stratospheric aerosol. Amer. Meteor. Soc. J. Meteor. 18:753–759, 1961.
260. Friend, J. P. The global sulfur cycle, pp. 177–201. In S. I. Rasool, Ed. Chemistry of the Lower Atmosphere. New York: Plenum Press, 1973.
261. Fuchs, N. A. The Mechanics of Aerosols. New York: Pergamon Press, 1964. 408 pp.
262. Fujii, T., and A. Hirata. Studies of the effects of air pollution on the fading of dyed fabrics. Taiki Osen Kenkyu (J. Jap. Soc. Air Pollut.) 5:164, 1970. (in Japanese)
263. Fujii, T., and C. Tsuda. Studies on the effects of air pollution on the fading of dyed fabrics. J. Jap. Soc. Air Pollut. 6(1)162, 1971. (in Japanese)
264. Fuller, E. C., and R. H. Christ. The rate of oxidation of sulfite ions by oxygen. J. Amer. Chem. Soc. 63:1644–1650, 1941.
265. Furiosi, N. J., S. C. Crane, and G. Freeman. Mixed sodium chloride aerosol and nitrogen dioxide in air. Biological effects on monkeys and rats. Arch. Environ. Health 27:405–408, 1973.
266. Garland, J. A., J. R. Branson, and L. C. Cox. A study of the contribution of pollution to visibility in a radiation fog. Atmos. Environ. 7:1079–1092, 1973.
267. Gartrell, G., Jr., and S. K. Friedlander. Relating particulate pollution to sources: The 1972 California aerosol characterization study. Atmos. Environ. 9:279–299, 1975.
268. Gerhard, E., and H. F. Johnstone. Photochemical oxidation of sulfur dioxide in air. Ind. Eng. Chem. 47:972–976, 1955.
269. Giacomelli-Maltoni, G., C. Merlandri, V. Prodi, and G. Tarroni. Deposition efficiency of monodisperse particles in human respiratory tract. Amer. Ind. Hyg. Assoc. J. 33:603–610, 1972.
270. Gibbons, E. V. The Corrosion Behavior of Some Architectural Metals in Canadian Atmospheres, Summary of Ten-Year Results of Group I. Division of Building Research Technical Paper No. 328. Ottawa: National Research Council of Canada, 1972. 81 pp.

271. Gibbons, E. V. The Corrosion Behavior of Some Architectural Metals in Canadian Atmospheres: Summary of Ten-Year Results of Group II. Division of Building Research Technical Paper No. 373. Ottawa: National Research Council of Canada, 1972.
272. Gibbons, R. A. The biochemical and physical properties of epithelial mucus. Amer. Rev. Resp. Dis. 83:568–569, 1961.
273. Gil, J., and E. R. Weibel. Extracellular lining of bronchioles after perfusion-fixation of rat lungs for electron microscopy. Anat. Rec. 169: 185–200, 1971.
274. Gilboa, A., and A. Silberberg. Characterization of epithelial mucus and its function in clearance by ciliary propulsion, pp. 49–63. In E. F. Aharonson, A. Ben-David, and M. A. Klingberg, Eds. Air Pollution and the Lung. Proceedings of the Twentieth Annual "OHOLO" Biological Conference, Máalot, Israel, March 16–19, 1975. New York: John Wiley & Sons, Inc., 1976.
275. Gillette, D. A., I. H. Blifford, Jr., and C. R. Fenster. Measurements of aerosol size distributions and vertical fluxes of aerosols on land subject to wind erosion. J. Appl. Meteor. 11:977–987, 1972.
276. Gladney, E. S., W. H. Zoller, A. G. Jones, and G. E. Gordon. Composition and size distributions of atmospheric particulate matter in Boston area. Environ. Sci. Tech. 8:551–557, 1974.
277. Glasser, M., and L. Greenburg. Air pollution, mortality, and weather. New York City, 1960–1964. Arch. Environ. Health 22:334–343, 1971.
278. Glasser, M., L. Greenburg, and F. Field. Mortality and morbidity during a period of high levels of air pollution. New York, November 23 to 25, 1966. Arch. Environ. Health 15:684–694, 1967.
279. Glasstone, S. Textbook of Physical Chemistry. (2nd ed.) New York: D. Van Nostrand Co., Inc., 1946. 1320 pp.
280. Goco, R. V., M. B. Kress, and O. C. Brantigan. Comparison of mucus glands in the tracheobronchial tree of man and animals. Ann. N. Y. Acad. Sci. 106:555–571, 1963.
281. Goetz, A. On the nature of the synergistic action of aerosols. Int. J. Air Water Pollut. 4:168–184, 1961.
282. Goldberg, H. E., A. A. Cohen, J. F. Finklea, J. H. Farmer, F. B. Benson, and G. J. Love. Frequency and severity of cardiopulmonary symptoms in adult panels: 1970–1971 New York studies, pp. 5-85 to 5-108. In U.S. Environmental Protection Agency, Office of Research and Development. Health Consequences of Sulfur Oxides: A Report from CHESS, 1970–1971. EPA-650/11-74-004. Washington, D.C.: U.S. Government Printing Office, 1974.
283. Goldberg, H. E., J. F. Finklea, C. J. Nelson, W. B. Steen, R. S. Chapman, D. H. Swanson, and A. A. Cohen. Prevalance of chronic respiratory disease symptoms in adults: 1970 survey of New York communities, pp. 5-33 to 5-47. In U.S. Environment. Health Consequences of Sulfur Oxides: A Report from CHESS, 1970–1971. EPA-650/1-74-004. Washington, D.C.: U.S. Government Printing Office, 1974.
284. Goodwin, J. E., W. Sage, and G. P. Tilly. Study of erosion by solid particles. Inst. Mech. Eng. Proc. 184(15):279–292, 1969–1970.
285. Gordon, C. C. Mount Storm Study. Report to Environmental Protection Agency under Contract 68-02-0229, November 17, 1972.
286. Gore, A. T., and C. W. Shaddick. Atmospheric pollution and mortality in the County of London. Brit. J. Prev. Soc. Med. 12:104–113, 1958.

287. Gorham, E. Bronchitis and the acidity of urban precipitation. Lancet 2:691, 1958.
288. Gorham, E. Pneumonia and atmospheric sulphate deposit. Lancet 2: 287–288, 1959.
289. Gormley, P. G., and M. Kennedy. Diffusion from a stream flowing through a cylindrical tube. Proc. Roy. Irish Acad. A52:163–169, 1949.
290. Gosselin, R. E. Physiological regulators of ciliary motion. Amer. Rev. Resp. Dis. 93 (Suppl.—Symposium on Structure, Function and Measurement of Respiratory Cilia. Duke University Medical Center, Durham, North Carolina, February 18–19, 1965):41–59, 1966.
291. Green, G. M. Integrated defense mechanisms in models of chronic pulmonary disease. Arch. Intern. Med. 126:500–503, 1970.
292. Green, G. M. Alveolobronchiolar transport mechanisms. Arch. Intern. Med. 131:109–114, 1973.
293. Green, H. L., and W. R. Lane. Particulate Clouds: Dusts, Smokes and Mists; Their Physics and Physical Chemistry and Industrial and Environmental Aspects. New York: D. Van Nostrand Co., 1957. 452 pp.
294. Greenburg, L., C. Erhardt, F. Field, J. I. Reed, and N. S. Seriff. Intermittent air pollution episode in New York City, 1962. Public Health Rep. 78:1061–1064, 1963.
295. Greenburg, L., F. Field, C. L. Erhardt, M. Glasser, and J. I. Reed. Air pollution, influenza, and mortality in New York City, January-February 1963. Arch. Environ. Health 15:430–438, 1967.
296. Greenburg, L., F. Field, J. I. Reed, and C. L. Erhardt. Air pollution and morbidity in New York City. J.A.M.A. 182·161–164, 1962.
297. Greenburg, L., M. B. Jacobs, B. M. Drolette, F. Field, and M. M. Braverman. Report of an air pollution incident in New York City, November 1953. Public Health Rep. 77:7–16, 1962.
298. Gross, P., and H. Brieger. Silicotic bronchiolitis obliterans—a focal clearance failure, pp. 105–110. In C. N. Davies, Ed. Inhaled Particles and Vapours II. Proceedings of an International Symposium organized by the British Occupational Hygiene Society, Cambridge, 28 September–1 October 1965. Oxford: Pergamon Press, 1967.
299. Gudbjerg, C. E., and G. Thomsen. Inflammatory changes in the bronchial glands in chronic bronchitis. Acta Radiol. 42:269–275, 1954.
300. Guderian, R. Kurzberichte: H. Pajenkamp. Einwirkung des Zementofenstaubes auf Pflanzen and Tiere. Staub 21:518–519, 1961.
301. Gupton, E. D., and P. E. Brown. Chest clearance of inhaled cobalt-60 oxide. Health Phys. 23:767–769, 1972.
302. Guttman, H. Effects of atmospheric factors on the corrosion of rolled zinc. pp. 223–239. In Metal Corrosion in the Atmosphere. A symposium presented at the Seventieth Annual Meeting, American Society for Testing and Materials, Boston, Mass., 25–30 June, 1967. ASTM Special Technical Publication No. 435. Philadelphia: American Society for Testing and Materials, 1968.
303. Guttman, H., and E. V. Gibbons. Corrosion Behavior of Metal-Coated Panels at Eight Canadian Locations: Summary of a Fourteen-Year Program. Division of Building Research Technical Paper 354. Ottawa: National Research Council of Canada, 1971.
304. Haagen-Smit, A. J., M. F. Brunelle, and J. W. Haagen-Smit. Ozone cracking in the Los Angeles area. Rubber Chem. Technol 32:1134–1142, 1959.

305. Hadfield, E. H. Damage to human nasal mucosa by wood dust, pp. 855–861. In W. H. Walton, Ed. Inhaled Particles III, Vol. II. Proceedings of an International Symposium organized by the British Occupational Hygiene Society in London, 14–23 September, 1970. Old Woking, Surrey, England: Unwin Brothers Limited, 1971.
306. Haenszel, W. Cancer mortality among the foreign-born in the United States. J. Nat. Cancer Inst. 26:37–132, 1961.
307. Haenszel, W., D. B. Loveland, and M. G. Sirken. Lung-cancer mortality as related to residence and smoking histories. I. White males. J. Nat. Cancer Inst. 28:947–1001, 1962.
308. Haenszel, W., and K. E. Taeuber. Lung-cancer mortality as related to residence and smoking histories. II. White females. J. Nat. Cancer Inst. 32:803–838, 1964.
309. Hagen, L. J., and N. P. Woodruff. Air pollution from duststorms in the Great Plains. Atmos. Environ. 7:323–332, 1973.
310. Hall, T. C., Jr. Photochemical Studies of Nitrogen-dioxide and Sulfur-dioxide. (Ph.D. thesis, University of California/Los Angeles, 1954)
311. Hamada, M., R. Fujikawa, C. Itakura, K. Kishitani, M. Koike, E. Tajima, and others. Durability of Asphaltic Materials. Parts 1–5. Division of Building Research Technical Translation 1146. (Translated by H. J. Kondo from Proc. Architect. Inst. Jap. 69: 53–72, 1961) Ottawa: National Research Council of Canada, 1964. 57 pp.
312. Hammond, E. C. Smoking in relation to death rates of one million men and women. Nat. Cancer Inst. Monogr. 19:127–204, 1966.
313. Hammond, E. C., and D. Horn. The relationship between human smoking habits and death rates: A follow-up study of 187,766 men. J.A.M.A. 155:1316–1328, 1954.
314. Hammond, E. C., and D. Horn. Smoking and death rates—report on forty-four months of follow-up of 187,783 men. II. Death rates by cause. J.A.M.A. 166:1294–1308, 1958.
315. Hänel, G. New results concerning the dependence of visibility on relative humidity and their significance in a model for visibility forecast. Beitr. Phys. Atmos. 44:137–167, 1971.
316. Hänel, G. Computation of the extinction of visible radiation by atmospheric aerosol particles as a function of the relative humidity, based upon measured properties. Aerosol Sci. 3:377–386, 1972.
317. Hänel, G. The ratio of the extinction coefficient to the mass of atmospheric aerosol particles as a function of the relative humidity. Aerosol Sci. 3:455–460, 1972.
318. Harrison, P. R., W. R. Matson, and J. W. Winchester. Time variations of lead, copper and cadmium concentrations in aerosols in Ann Arbor, Michigan. Atmos. Environ. 5:613–619, 1971.
319. Hatch, T. F., and P. Gross. Pulmonary Deposition and Retention of Inhaled Aerosols. New York: Academic Press, 1964. 192 pp.
320. Hayes, C. G., D. I. Hammer, C. M. Shy, V. Hasselblad, C. R. Sharp, J. P. Creason, and K. E. McClain. Prevalence of chronic respiratory disease symptoms in adults: 1970 survey of five Rocky Mountain communities, pp. 3-19 to 3-33. In U.S. Environmental Protection Agency, Office of Research and Development. Health Consequences of Sulfur Oxides: A Report from CHESS, 1970–1971. EPA-650/1-74-004. Washington, D.C.: U.S. Government Printing Office, 1974.

321. Haynie, F. H., and J. B. Upham. Effects of atmospheric sulfur dioxide on the corrosion of zinc. Mater. Prot. Perform. 9(8):35–40, 1970.
322. Haynie, F. H., and J. B. Upham. Effects of atmospheric pollutants on corrosion behavior of steels. Mater. Prot. Perform. 10:18–21, 1971.
323. Heinsohn, R. J., C. Birnie, Jr., and T. A. Cuscino. Fugitive Dust from Vehicles using Unpaved Roads. Center for Air Environment Studies (CAES) Publication No. 419-75. University Park, PA: The Pennsylvania State University, 1975. 29 pp.
324. Hermance, H. W. Combatting the effects of smog on wire-spring relays. Bell Lab. Rec. 44:48–52, 1966.
325. Hewitt, D. Mortality in the London boroughs, 1950–52, with special reference to respiratory disease. Brit. J. Prev. Soc. Med. 10:45–57, 1956.
326. Heyder, J., L. Armbruster, J. Gebhart, and W. Stahlhofen. Deposition of aerosol particles in the human respiratory tract, pp. 122–125. In V. Böhlau, Ed. Aerosole in Physik, Medizin and Technik. Bad Soden, West Germany: Gesellschaft für Aerosolforschung, 1973.
327. Heyder, J., J. Gebhart, G. Heigwer, C. Roth, and W. Stahlhofen. Experimental studies of the total deposition of aerosol particles in the human respiratory tract. Aerosol Sci. 4:191–208, 1973.
328. Hidy, G. M., Ed. Aerosols and Atmospheric Chemistry. Proceedings of the American Chemical Society. Los Angeles, California, March 28–April 2, 1971. New York: Academic Press, 1972. 348 pp.
329. Hidy, G. M. Some aspects of airborne particles and radiation in the atmosphere, pp. 1286–1298. In O. F. Nygaard, H. I. Adler, and W. K. Sinclair, Eds. Radiation Research. Biomedical, Chemical, and Physical Perspectives. Proceedings of the Fifth International Congress of Radiation Research, Seattle, Washington, July 14–20, 1974. New York: Academic Press, 1975.
330. Hidy, G. M., B. Appel, R. J. Charlson, W. E. Clark, D. Covert, D. Dockweiler, S. K. Friedlander, P. Friedman, R. Giauque, S. Green, S. Heisler, W. W. Ho, J. J. Huntzicker, R. B. Husar, K. Kubler, G. Lauer, P. K. Mueller, T. Novakov, R. Ragaini, L. W. Richards, T. B. Smith, E. R. Stephens, G. Sverdrup, S. Twiss, A. Waggoner, S. Wall, H. H. Wang, J. J. Wesolowski, K. T. Whitby, and W. White. Characterization of Aerosols in California (ACHEX). Final Report, Vols. I–IV. Submitted to the Air Resources Board, State of California, Sept. 30, 1974.
331. Hidy, G. M., B. R. Appel, R. J. Charlson, W. E. Clark, S. K. Friedlander, D. H. Hutchison, T. B. Smith, J. Suder, J. J. Wesolowski, and K. T. Whitby. Summary of the California aerosol characterization experiment. J. Air Pollut. Control Assoc. 25:1106–1114, 1975.
332. Hidy, G. M., and J. R. Brock. The Dynamics of Aerocolloidal Systems. New York: Pergamon Press, 1970. 379 pp.
333. Hidy, G. M., and J. R. Brock. An assessment of the global sources of tropospheric aerosols, pp. 1088–1097. In H. M. Englund and W. T. Beery, Eds. Proceedings of the Second International Clean Air Congress. Held at Washington, D.C., December 6–11, 1970. New York: Academic Press, 1971.
334. Higgins, I. T. T. Sulfur oxides and particulates, pp. 20–33. In Medical Aspects of Air Pollution. A Continuing Engineering Education Course developed by SAE Engineering Education Activity. Detroit, Michigan, January 14, 1971. New York: Society of Automotive Engineers, Inc., 1971.

335. Higgins, I. T. T. Trends in respiratory cancer mortality in the United States and in England and Wales. Arch. Environ. Health 28:121-129, 1974.
336. Higgins, I. T. T. Epidemiological evidence on the carcinogenic risk of air pollution, pp. 41–52. In C. Rosenfeld and W. Davis, Eds. Environmental Pollution and Carcinogenic Risks. INSERM symposium Series Vol. 52. IARC Scientific Publications 13. Paris: INSERM (Institut National de la Santé et de la Recherche Médicale), 1976.
337. Hilding, A. C. On cigarette smoking, bronchial carcinoma and ciliary action. II. Experimental study on the filtering action of cow's lungs, the deposition of tar in the bronchial tree and removal by ciliary action. New Eng. J. Med. 254:1155–1160, 1956.
338. Hilding, A. C. Ciliary streaming in the lower respiratory tract. Amer. J. Physiol. 191:404–410, 1957.
339. Hilding, A. C. Experimental studies on some little understood aspects of the physiology of the respiratory tract and their clinical importance. Trans. Amer. Acad. Ophthal. Otolaryng. 65:475–495, 1961.
340. Hindawi, I. J., and H. C. Ratsch. Growth Abnormalities of Christmas Trees Attributed to Sulfur Dioxide and Particulate Acid Aerosol. ACPA Paper No. 74-252, presented at the Air Pollution Control Association Meeting at Denver, Colorado, June 9–13, 1974.
341. Hinton, D. O., T. D. English, B. F. Parr, V. Hasselblad, R. C. Dickerson, and J. G. French. Human exposure to air pollutants in the Chicago-Northwest Indiana metropolitan region, 1950–1971, pp. 4-3 to 4-21. In U.S. Environmental Protection Agency, Office of Research and Development. Health Consequences of Sulfur Oxides: A Report from CHESS, 1970–1971. EPA-650/1-74-004. Washington, DC: U.S. Government Printing Office, 1974.
342. Hirayama, T. Smoking in Relation to the Death Rates of 265,118 Men and Women in Japan. A Report of Five Years of Follow-up. Presented at the American Cancer Society's Fourteenth Science Writers' Seminar, March 24–29, Clearwater Beach, FL, 1972. 15 pp.
343. Hitosugi, M. Epidemiological study of lung cancer with special reference to the effect of air pollution and smoking habits. Inst. Pub. Health Bull. 17:237–256, 1968.
344. Ho, W., G. M. Hidy, and R. M. Govan. Microwave measurements of the liquid water content of atmospheric aerosols. J. Appl. Meteorol. 13:871–879, 1974.
345. Hobbs, P. V. Comparison of ice nucleus concentrations measured with an acoustical counter and Millipore filters. J. Appl. Meteorol. 9:828–829, 1970.
346. Hobbs, P. V., and J. P. Locatell. Ice nuclei from a natural forest fire. J. Appl. Meteorol. 8:833–834, 1969.
347. Hobbs, P. V., and L. F. Radke. Cloud condensation nuclei from a simulated forest fire. Science 163:279–280, 1967.
348. Hobbs, P. V., L. F. Radke, and S. E. Shumway. Cloud condensation nuclei from industrial sources and their apparent influence on precipitation in Washington State. J. Atmos. Sci. 27:81–89, 1970.
349. Hocking, L. M., and P. R. Jonas. The collision efficiency of small drops. Q. J. Roy. Meteorol. Soc. 96:722–729, 1970.
350. Hodge, P. W. Large decrease in the clean air transmission of the air 1.7 km above Los Angeles. Nature 229:549, 1971.

351. Hodge, P. W., N. Laulainen, and R. J. Charlson. Astronomy and air pollution. Science 178:1123–1124, 1972.
352. Hodgson, T. A., Jr. Short-term effects of air pollution on mortality in New York City. Environ. Sci. Technol. 4:589–597, 1970.
353. Hoffman, A. J., T. C. Curran, T. B. McMullen, W. M. Cox, and W. F. Hunt, Jr. EPA's role in ambient air quality monitoring. Science 190:243–248, 1975.
354. Hogan, A. W. Antarctic aerosols. J. Appl. Meteorol. 14:550–559, 1975.
355. Holbrow, G. L. Atmospheric pollution: Its measurement and some effects on paint. J. Oil Colour Chem. Assoc. 45:701–718, 1962.
356. Holland, A. C., and G. Gagne. The scattering of polarized light by polydisperse systems of irregular particles. Appl. Opt. 9:1113–1121, 1970.
357. Holland, W. W., T. Halil, A. E. Bennett, and A. Elliott. Factors influencing the onset of chronic respiratory disease. Brit. Med. J. 2:205–208, 1969.
358. Holland, W. W., and D. D. Reid. The urban factor in chronic bronchitis. Lancet 1:445–448, 1965.
359. Holland, W. W., D. D. Reid, R. Seltser, and R. W. Stone. Respiratory disease in England and the United States. Studies of comparative prevalence. Arch. Environ. Health 10:338–343, 1965.
360. Holzworth, G. C. Some effects of air pollution on visibility in and near cities, pp. 69–88. In Public Health Service, Robert A. Taft Sanitary Engineering Center. Symposium. Air Over Cities. Held November 6–7, 1961, Cincinnati, Ohio. SEC Technical Report A62-5. Washington, D.C.: U.S. Government Printing Office, 1962.
361. Hoorn, B., and D. A. J. Tyrrel. A new virus cultivated in organ cultures of human ciliated epithelium. Arch. Ges. Virusforsch. 18:210–225, 1966.
362. Horsfield, K., G. Dart, D. E. Olson, F. F. Filley, and G. Cumming. Models of the human bronchial tree. J. Appl. Physiol. 31:207–217, 1971.
363. Horvath, H. On the brown color of atmospheric haze. Atmos. Environ. 5:333–344, 1971.
364. Horvath, H., and K. E. Noll. The relationship between atmospheric light scattering coefficient and visibility. Atmos. Environ. 3:543–550, 1969.
365. Hosking, E. S. Effects of Air Pollution on the Abrasion Resistance of Selected Fabrics. M.S. thesis. Los Angeles: Loma Linda University, 1970. 61 pp.
366. Hounam, R. F., A. Black, and M. Walsh. Deposition of aerosol particles in the nasopharyngeal region of the human respiratory tract. Nature 221:1254–1255, 1969.
367. House, D. E., J. F. Finklea, C. M. Shy, D. C. Calafiore, W. B. Riggan, J. W. Southwick, and L. J. Olsen. Prevalence of chronic respiratory disease symptoms in adults: 1970 survey of Salt Lake Basin communities, pp. 2-41 to 2-54. In U.S. Environmental Protection Agency, Office of Research and Development. Health Consequences of Sulfur Oxides: A Report from CHESS, 1970–1971. EPA-650/1-74-004. Washington, D.C.: U.S. Government Printing Office, 1974.
368. Howell, W. E. The growth of cloud drops in uniformly cooled air. J. Meteorol. 6:134–149, 1949.
369. Huff, F. A., and S. A. Changnon. Precipitation modification by major urban areas. Bull. Amer. Meteorolog. Soc. 54:1220–1232, 1973.
370. Hukui, S., and A. Yamamoto. Studies on the corrosion of metallic materials by polluted air. Netsu Kanri (Heat Management, Energy and Pollution Control) 21(6):28–44, 1969. (in Japanese)

371. Husar, R. B. Coagulation of Knudsen Aerosols. (Ph.D. thesis, University of Minnesota, 1971. 212 pp.)
372. Husar, R. B., N. V. Gillani, J. D. Husar, and D. E. Patterson. A Study of Long Range Transport from Visibility Observations, Trajectory Analysis and Local Air Pollution Monitoring Data. Presented at the 7th International Technical Meeting on Air Pollution Modeling and Its Application, NATO Committee on the Challenges of Modern Society. Held September 7–10, 1976 in Airlie, Virginia.
373. Husar, R. B., K. T. Whitby, and B. Y. H. Liu. Physical mechanisms governing the dynamics of Los Angeles smog aerosol. J. Colloid Interface Sci. 39:211–224, 1972.
374. Husar, R. B., and W. H. White. On the color of the Los Angeles smog. Atmos. Environ. 10:199–204, 1976.
375. Huschke, R. E., Ed. Glossary of Meteorology. Boston: American Meteorology Society, 1959. 638 pp.
376. IIT Research Institute. Particle Technology 1973. Proceedings of the First International Conference in Particle Technology. August 21–24, 1973, Chicago. Chicago: ITT Research Institute, 1973. 328 pp.
377. Inadvertent Climate Modification. Report of the Study of Man's Impact on Climate (SMIC). Cambridge: The MIT Press, 1971. 308 pp.
378. International Conference on Environmental Sensing and Assessment. Las Vegas, Nevada, September 14–19, 1975. New York: Institute of Electrical and Electronics Engineers, 1976.
379. Ipsen, J., M. Deane, and F. E. Ingenito. Relationships of acute respiratory disease to atmospheric pollution and meteorological conditions. Arch. Environ. Health 18:462–472, 1969.
380. Iravani, J. Clearance function of the respiratory ciliated epithelium in normal and bronchitic rats, pp. 143–146. In W. H. Walton, Ed. Inhaled Particles III, Vol. I. Proceedings of an International Symposium organized by the British Occupational Hygiene Society in London, 14–23 September, 1970. Old Woking, Surrey, England: Unwin Brothers Limited, 1971.
381. Iravani, J., and A. Van As. Mucus transport in the tracheobronchial tree of normal and bronchitic rats. J. Path. 106:81–93, 1972.
382. Jaenicke, R., and C. Junge. Studien zur oberen Grenzgrösse des natürlichen Aerosoles. Beitr. Phys. Atmos. 40:129–143, 1967.
383. Jaenicke, R., C. Junge, and H. J. Kanter. Messungen der Aerosolgrössenverteilung über dem Atlantik. "Meteor" Forschungsergebnisse, Reihe B. 7:1–54, 1971.
384. Jammet, H., J. Lafuma, J. C. Nenot, M. Chameaud, M. Perreau, M. LeBouffant, M. Lefevre, and M. Martin. Lung clearance: Silicosis and anthracosis, pp. 435–437. In H. A. Shapiro. Ed. Pneumoconiosis. Proceedings of the International Conference, Johannesburg, 1969. Capetown: Oxford University Press, 1970.
385. Jeffery, P., and L. Reid. Intra-epithelial nerves in normal rat airways: A quantitative electron microscopic study. J. Anat. 114:35–45, 1973.
386. Jellinek, H. H. G. Chain scission of polymers by small concentrations (1 to 5 ppm) of sulfur dioxide and nitrogen dioxide, respectively, in presence of air and near ultraviolet radiation. J. Air Pollut. Control Assoc. 20:672–674, 1970.
387. Jellinek, H. H. G., and J. F. Kryman. Photolytic oxidation of isotactic polystyrene in presence of sulfur dioxide. Part I. Chain scission as function

of temperature at constant oxygen and sulfur dioxide pressures and constant light intensity, pp. 85–90. In R. F. Reinisch, Ed. Photochemistry of Macromolecules. Proceedings of a Symposium held at the Pacific Conference on Chemistry and Spectroscopy, Anaheim, California, October 8–9, 1969. New York: Plenum Press, 1970.
388. Jellinek, H. H. G., and J. Pavlinec. Photolytic oxidation of isotactic polystyrene in presence of sulfur dioxide. Part II. Photolysis reaction as function of light intensity, sulfur dioxide and oxygen pressures, pp. 91–104. In R. F. Reinisch, Ed. Photochemistry of Macromolecules. Proceedings of a Symposium held at the Pacific Conference on Chemistry and Spectroscopy, Anaheim, California, October 8–9, 1969. New York: Plenum Press, 1970.
389. Jennings, O. E. Smoke injury to shade trees, pp. 44–48. In R. P. White, Ed. Tenth Annual Shade Tree Conference. Proceedings of Annual Meeting, Pittsburgh, August 30–31, 1934.
390. Jiusto, J. E. Aerosol and cloud microphysics measurements in Hawaii. Tellus 19:359–368, 1973.
391. John, W. Contact electrification applied to particulate matter-monitoring, pp. 649–667. In B. Y. H. Liu, Ed. Fine Particles. Aerosol Generation, Measurement, Sampling, and Analysis. Proceedings of a Symposium by the U.S. Environmental Protection Agency held in Minneapolis, Minnesota, May 28–30, 1975. New York: Academic Press, Inc., 1976.
392. Johnstone, H. F., and D. R. Coughanowr. Absorption of sulfur dioxide from air: Oxidation in drops containing dissolved catalysts. Ind. Eng. Chem. 50:1169–1172, 1958.
393. Johnstone, H. F., and A. J. Moll. Formation of sulfuric acid in fogs. Ind. Eng. Chem. 52:861–863, 1960.
394. Jonsson, B., and R. Sundberg. Has the Acidification by Atmospheric Pollution Caused a Growth Reduction in Swedish Forests? A Comparison of Growth between Regions with Different Soil Properties. Research Note 20. Stockholm: Royal College of Forestry, Department of Forest Yield Research, 1972. (in Swedish with summary in English)
395. Joseph, J. H., and A. Manes. Secular and seasonal variations of atmospheric turbidity at Jerusalem. J. Appl. Meteorol. 10:453–462, 1971.
396. Judge, J. Venice fights for life. Nat. Geographic 142:591–631, 1972.
397. Junge, C. Die Konstitution der atmosphärischen Aerosole. Ann. Meteorol. 5 (Beiheft):1-55, 1952.
398. Junge, C. Die Rolle der Aerosole and der gasförmigen Beimengungen der Luft im Spurenstoffhaushalt der Troposphäre. Tellus 5:1-26, 1953.
399. Junge, C. The size distribution and aging of natural aerosols as determined from electric and optical data on the atmosphere. Amer. Meteor. Soc. J. Meteor. 12:13–25, 1955.
400. Junge, C. E. The chemical composition of atmospheric aerosols, I: Measurements at Round Hill Field Station, June-July 1953. J. Meteorol. 11:323–333, 1954.
401. Junge, C. E. Vertical profiles of condensation nuclei in the stratosphere. Amer. Meteor. Soc. J. Meteor. 18:501–509, 1961.
402. Junge, C. E. Air Chemistry and Radioactivity. New York: Academic Press, 1963. 382 pp.
403. Junge, C. E. Our knowledge of the physico-chemistry of aerosols in the undisturbed marine environment. J. Geophys. Res. 77:5183–5200, 1972.

404. Junge, C. E., and T. G. Ryan. Study of the SO_2 oxidation in solution and its role in atmospheric chemistry. Q. J. Roy. Meteorol. Soc. 84:46–56, 1958.

405. Karuhn, R. F. The development of a new acoustic particle counter for particle size analysis, pp. 202–207. In IIT Research Institute. Particle Technology 1973. Proceedings of the First International Conference in Particle Technology. August 21–24, 1973, Chicago. Chicago: IIT Research Institute, 1973.

406. Katz, M. Photoelectric determination of atmospheric sulfur dioxide, employing dilute starch-iodine solutions. Anal. Chem. 22:1040–1047, 1950.

407. Katz, M., and S. B. Gale. Mechanism of photooxidation of sulfur dioxide in atmosphere, pp. 336–343. In H. M. Englund and W. T. Beery, Eds. Proceedings of the Second International Clean Air Congress held at Washington, D.C., December 6–11, 1970. Sponsored by The International Union of Air Pollution Prevention Associations. New York: Academic Press, 1971.

408. Keimowitz, R. I. Immunoglobulins in normal human tracheobronchial washings. A qualitative and quantitative study. J. Lab. Clin. Med. 63:54–59, 1964.

409. Keiser-Nielson, H. Mucin. Copenhagen: Dansk Videnskabs Forlag, 1953. 307 pp. (in Danish)

410. Keith, C. H., and A. B. Arons. The growth of sea-salt particles by condensation of atmospheric water vapor. J. Meteorol. 11:173–184, 1954.

411. Kelsey, J. L., E. W. Mood, and R. M. Acheson. Population mobility and epidemiology of chronic bronchitis in Connecticut. Arch. Environ. Health 16:853–861, 1968.

412. Kensler, C. J., and S. P. Battista. Chemical and physical factors affecting mammalian ciliary activity. Amer. Rev. Resp. Dis. 93: (Suppl.—Symposium on Structure, Function and Measurement of Respiratory Cilia. Duke University Medical Center, Durham, North Carolina, February 18–19, 1965):93–102, 1966.

413. Kertész-Sáringer, M., E. Mészáros, and T. Várkonyi. On the size distribution of benzo(*a*)pyrene containing particles in urban air. Atmos. Environ. 5:429–431, 1971.

414. Kilburn, K. H. Mucociliary clearance from bullfrog (*Rana cantesbiana*) lung. J. Appl. Physiol. 23:804–810, 1967.

415. Kilburn, K. H. A hypothesis for pulmonary clearance and its implications. Amer. Rev. Resp. Dis. 98:449–463, 1968.

416. King, M., A. Gilboa, F. A. Meyer, and A. Silberberg. On the transport of mucus and its rheologic stimulants in ciliated systems. Amer. Rev. Resp. Dis. 110:740–745, 1974.

417. Kloke, A., and K. Riebartsch. Verunreinigung von Kulturpflanzen mit Blei aus Kraftfahrzeugabgasen. Naturwissenschaften 51: 367–368, 1964.

418. Kocmond, W. C., W. D. Garrett, and E. J. Mack. Modification of laboratory fog with organic surface films. J. Geophys. Res. 77:3221–3231, 1972.

419. Kocmond, W. C., D. B. Kittelson, J. Y. Yang, and K. L. Demerjian. Determination of the Formation Mechanisms and Composition of Photochemical Aerosols. Annual Summary Report (Final). Prepared for the Environmental Protection Agency under Contract EPA-68-02-0557, 1973.

420. Kocmond, W. C., and E. J. Mack. The vertical distribution of cloud and Aitken nuclei downwind of urban pollution sources. J. Appl. Meteorol. 11:141–148, 1972.
421. Koenig, L. R. Some observations suggesting ice multiplication in the atmosphere. J. Atmos. Sci. 25:460–463, 1968.
422. Kondratyev, K. Ya. The Complete Atmospheric Energetics Experiments. Global Atmospheric Research Programme (GARP) Publication Series No. 12. Geneva: World Meteorological Organization, 1973. 43 pp.
423. Koschmieder, H. Theorie der horizontalen Sichtweite. Beitr. Phys. Freien Atmos. 12:33–55, 1925.
424. Koschmieder, H. Theorie der horizontalen Sichtweite II: Kontrast und Sichtweite. Beitr. Phys. Freien Atmos. 12:171–181, 1925.
425. Kotrappa, P., and M. E. Light. Design and performance of the Lovelace aerosol particle separator. Rev. Sci. Instrum. 43:1106–1112, 1972.
426. Krahl, V. E., and M. H. Bulmash. Studies on living ciliated epithelium. Amer. Rev. Resp. Dis. 99:711–718, 1969.
427. Kratky, B. A., E. T. Fukunaga, J. W. Hylin, and R. T. Nakano. Volcanic air pollution: Deleterious effects on tomatoes. J. Environ. Quality 3:138–140, 1974.
428. LaBelle, C. W., and H. Brieger. Synergistic effects of aerosols. II. Effects on rate of clearance from lung. A.M.A. Arch. Ind. Health 20:100–105, 1959.
429. LaBelle, C. W., J. E. Long, and E. E. Cristofano. Synergistic effects of aerosols. Particulates as carriers of toxic vapors. Arch. Ind. Health 11:297–304, 1955.
430. Lamb, D., and L. Reid. Histochemical types of acidic glycoprotein produced by mucous cells of the tracheobronchial glands in man. J. Path. 98:213–229, 1969.
431. Lamb, D., and L. Reid. The tracheobronchial submucosal glands in cystic fibrosis: A qualitative and quantitative histochemical study. Brit. J. Dis. Chest 66:239–247, 1972.
432. Lambert, P. M., and D. D. Reid. Smoking, air pollution, and bronchitis in Britain. Lancet 1:853–857, 1970.
433. Landahl, H. D. On the removal of air-borne droplets by the human respiratory tract: I. The lung. Bull. Math. Biophys. 12:43–56, 1950.
434. Landahl, H. D. Particle removal by the respiratory system. Note on the removal of airborne particulates by the human respiratory tract with particular reference to the role of diffusion. Bull. Math. Biophys. 25:29–39, 1963.
435. Landahl, H. D., T. N. Tracewell, and W. H. Lassen. On the retention of air-borne particulates in the human lung: II. A.M.A. Arch. Ind. Hyg. Occup. Med. 3:359–366, 1951.
436. Landahl, H. D., T. N. Tracewell, and W. H. Lassen. Retention of air-borne particulates in the human lung: III. A.M.A. Arch. Ind. Hyg. Occup. Med. 6:508–511, 1952.
437. Landsberg, H. E. The climate of towns, pp. 584–606. In W. L. Thomas, Jr., Ed. Man's Role in Changing the Face of the Earth. Chicago: University of Chicago Press, 1956.
438. Lapp, N. L., J. L. Hankinson, H. Amandus, and E. D. Palmes. Variability in the size of airspaces in normal human lungs as estimated by aerosols. Thorax 30:293–299, 1975.

439. Lapp, N. L., and A. Seaton. Lung mechanics in coal workers' pneumoconiosis. Ann. N. Y. Acad. Sci. 200:433–454, 1972.
440. Larson, T. V., D. S. Covert, R. Frank, and R. J. Charlson. Ammonia in the human airways: Neutralization of inspired acid sulfate aerosols. Science 197:161–163, 1977.
441. Laurenzi, G. A., S. Yin, and J. J. Guarneri. Adverse effect of oxygen on tracheal mucus flow. New Eng. J. Med. 279:333–339, 1968.
442. Lave, L. B., and E. P. Seskin. Air pollution and human health. The quantitative effect, with an estimate of the dollar benefit of pollution abatement, is considered. Science 169:723–733, 1970.
443. Lave, L. B., and E. P. Seskin. Air pollution, climate, and home heating: Their effects on U.S. mortality rates. Amer. J. Public Health 62:909–916, 1972.
444. Lawther, P. J. Effects of inhalation of sulphur dioxide on respiration and pulse-rate in normal subjects. Lancet 2:745–748, 1955.
445. Lawther, P. J. Climate, air pollution and chronic bronchitis. Proc. Roy. Soc. Med. 51:262–266, 1958.
446. Lawther, P. J. Compliance with the Clean Air Act: Medical aspects. J. Inst. Fuel 36:341–344, 1963.
447. Lawther, P. J., A. G. F. Brooks, P. W. Lord, and R. E. Waller. Day-to-day changes in ventilatory function in relation to the environment. Part I. Spirometric values. Environ. Res. 7:27–40, 1974.
448. Lawther, P. J., A. G. F. Brooks, P. W. Lord, and R. E. Waller. Day-to-day changes in ventilatory function in relation to the environment. Part II. Peak expiratory flow values. Environ. Res. 7:41–53, 1974.
449. Lawther, P. J., A. G. F. Brooks, P. W. Lord, and R. E. Waller. Day-to-day changes in ventilatory function in relation to the environment. Part III. Frequent measurements of peak flow. Environ. Res. 8:119–130, 1974.
450. Lawther, P. J., R. E. Waller, and M. Henderson. Air pollution and exacerbations of bronchitis. Thorax 25:525–539, 1970.
451. Lawther, P. J., R. E. Waller, and A. E. Martin. Clean air and health in London, pp. 71–78. In National Society for Clean Air. Clean Air Conference, Eastbourne, 21–24 October 1969. Part 1, Pre-prints of Papers. London: Spottiswoode, Ballantyne and Co. Ltd., 1969.
452. Le Bouffant, L. Influence de la nature des poussières et de la charge pulmonair sur l'epuration, pp. 227–237. In W. H. Walton, Ed. Inhaled Particles III, Vol. I. Proceedings of an International Symposium organized by the British Occupational Hygiene Society in London, 14–23 September, 1970. Old Woking, Surrey: Unwin Brothers, Ltd., 1971.
453. Lebowitz, M. D., P. Bendheim, G. Cristea, D. Markovitz, J. Misiaszek, M. Staniec, and D. Van Wyck. The effect of air pollution and weather on lung function in exercising children and adolescents. Amer. Rev. Resp. Dis. 109:262–273, 1974.
454. Lecrenier, A., and G. Piquer. Essais sur l'action des poussières de cimenterie sur la végétation. Bull. Hort. 74:56–58, 1956.
455. Lee, R. E., Jr. The size of suspended particulate matter in air. Science 178:567–575, 1972.
456. Lee, R. E., Jr., S. S. Goranson, R. E. Enrione, and G. B. Morgan. National Air Surveillance cascade impactor network. II. Size distribution measurements of trace metal components. Environ. Sci. Technol. 6:1025–1030, 1972.

457. Lee, R. E., Jr., R. K. Patterson, and J. Wagman. Particle-size distribution of metal components in urban air. Environ. Sci. Technol. 2:288–290, 1968.
458. Leh, H. O. Verunreinigungen von Kulturpflanzen mit Blei aus Kraftfahrzeugabgasen. Ges. Pflanzen 18:21–24, 1966.
459. Leonov, L. F., P. S. Prokhorov, T. A. Efanova, and I. A. Zolotárev. Passivation of condensation nuclei by cetyl alcohol vapor, pp. 12–20. In V. A. Fedoseev, Ed. Advances in Aerosol Physics. No. 2. New York: Halsted Press, 1972.
460. Lerman, S. Cement-Kiln Dust and the Bean Plant (*Phaseolus vulgaris* L. Black Valentine var.); In-depth Investigations into Plant Morphology, Physiology and Pathology, Ph.D. thesis, University of California, Riverside, 1972.
461. Lerman, S. L., and Darley, E. F. Particulates, pp. 141–158. In J. B. Mudd and T. T. Kozlowski, Eds. Responses of Plants to Air Pollution. New York: Academic Press, 1975.
462. Levy, H., II. Normal atmosphere: Large radical and formaldehyde concentrations predicted. Science 173:141–143, 1971.
463. Lewis, T. R., W. J. Moorman, W. F. Ludmann, and K. I. Campbell. Toxicity of long-term exposure to oxides of sulfur. Arch. Environ. Health 26:16–21, 1973.
464. Liebow, A. A. Aspects of transcapillary fluid and protein exchange in the lung, pp. 99–112. In A. P. Fishman and H. H. Hecht, Eds. The Pulmonary Circulation and Interstitial Space. Chicago: The University of Chicago Press, 1969.
465. Lightbody, A., M. E. Roberts, and C. J. Wessel. Plastics and rubber, pp. 537–595. In G. A. Greathouse and C. J. Wessel, Eds. Deterioration of Materials. Causes and Preventive Techniques. New York: Reinhold Publishing Corporation, 1954.
466. Likens, G. E., and F. H. Bormann. Acid rain. A serious regional environmental problem. Science 184:1176–1179, 1974.
467. Lin, C., M. Baker, and R. J. Charlson. Absorption coefficient of atmospheric aerosol: A method for measurement. Appl. Opt. 12: 1356–1363, 1973.
468. Lindberg, J. D., and L. S. Laude. Measurement of the absorption coefficient of atmospheric dust. Appl. Opt. 13:1923–1927, 1974.
469. Lippmann, M. Deposition and clearance of inhaled particles in the human nose. Ann. Otol. Rhinol. Laryngol. 79:519–528, 1970.
470. Lippmann, M. "Respirable" dust sampling. Amer. Ind. Hyg. Assoc. J. 31:138–159, 1970.
471. Lippmann, M. Regional deposition of particles in the human respiratory tract, pp. 213–232. In D. H. K. Lee, H. L. Falk, and S. D. Murphy, Eds. Handbook of Physiology, Section 9: Reactions to Environmental Agents. Bethesda, MD: The American Physiological Society, 1977.
472. Lippmann, M., and R. E. Albert. The effect of particle size on the regional deposition of inhaled aerosols in the human respiratory tract. Amer. Ind. Hyg. Assoc. J. 30:257–275, 1969.
473. Lippmann, M., R. E. Albert, and H. T. Peterson, Jr. The regional deposition of inhaled aerosols in man, pp. 105–120. In W. H. Walton, Ed. Inhaled Particles III, Vol. I. Proceedings of an International Symposium organized by the British Occupational Hygiene Society in London, 14–23 September, 1970. Old Woking, Surrey: Unwin Brothers, Ltd., 1971.

474. Lippmann, M., and B. Altshuler. Regional deposition of aerosols, pp. 25–48. In E. F. Aharonson, A. Ben-David, and M. A. Klingberg, Eds. Air Pollution and the Lung. Proceedings of the Twentieth Annual "OHOLO" Biological Conference, Máalot, Israel, March 16–19, 1975. New York: John Wiley & Sons, Inc., 1976.
475. Litt, M. Flow behavior of mucus. Ann. Otol. Rhinol. Laryngol. 80:330–335, 1971.
476. Litt, M. Rheological aspects of mucociliary flow. Ann. N.Y. Acad. Sci. 221:212–213, 1974.
477. Litvin, A. Clear coatings for exposed architectural concrete. J. Portland Cem. Assoc. 10:49–57, 1968.
478. Liu, B. Y. H., Ed. Fine Particles. Aerosol Generation, Measurement, Sampling, and Analysis. Proceedings of a Symposium by the U.S. Environmental Protection Agency held in Minneapolis, Minnesota, May 28–30, 1975. New York: Academic Press, Inc., 1976. 837 pp.
479. Liu, B. Y. H., and K. W. Lee. An aerosol generator of high stability. Amer. Ind. Hyg. Assoc. J. 36:861–865, 1975.
480. Liu, B. Y. H., and D. Y. H. Pui. On the performance of the electrical aerosol analyzer. J. Aerosol Sci. 6:249–264, 1975.
481. Liu, B. Y. H., and D. Y. H. Pui. Electrical Behavior of Aerosols. Presented at the 82nd National Meeting of American Institute of Chemical Engineers, Atlantic City, August, 1976.
482. Lodge, J. P., Jr., and E. R. Frank. Chemical identification of some atmospheric components in the Aitken nucleus size range. J. Rech. Atmospheriques 2:139–140, 1966.
483. Logan, W. P. D. Mortality in the London fog incident. Lancet 1:336–338, 1953.
484. Loo, B. W., J. M. Jaklevic, and F. S. Goulding. Dichotomous virtual impactors for large scale monitoring of airborne particulate matter, pp. 311–350. In B. Y. H. Liu, Ed. Fine Particles. Aerosol Generation, Measurement, Sampling, and Analysis. Proceedings of a Symposium by the U.S. Environmental Protection Agency held in Minneapolis, Minnesota, May 28–30, 1975. New York: Academic Press, Inc., 1976.
485. Lourenço, R. V., M. F. Klimek, and C. J. Borowski. Deposition and clearance of 2 μ particles in the tracheobronchial tree of normal subjects—smokers and nonsmokers. J. Clin. Invest. 50:1411–1419, 1971.
486. Lourenço, R. V., R. Loddenkemper, and R. W. Carton. Patterns of distribution and clearance of aerosols in patients with bronchiectasis. Amer. Rev. Resp. Dis. 106:857–866, 1972.
487. Love, G. J., A. A. Cohen, J. F. Finklea, J. G. French, G. R. Lowrimore, W. C. Nelson, and P. B. Ramsey. Prospective surveys of acute respiratory disease in volunteer families: 1970–1971 New York studies, pp. 5–49 to 5–69. In U.S. Environmental Protection Agency, Office of Research and Development. Health Consequences of Sulfur Oxides: A Report from CHESS, 1970–1971. EPA-650/1-74-004. Washington, D.C.: U.S. Government Printing Office, 1974.
488. Love, R. G., and D. C. F. Muir. Aerosol deposition and airway obstruction. Amer. Rev. Resp. Dis. 114:891–897, 1976.
489. Lovejoy, F. W., Jr., H. Constantine, J. Flatley, N. Kaltreider, and L. Dautrebande. Measurement of gas trapped in the lungs during acute

changes in airway resistance in normal subjects and in patients with chronic pulmonary disease. Amer. J. Med. 30:884–892, 1961.

490. Lowe, C. R., H. Campbell, and T. Khosla. Bronchitis in two integrated steel works. III. Respiratory symptoms and ventilatory capacity related to atmospheric pollution. Brit. J. Ind. Med. 27:121–129, 1970.

491. Lowe, C. R., P. L. Pelmear, H. Campbell, R. A. N. Hitchens, T. Khosla, and T. C. King. Bronchitis in two integrated steel works. I. Ventilatory capacity, age, and physique of non-bronchitic men. Brit. J. Prev. Soc. Med. 22:1-11, 1968.

492. Lucas, A. M., and L. C. Douglas. Principles underlying ciliary activity in the respiratory tract. II. A comparison of nasal clearance in man, monkey and other mammals. Arch. Otolaryng. 20:518–541, 1934.

493. Luchsinger, P. C., B. LaGarde, and J. E. Kilfeather. Particle clearance from the human tracheobronchial tree. Amer. Rev. Resp. Dis. 97: 1046–1050, 1968.

494. Ludwig, J. H., G. B. Morgan, and T. B. McMullen. Trends in urban air quality, pp. 321–338. In W. H. Matthew, W. W. Kellogg, and G. O. Robinson, Eds. Man's Impact on the Climate. Cambridge: The MIT Press, 1971.

495. Lundgren, D. A. Mass Distribution of Large Atmospheric Particles. Ph.D. Thesis, University of Minnesota, Minneapolis, Minn., 1973. 161 pp.

496. Lundgren, D. A., L. D. Carter, and P. S. Daley. Aerosol mass measurement using piezoelectric crystal sensors, pp. 485–510. In B. Y. H. Liu, Ed. Fine Particles. Aerosol Generation, Measurement, Sampling, and Analysis. Proceedings of a Symposium by the U.S. Environmental Protection Agency held in Minneapolis, Minnesota, May 28–30, 1975. New York: Academic Press, Inc., 1976.

497. Lunn, J. E., J. Knowelden, and A. J. Handyside. Patterns of respiratory illness in Sheffield infant school children. Brit. J. Prev. Soc. Med. 21:7–16, 1967.

498. Lunn, J. E., J. Knowelden, and J. W. Roe. Patterns of respiratory illness in Sheffield junior school children. A follow-up study. Brit. J. Prev. Soc. Med. 24:223–228, 1970.

499. Lyons, W. A., and R. B. Husar. SMS/GOES visible images detect a synoptic-scale air pollution episode. Mon. Weather Rev. 104:1623–1626, 1976.

500. Lyons, W. A., and S. R. Pease. Detection of particulate air pollution plumes from major point sources using ERTS-1 imagery. Bull. Amer. Meteor. Soc. 54:1163–1170, 1973.

501. MacFarland, H. N., C. E. Ulrich, A. Martin, A. Krumm, W. M. Busey, and Y. Alarie. Chronic exposure of cynomolgus monkeys to fly ash, pp. 313–327. In W. H. Walton, Ed. Inhaled Particles III, Vol. 1. Proceedings of an International Symposium organized by the British Occupational Hygiene Society in London, 14–23 September 1970. Old Woking, Surrey: Unwin Brothers Limited, 1971.

502. Macias, E. S., and R. B. Husar. A review of atmospheric particulate mass measurement via the beta attenuation technique, pp. 535–564. In B. Y. H. Liu, Ed. Fine Particles. Aerosol Generation, Measurement, Sampling, and Analysis. Proceedings of a Symposium by the U.S. Environmental Protection Agency held in Minneapolis, Minnesota, May 28–30, 1975. New York: Academic Press, Inc., 1976.

503. Macklin, C. C. Pulmonary sumps, dust accumulations, alveolar fluid and lymph vessels. Acta Anat. 23:1–33, 1955.
504. Manos, N. E. Comparative Mortality among Metropolitan Areas of the United States, 1949–1951, 102 Causes of Death. Ratios of Observed to Expected Mortality for 163 Metropolitan Areas, Mortality Rates and Ratios by Degree of Urbanization for the Total United States, and other Background Information Necessary to the Interpretation of these Rates and Ratios. PHS Publication 562. Washington, D.C.: U.S. Government Printing Office, 1957. 143 pp.
505. Martin, M. G., and P. E. Morrow. Effect of changing inspired O_2 and CO_2 levels on trachel mucociliary transport rate. J. Appl. Physiol. 27:385–388, 1969.
506. Marple, V. A., and K. Willeke. Inertial impactors: Theory, design and use, pp. 412–446. In B. Y. H. Liu, Ed. Fine Particles. Aerosol Generation, Measurement, Sampling, and Analysis. Proceedings of a Symposium by the U.S. Environmental Protection Agency held in Minneapolis, Minnesota, May 28–30, 1975. New York: Academic Press, Inc., 1976.
507. Märtens, A., and W. Jacobi. Die in-vivo-Bestimmung der Aerosolteilchendeposition im Atemtrakt bei Mund- bzw. Nasenatmung, pp. 117–121. In V. Böhlau, Ed. Aerosole in Physik, Medizin and Technik. Kongressbericht der Aerosoltagung im Taunus-Sanatorium am 17. und 18. Oktober 1973 der LVA Wttbg. Bad Soden, West Germany: Gesellschaft für Aerosolforschung, 1973.
508. Martin, A. E. Mortality and morbidity statistics and air pollution. Proc. Roy. Soc. Med. 57:969–975, 1964.
509. Martin, A. E., and W. H. Bradley. Mortality, fog, and atmospheric pollution. An investigation during the winter of 1958–59. Monthly Bull. Min. Health Pub. Health Lab. Serv. 19:56–73, 1960.
510. Martin, S. W., and R. A. Willoughby. Organic dusts, sulfur dioxide, and the respiratory tract of swine. Arch. Environ. Health 25:158–165, 1972.
511. Martinez-Tello, F. J., D. G. Braun, and W. A. Blanc. Immunoglobulin production in bronchial mucosa and bronchial lymph nodes, particularly in cystic fibrosis of the pancreas. J. Immunol. 101:989–1003, 1968.
512. Massachusetts Institute of Technology. Man's Impact on the Global Environment. Assessment and Recommendations for Action. Report of the Study of Critical Environmental Problems (SCEP). Cambridge, Mass.: MIT Press, 1970. 319 pp.
513. Matsuba, K., and W. M. Thurlbeck. The number and dimensions of small airways in nonemphysematous lungs. Amer. Rev. Resp. Dis. 104:516–524, 1971.
514. Matteson, M., W. Stöber, and H. Luther. Kinetics of the oxidation of sulfur dioxide by aerosols of manganese sulfate. Ind. Eng. Chem. Fund. 8:677–687, 1969.
515. Matthews, L. W., S. Spector, J. Lemm, and J. L. Potter. Studies on pulmonary secretions. I. The over-all chemical composition of pulmonary secretions from patients with cystic fibrosis, bronchiectasis, and laryngectomy. Amer. Rev. Resp. Dis. 88:199–204, 1963.
516. Matthys, H., M. Müller, and N. Konietzko. Quantitative and selective bronchial clearance studies using $^{99m}T_c$-sulfate particles. Scand. J. Resp. Dis. Suppl. 85:33–37, 1974.
517. May, K. R. The cascade impactor: An instrument for sampling coarse aerosols. J. Sci. Instrum. 22:187–195, 1945.

518. McBurney, J. W. Effect of atmosphere on masonry and related materials, pp. 45–56. In Symposium on Some Approaches to Durability in Structures. ASTM Special Technical Publication No. 236. Philadelphia: American Society for Testing and Materials, 1959.
519. McCarroll, J., and W. Bradley. Excess mortality as an indicator of health effects of air pollution. Amer. J. Public Health 56:1933–1942, 1966.
520. McCarroll, J., E. J. Cassell, D. W. Wolter, J. D. Mountain, J. R. Diamond, and I. M. Mountain. Health and urban environment. V. Air pollution and illness in a normal urban population. Arch. Environ. Health 14:178–184, 1967.
521. McCarthy, C., and L. Reid. Intracellular mucopolysaccharides in the normal human bronchial tree. Q. J. Exp. Physiol. 49:85–94, 1964.
522. McCarthy, D. S., R. Spencer, R. Greene, and J. Milic-Emili. Measurement of "closing volume" as a simple and sensitive test for early detection of small airway disease. Amer. J. Med. 52:747–753, 1972.
523. McCormick, R. A., and J. H. Ludwig. Climate modification by atmospheric aerosols. Science 156:1358–1360, 1967.
524. McCune, D. C., A. E. Hitchcock, J. S. Jacobson, and L. H. Weinstein. Fluoride accumulation and growth of plants exposed to particulate cryolite in the atmosphere. Contrib. Boyce Thompson Inst. 23:1-12, 1965.
525. McCune, D. C., D. H. Silberman, and L. H. Weinstein. Effects of relative humidity and free water on the phytotoxicity of hydrogen fluoride and cryolite, pp. 116–119. In Proceedings of the Fourth International Clean Air Congress, Tokyo, 16–20 May 1977. Tokyo: Japanese Union of Air Pollution Prevention Associations, 1977.
526. McJilton, C., R. Frank, and R. Charlson. Role of relative humidity in the synergistic effect of a sulfur dioxide-aerosol mixture on the lung. Science 182:503–504, 1973.
527. McJilton, C., J. Thielke, and R. Frank. Ozone Uptake Model for the Respiratory System. Paper presented at American Industrial Hygiene Association Conference, San Francisco, California, May 14–19, 1972. (Abstr.: Amer. Ind. Hyg. Assoc. J. 33:20, 1972.)
528. McKinney, N., and H. W. Hermance. Stress corrosion cracking rates of a nickel-brass alloy under applied potential, pp. 274–291. In American Society for Testing and Materials. Symposium on Stress Corrosion Testing, presented at the sixty-ninth annual meeting, Atlantic City, 26 June–1 July 1966. ASTM Special Technical Publication No. 425. Philadelphia: American Society for Testing and Materials, 1967.
529. McLaughlin, R. F., Jr., W. S. Tyler, and R. O. Canada. Subgross pulmonary anatomy of the rabbit, rat, and guinea pig, with additional notes on the human lung. Amer. Rev. Resp. Dis. 94:380–387, 1966.
530. McLeod, W., and R. R. Rogers. Corrosion of metals by aqueous solutions of the atmospheric pollutant sulfurous acid. Electrochem. Technol. 6:231–235, 1968.
531. McLeod, W., and R. R. Rogers. The nature of corrosion of zinc by sulfurous acid at ordinary temperatures. Corrosion 25:74–76, 1969.
532. Mead, J. The lung's "quiet zone." New Eng. J. Med. 282:1318–1319, 1970.
533. Mercer, T. T. On the role of particle size in the dissolution of lung burdens. Health Phys. 13:1211–1221, 1967.
534. Mercer, T. T. Aerosol Technology in Hazard Evaluation. New York: Academic Press, 1973. 394 pp.

535. Mercer, T. T., P. E. Morrow, and W. Stöber, Eds. Assessment of Airborne Particles. Fundamentals, Applications, and Implications to Inhalation Toxicity. Springfield, Ill.: Charles C Thomas, 1972. 540 pp.
536. Meyrick, B., J. M. Sturgess, and L. Reid. A reconstruction of the duct system and secretory tubules of the human bronchial submucosal gland. Thorax 24:729–736, 1969.
537. Middleton, J. T., E. F. Darley, and R. F. Brewer. Damage to vegetation from polluted atmospheres. J. Air Pollut. Control Assoc. 8:9–15, 1958.
538. Middleton, W. E. K. Vision Through the Atmosphere. Toronto: University of Toronto Press, 1952. 250 pp.
539. Milburn, R. H., W. L. Crider, and S. D. Morton. The retention of hygroscopic dusts in the human lungs. A.M.A. Arch. Ind. Health 15:59–62, 1957.
540. Miller, C. E. Flow induced by mechanical analogues of mucociliary systems. Ann. N. Y. Acad. Sci. 130:880–890, 1966.
541. Miller, C. E. The kinematics and dynamics of ciliary fluid systems. J. Exp. Biol. 49:617–629, 1968.
542. Miller, C. E. Streamlines, streak lines, and particle path lines associated with a mechanically-induced flow homorphic with the mammalian mucociliary system. Biorheology 6:127–135, 1969.
543. Miller, M. S., S. K. Friedlander, and G. M. Hidy. A chemical element balance for the Pasadena aerosol, pp. 301–312. In G. M. Hidy, Ed. Aerosols and Atmospheric Chemistry. Proceedings of the American Chemical Society. Los Angeles, California, March 28–April 2, 1971. New York: Academic Press, 1972.
544. Miller. P. M., and S. Rich. Soot damage to greenhouse plants. Plant Dis. Rep. 51:712, 1967.
545. Miller, W. S. Muscle in the pleura, pp. 155–156. In The Lung. (2nd ed.) Springfield, Ill.: Charles C Thomas, 1947.
546. Ministry of Health (Great Britain). Mortality and Morbidity During the London Fog of December 1952. Report by a Committee of Departmental Officers and Expert Advisers appointed by the Minister of Health. Report No. 95. London: Her Majesty's Stationery Office, 1954. 61 pp.
547. Ministry of Pensions and National Insurance. Report on an Enquiry into the Incidence of Incapacity for Work. Part II. Incidence of Incapacity for Work in Different Areas and Occupations. London: Her Majesty's Stationery Office, 1965. 163 pp.
548. Mitchell, J. M., Jr. Summary of the problem of air pollution effects on climate, pp. 164–175. In W. H. Matthews, W. W. Kellogg, and G. O. Robinson, Eds. Man's Impact on the Climate. Cambridge: The MIT Press, 1971.
549. Mitchell, J. M., Jr. The effect of atmospheric particles on radiation and temperature, pp. 295–301. In W. H. Matthews, W. W. Kellogg, and G. O. Robinson, Eds. Man's Impact on the Climate. Cambridge: The MIT Press, 1971.
550. Moore, D. G., and A. Potter. Effect of Exposure Site on Weather Resistance of Porcelain Enamels Exposed for Three Years. National Bureau of Standards Monograph 44. Washington, D.C.: U.S. Government Printing Office, 1962. 13 pp.
551. Morgan, A., A Black, J. C. Evans, E. H. Hadfield, R. G. Macbeth, and M. Walsh. Impairment of nasal mucociliary clearance in woodworkers in the furniture industry, pp. 335–338. In Proceedings, First International

Congress on Aerosols in Medicine, 19–21 September 1973, Baden/Wien, Austria. Baden/Wien: Internationale Gesellschaft für Aerosole in der Medizin, 1973.

552. Mork, T. A comparative study of respiratory disease in England and Wales and Norway. Acta Med. Scand. 172 (Suppl. 384):1–100, 1962.
553. Morrow, P. E. Lymphatic drainage of the lung in dust clearance. Ann. N.Y. Acad. Sci. 200:46–65, 1972.
554. Morrow, P. E. Alveolar clearance of aerosols. Arch. Int. Med. 131:101–108, 1973.
555. Morrow, P. E. An evaluation of recent NO_x toxicity data and an attempt to derive an ambient air standard for NO_x by established toxicological procedures. Environ. Res. 10:92–112, 1975.
556. Morrow, P. E., F. R. Gibb, H. Davies, and M. Fisher. Dust removal from the lung parenchyma: An investigation of clearance simulants. Toxicol. Appl. Pharmacol. 12:372–396, 1968.
557. Morrow, P. E., F. R. Gibb, and K. M. Gazioglu. A study of particulate clearance from the human lungs. Amer. Rev. Resp. Dis. 96:1209–1221, 1967.
558. Morrow, P. E., F. R. Gibb, and K. Gazioglu. The clearance of dust from the lower respiratory tract of man. An experimental study, pp. 351–358. In C. N. Davies, Ed. Inhaled Particles and Vapours II. Proceedings of an International Symposium organized by the British Occupational Hygiene Society, Cambridge, 28 September–1 October 1965. Oxford: Pergamon Press, 1967.
559. Morrow, P. E., F. R. Gibb, and L. Johnson. Clearance of insoluble dust from the lower respiratory tract. Health Phys. 10:543–555, 1964.
560. Mossop, S. C. Comparisons between concentration of ice crystals in cloud and the concentration of ice nuclei. J. Rech. Atmospheriques 3:119–124, 1968.
561. Motto, H. L., R. H. Daines, D. M. Chilko, and C. K. Motto. Lead in soils and plants: Its relationship to traffic volume and proximity to highways. Environ. Sci. Technol. 4:231–237, 1970.
562. Muir, D. C. F., and C. N. Davies. The deposition of 0.5-μ diameter aerosols in the lungs of man. Ann. Occup. Hyg. 10:161–174, 1967.
563. Mulcahy, M. F. R., J. R. Steven, and J. C. Ward. The kinetics of reaction between oxygen atoms and sulfur dioxides: An investigation by electron spin resonance spectrometry. J. Phys. Chem. 71:2124–2131, 1967.
564. Munn, R. E., and B. Bolin. Global air pollution—meteorological aspects. A survey. Atmos. Environ. 5:363–402, 1971.
565. Nadel, J. A. Aerosol effects on smooth muscle and airway visualization technique. Arch. Intern. Med. 131:83–87, 1973.
566. Nadel, J. A., H. Salem, B. Tamplin, and Y. Tokiwa. Mechanism of bronchoconstriction during inhalation of sulfur dioxide. J. Appl. Physiol. 20:164–167, 1965.
567. Nadel, J. A., and J. G. Widdicombe. Reflex effects of upper airway irritation on total lung resistance and blood pressure. J. Appl. Physiol. 17:861–865, 1962.
568. Nagaishi, C. Functional Anatomy and Histology of the Lung. Baltimore: University Park Press, 1972. 295 pp.
569. Nakamura, K. Response of pulmonary airway resistance by interaction of aerosols and gases in different physical and chemical nature. Jap. J. Hyg. 19:322–333, 1964. (in Japanese). (Translated by EPA; available from the

Air Pollution Technical Center, Research Triangle Park, N.C. as APTIC No. 11425.)

570. National Academy of Sciences, Committee on Geological Sciences. Enter pollutants, pp. 29–30. In The Earth and Human Affairs. San Francisco: Canfield Press, 1972.

571. National Research Council, Committee on Biologic Effects of Atmospheric Pollutants. Lead: Airborne Lead in Perspective. Washington, D.C.: National Academy of Sciences, 1972. 330 pp.

572. National Research Council, Committee on Biologic Effects of Atmospheric Pollutants. Particulate Polycyclic Organic Matter. Washington, D.C.: National Academy of Sciences, 1972. 361 pp.

573. National Research Council, Subcommittee on Nitrogen Oxides, Committee on Medical and Biologic Effects of Environmental Pollutants. Nitrogen Oxides. Washington, D.C.: National Academy of Sciences, 1977. 330 pp.

574. Natusch, D. F. S., and J. R. Wallace. Urban aerosol toxicity: The influence of particle size. Science 186:695–699, 1974.

575. Natusch, D. F. S., J. R. Wallace, and C. A. Evans, Jr. Toxic trace elements: Preferential concentration in respirable particles. Science 183:202–204, 1974.

576. Negus, V. E. The function of mucus. Acta Otolaryng. 56:204–213, 1963.

577. Nelson, W. C., J. F. Finklea, D. E. House, D. C. Calafiore, M. B. Hertz, and D. H. Swanson. Frequency of acute lower respiratory disease in children: Retrospective survey of Salt Lake Basin communities, 1967–1970, pp. 2-55 to 2-73. In U.S. Environmental Protection Agency, Office of Research and Development. Health Consequences of Sulfur Oxides: A Report from CHESS, 1970–1971. EPA-650/1-74-004. Washington, D.C.: U.S. Government Printing Office, 1974.

578. Nenot, J. C. Etude de l'irradiation sur l'epuration pulmonaire, pp. 239–246. In W. H. Walton, Ed. Inhaled Particles III, Vol. I. Proceedings of an International Symposium organized by the British Occupational Hygiene Society in London, 14–23 September, 1970. Old Woking, Surrey: Unwin Brothers, Ltd., 1971.

579. Niewoehner, D. E., and J. Kleinerman. Morphologic basis of pulmonary resistance in the human lung and effects of aging. J. Appl. Physiol. 316:412–418, 1974.

580. Noll, K. E., P. K. Mueller, and M. Imada. Visibility and aerosol concentration in urban air. Atmos. Environ. 2:465–475, 1968.

581. Norris, R. M., and J. M. Bishop. The effect of calcium carbonate dust on ventilation and respiratory gas exchange in normal subjects and in patients with asthma and chronic bronchitis. Clin. Sci. 30:103–115, 1966.

582. Novakov, T., S. G. Chang, and A. B. Harker. Sulfates as pollution particulates: Catalytic formation on carbon (soot) particles. Science 186:259–261, 1974.

583. Nuessle, A. C. Soil and stain resistance, pp. 465–505. In H. F. Mark, N. S. Wooding, and S. M. Atlas, Eds. Chemical Aftertreatment of Textiles. New York: Wiley-Interscience, 1971.

584. Ogden, T. L. The effect of rainfall on a large steelworks. J. Appl. Meteorol. 8:585–591, 1969.

585. Olsen, H. C., and J. C. Gilson. Respiratory symptoms, bronchitis, and ventilatory capacity in man. An Anglo-Danish comparison, with special reference to differences in smoking habits. Brit. Med. J. 1:450–456, 1960.

586. Olsen, D. E., M. F. Sudlow, K. Hosfield, and G. F. Filley. Convective patterns of flow during inspiration. Arch. Intern. Med. 131:51–57, 1973.
587. Oma, K., T. Sugaro, T. Ueki, and Y. Hirai. Studies on atmospheric corrosion of steels related to meteorological factors in Japan. II. Multiple correlation of meteorological and atmospheric pollution substances on corrosion of steel. Division of Building Research Technical Translation 1281. (Translated from Boshoku Gijutsu (Corros. Eng.) 14:16–19, 1965) Ottawa: National Research Council of Canada, 1965.
588. Orr, C., Jr., F. K. Hurd, and W. J. Corbett. Aerosol size and relative humidity. J. Colloid Sci. 13:472–482, 1958.
589. Pack, M. R., A. C. Hill, M. D. Thomas, and L. G. Transtrum. Determination of gaseous and particulate inorganic fluorides in the atmosphere, pp. 27–44. In American Society for Testing Materials Special Technical Publication No. 281. Symposium on Air Pollution Control, San Francisco, California, October 16, 1959. Philadelphia: American Society for Testing Materials, 1960.
590. Page, A. L., and T. J. Ganje. Accumulations of lead in soils for regions of high and low motor vehicle traffic density. Environ. Sci. Technol. 4:140–142, 1970.
591. Page, A. L., T. J. Ganje, and M. S. Joshi. Lead quantities in plants, soil, and air near some major highways in southern California. Hilgardia 41:1–31, 1971.
592. Pajenkamp, H. Einwirkung des Zementofenstaubes auf Pflanzen und Tiere. Zement-Kalk-Gips 14:88–95, 1961.
593. Palmer, J. D. Corrosion control involves pollution too. Can. Chem. Process. 53(10):78–81, 1969.
594. Palmer, K. N. V., D. Ballantyne, M. L. Diament, and W. F. D. Hamilton. The rheology of bronchitic sputum. Brit. J. Dis. Chest 64:185–191, 1970.
595. Palmes, E. D., and M. Lippmann. Influence of respiratory air space dimensions on aerosol deposition, pp. 127–136. In W. H. Walton, Ed. Inhaled Particles IV. London: Pergamon Press Ltd, 1977.
596. Panda, S. C., R. C. Nayak, and I. C. Mahapatra. A note on effect of cement dust on the soil fertility and crop growth around the factory area of Rajgangpur (Orissa). Indian J. Agron. 18:389–391, 1973.
597. Parish, S. B. The effect of cement dust on citrus trees. Plant World 13:288–291, 1910.
598. Patterson, R. K., and J. Wagman, Mass and composition of an urban aerosol as a function of particle size for several visibility levels. J. Aerosol Sci. 8:269–279, 1977.
599. Pattle, R. E. The retention of gases and particles in the human nose, pp. 302–309. In C. N. Davies, Ed. Inhaled Particles and Vapours. Oxford: Pergamon Press, 1961.
600. Pavia, D., M. D. Short, and M. L. Thomson. No demonstrable long term effects of cigarette smoking on the mucociliary mechanism of the human lung. Nature 226:1228–1231, 1970.
601. Pavia, D., and M. L. Thomson. Inhibition of mucociliary clearance from the human lung by hyoscine. Lancet 1:449–450, 1971.
602. Payne, W., L. J. Stief, and D. D. Davis. A kinetics study of the reaction of HO_2 with SO_2 and NO. J. Amer. Chem. Soc. 95:7614–7619, 1973.
603. Pedace, E. A., and E. B. Sansone. The relationship between "soiling index" and suspended particulate matter concentrations. J. Air Pollut. Control Assoc. 22:348–351, 1972.

604. Pedley, T. J., R. C. Schroter, and M. F. Sudlow. Flow and pressure drop in systems of repeatedly branching tubes. J. Fluid Mech. 46:365–383, 1971.
605. Peirce, G. J. An effect of cement dust on orange trees. Plant World 13:283–288, 1910.
606. Pemberton, J., and C. Goldberg. Air pollution and bronchitis. Brit. Med. J. 2:567–570, 1954.
607. Penkett, S. A. Oxidation of SO_2 and other atmospheric gases by ozone in aqueous solution. Nature (Phys. Sci.) 240:105–106, 1972.
608. Penkett, S. A., and J. A. Garland. Oxidation of sulphur dioxide in artificial fogs by ozone. Tellus 26(1-2):284–290, 1974.
609. Petrilli, F. L., G. Agnese, and S. Kanitz. Epidemiologic studies of air pollution effects in Genoa, Italy. Arch. Environ. Health 12:733–740, 1966.
610. Pierce, R. C., and M. Katz. Dependency of polynuclear aromatic hydrocarbon content on size distribution of atmospheric aerosols. Environ. Sci. Tech. 9:347–353, 1975.
611. Pilat, M. J. Application of gas-aerosol adsorption data to the selection of air quality standards. J. Air Pollut. Control Assoc. 18:751–753, 1968.
612. Podzimek, J., W. A. Sedlacek, and J. B. Haberl. Aircraft measurements of Aitken nuclei in the lower stratosphere. Technical report to the Office of Navel Research, Contract NOOO14-75-C-0413 and to the U.S. Department of Transportation, CIAP, from the Graduate Center for Cloud Physics Research. Rolla, MO: University of Missouri, 1975.
613. Policard, A., and P. Galy. Les Bronches. Structures et Méchanisms a l'État Normal et Pathologique. Paris: Masson et Cie, 1945. 191 pp.
614. Porch, W. M., D. S. Ensor, R. J. Charlson, and J. Heintzenberg. Blue moon: Is this a property of background aerosol? Appl. Opt. 12:34–36, 1973.
615. Porstendörfer, J. Untersuchungen zur Frage des Wachstums von inhalierten Aerosolteilchen im Atemtrakt. Aerosol Sci. 2:73–79, 1971.
616. Potter, J. L., L. W. Matthews, J. Lemm, and S. Spector. Human pulmonary secretions in health and disease. VI. Studies on mucus in relation to human disease. Ann. N.Y. Acad. Sci. (Art. 2) 106:692–697, 1963.
617. Potter, J. L., L. W. Matthews, S. Spector, and J. Lemm. Studies on pulmonary secretions. II. Osmolality and the ionic environment of pulmonary secretions from patients with cystic fibrosis, bronchiectasis, and laryngectomy. Amer. Rev. Resp. Dis. 96:83–87, 1967.
618. Prager, M. J., E. R. Stephens, and W. E. Scott. Aerosol formation from gaseous air pollutants. I/EC Ind. Eng. Chem. 52:521–524, 1960.
619. Pressley, H. Some effects of attack on refractories by the oxides of sodium, sulphur, and vanadium. Trans. Brit. Ceram. Soc. 69:205–210, 1970.
620. Prindle, R. A., G. W. Wright, R. O. McCaldin, S. C. Marcus, T. C. Lloyd, and W. E. Bye. Comparison of pulmonary function and other parameters in two communities with widely different air pollution levels. Amer. J. Public Health 53:200–217, 1963.
621. Proctor, D. F., Ib. Andersen, and G. Lundqvist. Clearance of inhaled particles from the human nose. Arch. Intern. Med. 131:132–139, 1973.
622. Prospero, J. M., and E. Bonatti. Continental dust in the atmosphere of the eastern equatorial Pacific. J. Geophys. Res. 74:3362–3371, 1969.
623. Prospero, J. M., and T. N. Carlson. Vertical and areal distribution of Saharan dust over the western equatorial North Atlantic. J. Geophys. Res. 77:5255–5265, 1972.

624. Pueschel, R. F., R. J. Charlson, and N. C. Ahlquist. On the anomalous deliquescence of sea-spray aerosols. J. Appl. Meteorol. 8:995–998, 1969.
625. Pueschel, R. F., and G. Langer. Sugar cane fires as a source of ice nuclei in Hawaii. J. Appl. Meteorol. 12:549–551, 1973.
626. Radke, L. F., and F. M. Turner. An improved automatic cloud condensation nucleus counter. J. Appl. Meteorol. 11:407–409, 1972.
627. Rao, A. K. An Experimental Study of Inertial Impactors. Ph.D. Thesis, University of Minnesota, Minn., 1975. 212 pp.
628. Rasmussen, R. A., and F. W. Went. Volatile organic material of plant origin in the atmosphere. Proc. Nat. Acad. Sci. U.S.A. 53:215–220, 1965.
629. Rasool, S. I., and S. H. Schneider. Atmospheric carbon dioxide and aerosols: Effects of large increases in global climate. Science 173:138–141, 1971.
630. Ray, F. K., P. B. Mack, F. Bonnet, and A. H. Wachter. A comparison of the effect on rayon fabrics of various gases under controlled conditions. Amer. Dyest. Rep. 37:391–396, 1948.
Rayleigh, Lord. *See* Strutt, J. W.
631. Raymond, V., and R. Nussbaum. A propos des poussières de cimenteries et de leurs effets sur l'homme, les plants et les animaux. Pollut. Atmos. 8:284–294, 1966.
632. Registrar-General, Great Britain. Appendix, pp. 71–362. In Third Annual Report of the Registrar-General of Births, Deaths and Marriages in England. London: Her Majesty's Stationery Office, 1841.
633. Reid, D. D. Environmental factors in respiratory disease. Lancet 1:1289–1294, 1958.
634. Reid, D. D. Air pollution as a cause of chronic bronchitis. Proc. Roy. Soc. Med. 57:965–968, 1964.
635. Reid, D. D. The beginnings of bronchitis. Proc. Roy. Soc. Med. 62:311–316, 1969.
636. Reid, D. D., D. O. Anderson, B. G. Ferris, and C. M. Fletcher. An Anglo-American comparison of the prevalence of bronchitis. Brit. Med. J. 2:1487–1491, 1964.
637. Reid, L. Measurement of the bronchial mucous gland layer: A diagnostic yardstick in chronic bronchitis. Thorax 15: 132–141, 1960.
638. Reid, L. Bronchial mucus production in health and disease, pp. 87–108. In A. A. Liebow, and D. E. Smith, Eds. The Lung. Baltimore: The Williams and Wilkins Company, 1968.
639. Reid, L. Evaluation of model systems for study of airway epithelium, cilia, and mucus. Arch. Intern. Med. 126:428–434, 1970.
640. Reid, L. The bronchitic component in airways obstruction. Bull. Physiopathol. Resp. 9:913–923, 1973.
641. Renzetti, N. A., and G. J. Doyle. Photochemical aerosol formation in sulfur dioxide-hydrocarbon systems. Int. J. Air Pollut. 2:327–345, 1960.
642. Rhodin, J. G. Ultrastructure and function of the human tracheal mucosa. Amer. Rev. Resp. Dis. 93 (Suppl.—Symposium on Structure, Function and Measurement of Respiratory Cilia. Duke University Medical Center, Durham, North Carolina, February 18–19, 1965):41–59, 1966.
643. Rhodin, J. G., and T. Dalhamn. Electron microscopy of the tracheal ciliated mucosa in rat. Z. Zellforsch. Mikrosk. Anat. 44:345–412, 1956.
644. Ricks, G. R., and R. J. H. Williams. Effects of atmospheric pollution on deciduous woodland. Part 2: Effects of particulate matter upon stomatal

diffusion resistance in leaves of *Quercus petraea* (mattuschka) Leibl. Environ. Pollut. 6:87–109, 1974.

645. Robinson, E. Effect on the physical properties of the atmosphere, pp. 349–400. In A. C. Stern, Ed. Air Pollution. Vol. I. Air Pollution and Its Effects. New York: Academic Press, 1968.

646. Robinson, E., and R. C. Robbins. Sources, Abundance, and Fate of Gaseous Atmospheric Pollutants. Prepared for the American Petroleum Institute by the Stanford Research Institute. SRI Project PR-6755. Menlo Park, California: Stanford Research Institute, 1968. 123 pp.

647. Rodhe, H., and J. Grandell. On the removal time of aerosol particles from the atmosphere by precipitation scavenging. Tellus 24:442–454, 1972.

648. Roosen, R. G., R. J. Angione, and C. H. Klemcke. Worldwide variations in atmospheric transmission: 1. Baseline results from Smithsonian observations. Bull. Amer. Meteorol. Soc. 54:307–316, 1973.

649. Rosenbaum, S. Home localities of national servicement with respiratory disease. Brit. J. Prev. Soc. Med. 15:61–67, 1961.

650. Ross, S. M. A Wavy Wall Analytical Model of Mucociliary Pumping. Ph.D. thesis, The Johns Hopkins University, 1971. 305 pp.

651. Ross, S. M., and S. Corrsin. Results of an analytical model of mucociliary pumping. J. Appl. Physiol. 37:333–340, 1974.

652. Roussel, P., P. Degand, A. Randoux, and R. Havez. Activités fonctionnelles des mucines bronchiques, pp. 15–28. In Progress in Respiration Research, Vol. 6. Chronic Inflammation of the Bronchi. Basel: S. Karger, 1971.

653. Rudolf, G., and J. Heyder. Deposition of aerosol particles in the human nose. In V. Böhlau, Ed. Aerosole in Naturwissenschaft, Medizin and Technik. Proceedings of a conference held in Bad Soden, October 16–19, 1974. Bad Soden, West Germany: Gesellschaft für Aerosolforschung, 1974.

654. Rundo, J. A case of accident inhalation of irradiated uranium. Brit. J. Radiol. 38:39–50, 1965.

655. Rushmer, M. A., and M. D. Burdick. Weather Resistance of Porcelain Enamels. Effect of Exposure Site and Other Variables After Seven Years. United States Department of Commerce, National Bureau of Standards. Building Science Series 4. Washington, D.C.: U.S. Government Printing Office, 1966. 16 pp.

656. Russell, W. T. The relative influence of fog and low temperature on the mortality from respiratory disease. Lancet 211:1128–1130, 1926.

657. Sadé, and N. Eliezer. Secretory otitis media and the nature of the mucociliary system. Acta Otolarngol. 70:351–357, 1970.

658. Sadé, J., N. Eliezer, A. Silberberg, and A. C. Nevo. The role of mucus in transport by cilia. Amer. Rev. Resp. Dis. 102:48–52, 1970.

659. Sakakura, Y., Y. Sasaki, Y. Togo, H. N. Wagner, Jr., R. B. Hornick, A. R. Schwartz, and D. F. Proctor. Mucociliary function during experimentally induced rhinovirus infection in man. Ann. Otol. Rhinol. Laryngol. 82:203–211, 1973.

660. Salmon, R. L. Systems Analysis of the Effects of Air Pollution on Materials. Final Report for Economic Effects Research Division, National Air Pollution Control Administration. Kansas City: Midwest Research Institute, 1970. 192 pp.

661. Salvin, V. S. Effect of atmospheric contaminants on lightfastness testing. Amer. Dyest. Rep. 47:P450-P451, 1958.
662. Samuels, H. J., S. Twiss, and E. W. Wong. Visibility, Light Scattering and Mass Concentration of Particulate Matter. Report of the California Tri-City Aerosol Sampling Project. Sacramento: State of California Air Resources Board, 1973. 60 pp.
663. Santa Cruz, R., J. Landa, J. Hirsch, and M. A. Sackner. Tracheal mucous velocity in normal man and patients with obstructive lung disease; effects of terbutaline. Amer. Rev. Resp. Dis. 109:458–463, 1974.
664. Satir, P. How cilia move. Sci. Amer. 231(4):44–52, 1974.
665. Saxena, V. K., D. J. Alofs, and A. C. Tebelak. A comparative study of Aitken nuclei counters. J. Rech. Atmos. 6:495–505, 1972.
666. Schaefer, V. J. The Air Quality Patterns of Aerosols on the Global Scale. Part 1 and Part 2. Final Report. Prepared for the National Science Foundation. ASRC-SUNY Publ. No. 406. Albany, N.Y.: Atmospheric Sciences Research Center, State University of New York, 1976. 110 pp and 262 pp.
667. Schaefer, V. J. The inadvertent modification of the atmosphere by air pollution. Bull. Amer. Meteorol. Soc. 50:199–206, 1969.
668. Schaffer, R. J. The Weathering of Natural Building Stones. London: His Majesty's Stationery Office, 1932. 149 pp.
669. Scheffer, F., E. Przemeck, and W. Wilms. Untersuchungen über den Einfluss von Zementofen-Flugstaub auf Boden und Pflanze. Staub 21:251–254, 1961.
670. Schiller, E. Inhalation, retention and elimination of dusts from dogs' and rats' lungs with special reference to the alveolar phagocytes and bronchial epithelium, pp. 342–347. In C. N. Davies, Ed. Inhaled Particles and Vapours. Proceedings of an International Symposium organized by the British Occupational Hygiene Society, Oxford, 29 March–1 April 1960. Oxford: Pergamon Press, 1961.
671. Schimmel, H., and L. Greenburg. A study of the relation of pollution to mortality. New York City, 1963–1968. J. Air Pollut. Control Assoc. 22:607–616, 1972.
672. Schimmel, H., and T. J. Murawski. SO_2—Harmful pollutant or air quality indicator? J. Air Pollut. Control Assoc. 25:739–740, 1975.
673. Schlesinger, R. B., V. R. Cohen, and M. Lippmann. Studies of intrabronchial particle deposition using hollow bronchial casts. pp. 116–127. In E. Karbe and J. F. Park, Eds. Experimental Lung Cancer. Carcinogenesis and Bioassays. New York: Springer-Verlag, 1974.
674. Schlesinger, R. B., and M. Lippmann. Particle deposition in casts of the human upper tracheobronchial tree. Amer. Ind. Hyg. Assoc. J. 33:237–251, 1972.
675. Schlipköter, H. W. Die gesundheitliche Bedeutung des atmosphärischen Staubes. Staub 25:378–384, 1965.
676. Schneider, S. H. A comment on "climate: the influence of aerosols." J. Appl. Meteorol. 10:840–841, 1971.
677. Schönbeck, H. Beobachtungen zur Frage des Einflusses von industriellen Immissionen auf die Krankheitsbereitschaft der Pflanze, pp. 89–98. In (Nordrhein-Westfalen) Berichte aus der Landesanstalt für Bodennutzungsschutz, Bochum, 1960.
678. Schrenk, H. H., H. Heimann, G. D. Clayton, W. M. Gafafer, and W. Wexler. Air Pollution in Donora, Pennsylvania. Epidemiology of the

Unusual Smog Episode of October 1948. Preliminary Report. Public Health Bulletin No. 306. Washington, D.C.: U.S. Government Printing Office, 1949. 173 pp.

679. Schroeter, L. C. Kinetics of air oxidation of sulfurous acid salts. J. Pharmaceut. Sci. 52:559–563, 1963.
680. Schuck, E. A., and J. K. Locke. Relationship of automotive lead particulates to certain consumer crops. Environ. Sci. Technol. 4:324–330, 1970.
681. Schuetzle, D., A. L. Crittenden, and R. J. Charlson. Application of computer controlled high resolution mass spectrometry to the analysis of air pollutants. J. Air Pollut. Control Assoc. 23:704–709, 1973.
682. Schütz, L., and R. Jaenicke. Particle number and mass distributions above 10^{-4} cm radius in sand and aerosols of the Sahara desert. J. Appl. Meteor. 13:863–870, 1974.
683. Scott, J. A. Fog and deaths in London, December 1952. Public Health Rep. 68:474–479, 1953.
684. Scott, J. A. Fog and atmospheric pollution in London, winter 1958–59. Med. Officer 102:191–193, 1959.
685. Scott, J. A. The London fog of December, 1962. Med. Officer 109: 250–252, 1963.
686. Scott, J. A., I. Taylor, A. T. Gore, and C. W. Shaddick. Mortality in London in the winter of 1962–63. Med. Off. 111:327–330, 1964.
687. Seabright, L. H., and R. J. Fabian. The many faces of corrosion. Mater. Des. Eng. 57(1):85–91, 1963.
688. Sehmel, G. A. Particle resuspension from an asphalt road caused by car and truck traffic. Atmos. Environ. 7:291–309, 1973.
689. Sellick, H., and J. G. Widdicombe. Stimulation of lung irritant receptors by cigarette smoke, carbon dust and histamine aerosol. J. Appl. Physiol. 31:15–19, 1971.
690. Semonin, R. G. Urban-induced weather modification, pp. 40–41. In 1974 Earth Environment and Resources Conference Digest of Technical Papers. New York: Lewis Winner, 1974.
691. Shephard, R. J., M. L. Thomson, G. C. Carey, and J. J. Phair. Field testing of pulmonary dynamics. J. Appl. Physiol. 13:189–193, 1958.
692. Shephard, R. J., M. E. Turner, G. C. R. Carey, and J. J. Phair. Correlation of pulmonary function and domestic microenvironment. J. Appl. Physiol. 15:70–76, 1960.
693. Shriner, D. S. Effects of Simulated Rain Acidified with Sulfuric Acid on Host-Parasite Interactions. Ph.D. Thesis. Raleigh: North Carolina State University, 1974. 85 pp.
694. Shy, C. M. Human health consequences of nitrogen dioxide: A review, pp. 363–405. In Assembly of Life Sciences-National Academy of Sciences-National Research Council. Proceedings of the Conference on Health Effects of Air Pollutants, Washington, D.C., October 3–5, 1973. Washington, D.C.: U.S. Government Printing Office, 1973.
695. Shy, C. M., V. Hasselblad, R. M. Burton, C. J. Nelson, and A. A. Cohen. Air pollution effects of ventilatory function of U.S. schoolchildren. Results of studies in Cincinnati, Chattanooga, and New York. Arch. Environ. Health 27:124–128, 1973.
696. Shy, C. M., V. Hasselblad, J. F. Finklea, R. M. Burton, M. Pravda, R. S. Chapman, and A. A. Cohen. Ventilatory function in school children: 1970–1971 New York studies, pp. 5-109 to 5-119. In U.S. Environmental Protection Agency, Office of Research and Development. Health

Consequences of Sulfur Oxides: A Report from CHESS, 1970–1971. EPA-650/1-74-004. Washington, D.C.: U.S. Government Printing Office, 1974.

697. Shy, C. M., C. J. Nelson, F. B. Benson, W. B. Riggan, V. A. Newill, and R. S. Chapman. Ventilatory function in school children: 1967–1968 testing in Cincinnati neighborhoods, pp. 6-3 to 6-14. In U.S. Environmental Protection Agency, Office of Research and Development. Health Consequences of Sulfur Oxides: A Report from CHESS, 1970–1971. EPA-650/1-74-004. Washington, D.C.: U.S. Government Printing Office, 1974.

698. Sidebottom, H. W., C. C. Badcock, G. E. Jackson, J. G. Calvert, G. W. Reinhardt, and E. K. Damon. Photooxidation of sulfur dioxide. Environ. Sci. Technol. 6:72–79, 1972.

699. Sievers, F. J. Crop injury resulting from magnesium oxide dust. Phytopathology 14:108–113, 1924.

700. Silvers, G. W., J. C. Maisel, T. L. Petty, G. F. Filley, and R. S. Mitchell. Flow limitation during forced expiration in excised human lungs. J. Appl. Physiol. 36:737–744, 1974.

701. Sim, V. M., and R. E. Pattle. Effect of possible smog irritants on human subjects. J.A.M.A. 165:1908–1913, 1957.

702. Simmons, W. A., and W. Young. Correlation of the Integrating Nephelometer to High Volume Air Sampler. Commonwealth of Massachusetts Department of Public Health, Bureau of Air Use Management, Boston, Mass., 1970, 6 pp. (Available from Commonwealth of Massachusetts, Department of Environmental Quality Engineering Division of Air Quality Control, 600 Washington Street, Boston, MA 02111)

703. Simpson, J. W., and P. J. Horrobin, Eds. The Weathering and Performance of Building Materials. New York: Wiley-Interscience, 1970. 286 pp.

704. Sleigh, M. A. Further observations on co-ordination and the determination of frequency in the peristomial cilia of *Stentor*. J. Exp. Biol. 34:106–115, 1957.

705. Sleigh, M. A. Some aspects of the comparative physiology of cilia. Amer. Rev. Resp. Dis. 93 (Suppl):16–31, 1966.

706. Slinn, W. G. N., and J. M. Hales. A reevaluation of the role of thermophoresis as a mechanism of in- and below-cloud scavenging. J. Atmos. Sci. 28:1465–1471, 1971.

707. Smith, F. B. The turbulent spread of a falling cluster, pp. 193–210. In H. E. Landsberg and J. Van Mieghem, Eds. Advances in Geophysics, Vol. 6. New York: Academic Press, 1959.

708. Smith, J. P., and P. Urone. Static studies of sulfur dioxide reactions. Effects of NO_2, C_3H_6, and H_2O. Environ. Sci. Technol. 8:742–746, 1974.

709. Snell, R. E., and P. C. Luchsinger. Effects of sulfur dioxide on expiratory flow rateş and total respiratory resistance in normal human subjects. Arch. Environ. Health 18:693–698, 1969.

710. Sorokin, S. P. The respiratory system, pp. 661–711. In R. O. Greep and L. Weiss, Eds. Histology. (3rd ed.) New York: McGraw-Hill Book Company, 1973.

711. Soulage, G. Contribution des fumées industrielles a l'enrichissement de l'atmosphère en noyaux glaçogènes. I. Fumées d'aciéries électriques. Bull. Observ. Puy Dome No. 4:121–124, 1958.

712. Spector, W. G., and D. A. Willoughby. The inflammatory response. Bact. Rev. 27:117–154, 1963.

713. Spedding, D. J. Sulphur dioxide uptake by limestone. Atmos. Environ. 3:683, 685–686, 1969. (short communication)
714. Speizer, F. E. An epidemiological appraisal of the effects of ambient air on health: Particulates and oxides of sulfur. J. Air Pollut. Control Assoc. 19:647–655, 1969.
715. Speizer, F. E., and N. R. Frank. The uptake and release of SO_2 by the human nose. Arch. Environ. Health 12:725–728, 1966.
716. Spence, J. W., and F. H. Haynie. Paint Technology and Air Pollution: A Survey and Economic Assessment, pp. 20–24. Environmental Protection Agency, Office of Air Programs. Washington, D.C.: U.S. Government Printing Office, 1972.
717. Spicer, W. S., Jr., P. B. Storey, W. K. C. Morgan, H. D. Kerr, and N. E. Standiford. Variation in respiratory function in selected patients and its relation to air pollution. Amer. Rev. Resp. Dis. 86:705–712, 1962.
718. Spiegelman, J. R., G. D. Hanson, A. Lazarus, R. J. Bennett, M. Lippmann, and R. E. Albert. Effect of acute sulfur dioxide exposure on bronchial clearance in the donkey. Arch. Environ. Health 17:321–326, 1968.
719. Squires, P. The microstructure and colloidal stability of warm clouds. Part I. The relation between structure and stability. Tellus 10:256–261, 1958.
720. Squires, P. An estimate of the anthropogenic production of cloud nuclei. J. Rech. Atmos. 2:297–307, 1966.
721. Stedman, H. F. Natural and artificial weathering performance of rigid polyvinyl chloride (PVC) and other plastic materials, pp. 751–760. In Performance Concept in Buildings. Vol. 1: Invited Papers. Proceedings of a Symposium. National Bureau of Standards Special Publication 361. Washington, D.C.: U.S. Government Printing Office, 1972.
722. Steinhübel, G. Changes of starch reserves of the leaves of Ilex aquifolium after artificial spoiling by immovable dust. Biologia (Bratislava) 18:23–33, 1963. (in Slovak) (Abstract in German, pp. 32–33)
723. Stell, P. M. Smoking and laryngeal cancer. Lancet 1(7751):617–618, 1972.
724. Sterling, T. D., J. J. Phair, S. V. Pollack, D. A. Schumsky, and I. DeGroot. Urban morbidity and air pollution. A first report. Arch. Environ. Health 13:158–170, 1966.
725. Sterling, T. D., S. V. Pollack, and J. J. Phair. Urban hospital morbidity and air pollution. A second report. Arch. Environ. Health 15:362–374, 1967.
726. Stevens, R. K., A. R. McFarland, and J. B. Wedding. Comparison of the virtual dichotomous sampler with the hi-volume sampler. American Chemical Society, Division of Environmental Chemistry. Paper No. 25 in Preprints of Papers Presented at the 171st National Meeting, New York, N.Y., April 4–9, 1976. 16(1):84–85, 1976.
727. Stevenson, C. M. An improved Millipore filter technique for measuring the concentrations of freezing nuclei in the atmosphere. Q. J. Roy. Meteorol. Soc. 94:35–43, 1968.
728. Stevenson, H. J. R., D. E. Sanderson, and A. P. Altshuller. Formation of photochemical aerosols. Air Water Pollut. Int. J. 9:367–375, 1965.
729. Stöber, W. Design, performance and applications of spiral duct aerosol centrifuges, pp. 351–397. In B. Y. H. Liu, Ed. Fine Particles. Aerosol Generation, Measurement, Sampling, and Analysis. Proceedings of a Symposium by the U.S. Environmental Protection Agency held in Minneapolis, Minnesota, May 28–30, 1975. New York: Academic Press, Inc. 1976.

730. Stöber, W., and H. Flachsbart. Size-separating precipitation in a spinning spiral duct. Environ. Sci. Technol. 3:1280–1296, 1969.
731. Stöber, W., and F. J. Mönig. Measurements with a prototype mass distribution monitor for particulate air pollution. In Session 5, Monitoring and Evaluation of Atmospheric Particulate Matter. In International Conference on Environmental Sensing and Assessment, held in Las Vegas, Nevada, September 14–17, 1975. New York: Institute of Electrical and Electronics Engineers, 1976.
732. Stocks, P. Air pollution and cancer mortality in Liverpool Hospital region and North Wales. Int. J. Air Pollut. 1:1–13, 1958.
733. Stocks, P. Cancer and bronchitis mortality in relation to atmospheric deposit and smoke. Brit. Med. J. 1:74–79, 1959.
734. Stocks, P. On the relations between atmospheric pollution in urban and rural localities and mortality from cancer, bronchitis and pneumonia, with particular reference to 3:4 benzopyrene, beryllium, molybdenum, vanadium and arsenic. Brit. J. Cancer 14:397–418, 1960.
735. Stocks, P. Lung cancer and bronchitis in relation to cigarette smoking and fuel consumption in 20 countries. Brit. J. Prev. Soc. Med. 21:181–185, 1967.
736. Stocks, P., and J. M. Campbell. Lung cancer death rates among nonsmokers and pipe and cigarette smokers. An evaluation in relation to air pollution by benzpyrene and other substances. Brit. Med. J. 2:923–939, 1955.
737. Stokinger, H. E., L. T. Steadman, H. B. Wilson, G. E. Sylvester, S. Dziuba, and C. W. LaBelle. Lobar deposition and retention of inhaled insoluble particulates. A.M.A. Arch. Ind. Hyg. Occup. Med. 4:346–353, 1951.
738. Storebø, P. B. Formation of radioactivity size distributions in nuclear bomb debris. J. Aerosol Sci. 5:557–577, 1974.
739. Storozhilova, A. I. Passivation of hygroscopic condensation nuclei by surfactant absorption from gas phase, pp. 47–52. In V. A. Fedoseev, Ed. Advances in Aerosol Physics. No. 1. New York: Halsted Press, 1972.
740. Strutt, J. W. (Lord Rayleigh). On the scattering of light by small particles. Phil. Mag. 41:447–454, 1871.
741. Stuart, B. O. Deposition of inhaled aerosols. Arch. Int. Med. 131:60–73, 1973.
742. Sturgess, J., and L. Reid. Secretory activity of the human bronchial mucous glands *in vitro*. Exp. Molec. Pathol. 16:362–381, 1972.
743. Sturgess, J. M. The Control of the Bronchial Mucous Glands and Their Secretion. Ph.D. thesis, University of London, 1970.
744. Sturgess, J. M., A. J. Palfrey, and L. Reid. The viscosity of bronchial secretion. Clin. Sci. 38:145–156, 1970.
745. Sturgess, J. M., and L. Reid. A new pattern of sputum viscosity, pp. 368–385. In J. Lawson, Ed. Proceedings of the Fifth International Cystic Fibrosis Conference, Churchill College, Cambridge, England. September 22nd–26th, 1969. London: Cystic Fibrosis Research Trust, 1969.
746. Sverdrup, G. M., and K. T. Whitby. On the variation of the aerosol volume to light scattering coefficient. Particle Technology Laboratory Publication No. 273. Minneapolis: Mechanical Engineering Dept., University of Minnesota, 1976.
747. Sverdrup, G. M., K. T. Whitby, and W. E. Clark. Characterization of California aerosols—II. Aerosol size distribution measurements in the Mojave Desert. Atmos. Environ. 9:483–494, 1975.

748. Tajiri, Y. Effect of air pollution on metals. Kuki Seijo (Air Clean. J. Jap. Air Clean. Assoc.) 10(5):90–99, 1972. (in Japanese)

749. Task Group on Lung Dynamics (Committee II—ICRP). Deposition and retention models for internal dosimetry of the human respiratory tract. Health Phys. 12:173–207, 1966.

750. Taylor, C. E., A. C. Guyton, and V. S. Bishop. Permeability of the alveolar membrane to solutes. Circ. Res. 16:353–362, 1965.

751. Telford, J. W. Freezing nuclei from industrial processes. J. Meteorol. 17:676–681, 1960.

752. Thomas, M. D., R. H. Hendricks, F. D. Gunn, and T. Critchlow. Prolonged exposure of guinea pigs to sulfuric acid aerosol. A.M.A. Arch. Ind. Health 17:70–80, 1958.

753. Thomas, M. D., R. H. Hendricks, and G. R. Hill. Some impurities in the air and their effects on plants, pp. 41–47. In L. C. McCabe, Ed. Air Pollution. Proceedings of the United States Technical Conference on Air Pollution, Washington, D.C., 1950. New York: McGraw-Hill Book Company, Inc. 1952.

754. Thomson, M. L., and D. Pavia. Long-term tobacco smoking and mucociliary clearance from the human lung in health and respiratory impairment. Arch. Environ. Health 26:86–89, 1973.

755. Timbrell, V. An aerosol spectrometer and its applications, pp. 290–330. In T. T. Mercer, P. E. Morrow, and W. Stöber, Eds. Assessment of Airborne Particles. Fundamentals, Applications, and Implications to Inhalation Toxicity. A Proceedings Publication of the Third Rochester International Conference on Environmental Toxicity. Springfield: Charles C Thomas, 1972.

756. Timbrell, V. Inhalation and biological effects of asbestos, pp. 429–441. In T. T. Mercer, P. E. Morrow, and W. Stöber, Eds. Assessment of Airborne Particles. Fundamentals, Applications, and Implications to Inhalation Toxicity. A Proceedings Publication of the Third Rochester International Conference on Environmental Toxicity. Springfield: Charles C Thomas, 1972.

757. Tomori, Z., and J. G. Widdicombe. Muscular, bronchomotor and cardiovascular reflexes elicited by mechanical stimulation of the respiratory tract. J. Physiol. 200:25–49, 1969.

758. Toremalm, N. G. The daily amount of tracheo-bronchial secretions in man. A method for continuous tracheal aspiration in laryngectomized and tracheotomized patients. Acta Otolaryng. Suppl. 158:43–53, 1960.

759. Toyama, T. Studies on aerosols: I. Synergistic responses of pulmonary resistance on inhaling sodium chloride and SO_2 in man. Jap. J. Ind. Med. 4:18–24, 1962.

760. Toyama, T., and K. Nakamura. Synergistic response of hydrogen peroxide aerosols and sulfur dioxide to pulmonary airway resistance. Ind. Health (Japan) 2:34–35, 1964.

761. Tucker, A. D., J. H. Wyatt, and D. Undery. Clearance of inhaled particles from alveoli by normal interstitial drainage pathways. J. Appl. Physiol. 35:719–732, 1973.

762. Twomey, S. The nuclei of natural cloud formation. Part II. The supersaturation in natural clouds and the variation of cloud droplet concentration. Geofis. Pura Appl. 43:243–249, 1959.

763. Twomey, S. The influence of droplet concentration on rain formulation and stability in clouds. Bull. Observ. Puy Dome No. 2:33–41, 1959.

764. Twomey, S., and P. Squires. The influence of cloud nucleus population on the microstructure and stability of convective clouds. Tellus 11:408–411, 1959.
765. Twomey, S., and J. Warner. Comparison of measurements of cloud droplets and cloud nuclei. J. Atmos. Sci. 24:702–703, 1967.
766. Urone, P., H. Lutsep, C. M. Noyes, and J. F. Parcher. Static studies of sulfur dioxide reactions in air. Environ. Sci. Technol. 2:611–618, 1968.
767. Urone, P., W. H. Schroeder, and S. R. Miller. Reactions of sulfur dioxide in air, pp. 370–375. In H. M. Englund and W. T. Beery, Eds. Proceedings of the Second International Clean Air Congress, Held at Washington, D.C., December 6–11, 1970. Sponsored by The International Union of Air Pollution Prevention Associations. New York: Academic Press, 1971.
768. U.S. Department of Health, Education, and Welfare. Air Quality Criteria for Particulate Matter. National Air Pollution Control Administration Publ. No. AP-49. Washington, D.C.: U.S. Government Printing Office, 1969. 211 pp.
769. U.S. Department of Health, Education, and Welfare. Air Quality Criteria for Sulfur Oxides. National Air Pollution Control Administration Publ. No. AP-50. Washington, D.C.: U.S. Government Printing Office, 1970. 178 pp.
770. U.S. Environmental Protection Agency, Office of Air and Waste Management, Office of Air Quality Planning and Standards. Position Paper on Regulation of Atmospheric Sulfates. Report No. EPA-450/2-75-007. Research Triangle Park, NC: U.S. Environmental Protection Agency, 1975.
771. Venice preserved. Time 101:54, June 11, 1973.
772. Vohra, K. G., K. N. Vasudevan, and P. V. N. Nair. Mechanisms of nucleus-forming reactions in the atmosphere. J. Geophys. Res. 75: 2951–2960, 1970.
773. von Hayek, H. The Human Lung. (Translated by V. E. Krahl) New York: Hafner Publishing Company, Inc., 1960. 372 pp.
774. Waggoner, A. P., and R. J. Charlson. Simulating the color of polluted air. Appl. Opt. 10:957–958, 1971.
775. Waggoner, A. P., and R. J. Charlson. Measurements of aerosol optical parameters, pp. 511–533. In B. Y. H. Liu, Ed. Fine Particles. Aerosol Generation, Measurement, Sampling, and Analysis. Proceedings of a symposium by the U.S. Environmental Protection Agency held in Minneapolis, Minnesota, May 28–30, 1975. New York: Academic Press, Inc., 1976.
776. Waggoner, A. P., A. H. Vanderpol, R. J. Charlson, S. Larsen, L. Granat, and C. Trägårdh. Sulphate-light scattering ratio as an index of the role of sulphur in tropospheric optics. Nature 261:120–122, 1976.
777. Wagman, J., R. E. Lee, Jr., and C. J. Axt. Influence of some atmospheric variables on the concentration and particle size distribution of sulfate in urban air. Atmos. Environ. 1:479–489, 1967.
778. Wagner, H. N., Jr., V. Lopez-Majano, and J. K. Langan. Clearance of particulate matter from the tracheobronchial tree in patients with tuberculosis. Nature 205:252–254, 1965.
779. Wahdan, M. H. Atmospheric Pollution and Other Environmental Factors in Respiratory Disease in Children. Ph.D. thesis, London: University of London, 1963. 240 pp.

780. Waller, R. E. Acid droplets in town air. Air Water Pollut. 7:773–778, 1963.
781. Waller, R. E., and P. J. Lawther. Some observations on London fog. Brit. Med. J. 2:1356–1358, 1955.
782. Waller, R. E., and P. J. Lawther. Further observations on London fog. Brit. Med. J. 2:1473–1475, 1957.
783. Walter, H. Coagulation and size distribution of condensation aerosols. J. Aerosol Sci. 4:1–15, 1973.
784. Wang, W-C., and G. A. Domoto. The radiative effect of aerosol in the earth's atmosphere. J. Appl. Meteorol. 13:521–534, 1974.
785. Warner, C. G., G. M. Davies, J. G. Jones, and C. R. Lowe. Bronchitis in two integrated steel works. II. Sulfur dioxide and particulate atmospheric pollution in and around two works. Ann. Occup. Hyg. 12:150–170, 1969.
786. Warner, J. A reduction in rainfall associated with smoke from sugar-cane fires—an inadvertent weather modification? J. Appl. Meteorol. 7:247–251, 1968.
787. Warner, J. The microstructure of cumulus cloud. Part II. The effect on droplet size distribution of the cloud nucleus spectrum and updraft velocity. J. Atmos. Sci. 26:1272–1282, 1969.
788. Warner, J., and S. Twomey. The production of cloud nuclei by cane fires and the effect on cloud droplet concentration. J. Atmos. Sci. 24:704–706, 1967.
789. Warren, H. V., and R. E. Delavault. Lead in some food crops and trees. J. Sci. Food Agricul. 13:96–98, 1962.
790. Watanbe, H. Air pollution and its health effects in Osaka, Japan. (Preprint) Presented at the 58th Annual Meeting, Air Pollution Control Association, Toronto, Canada, June 20–24, 1965.
791. Weibel, E. R. Morphometry of the Human Lung. New York: Academic Press, 1963. 151 pp.
792. Weibel, E. R. Geometric and dimensional airway models of conductive, transitory and respiratory zones of the human lung, pp. 136–140. In Morphometry of the Human Lung. New York: Academic Press, 1963.
793. Wentzel, K. F. Literaturberichte: H. Pajenkamp. Einwirkung des Zementofenstaubes auf Pflanzen und Tiere. Z. Pflanzenkr. 69:478, 1962.
794. Wesolowski, J. J., A. Alcocer, and B. R. Appel. The Validation of the Lundgren Impactor. AIHL Report No. 138-B. Berkeley: Laboratory Services Branch, California State Department of Health, September 1975.
795. Wessel, C. J. Textiles and cordage, pp. 408–506. In G. A. Greathouse and C. J. Wessel, Eds. Deterioration of Materials. Causes and Preventive Techniques. New York: Reinhold Publishing Corporation, 1954.
796. Whitby, K. T. Second International Workshop on Cloud and Ice Nuclei held at Fort Collins, Colorado, 5–19 Aug. 1970. Particle Technology Laboratory Report No. 173. Environmental Division, Mechanical Engineering Department, University of Minnesota, 1973.
797. Whitby, K. T. Modeling of multimodal aerosol distributions. In V. Böhlau, Ed. Aerosole in Naturwissenschaft, Medizin und Technik. Proceedings of a conference held in Bad Soden, October 16–19, 1974. Bad Soden, West Germany: Gesellschaft für Aerosolforschung, 1974.
798. Whitby, K. T. Modeling of Atmospheric Aerosol Particle Size Distributions. A Progress Report on EPA Research Grant No. R800971, Sampling and Analysis of Atmospheric Aerosols. Particle Technology Laboratory

Report No. 253. Environmental Division, Mechanical Engineering Department, University of Minnesota, 1975.

799. Whitby, K. T. Electrical measuremnt of aerosols, pp. 584–624. In B. Y. H. Liu, Ed. Fine Particles. Aerosol Generation, Measurement, Sampling, and Analysis. Proceedings of a Symposium by the U.S. Environmental Protection Agency held in Minneapolis, Minnesota, May 28–30, 1975. New York: Academic Press, Inc., 1976.

800. Whitby, K. T., A. B. Algren, and R. C. Jordan. The dust spot method for evaluating air cleaners. Heating-Piping Air Cond. 28(11):151–157, 1956.

801. Whitby, K. T., and B. Cantrell. Atmospheric aerosols—characteristics and measurements. In Session 29, Fine Particles. In International Conference on Environmental Sensing and Assessment, held in Las Vegas, Nevada, September 14–17, 1975. New York: Institute of Electrical and Electronics Engineers, 1976.

802. Whitby, K. T., R. E. Charlson, W. E. Wilson, and R. K. Stevens. The size of suspended particle matter in air. Science 183:1098–1099, 1974. (Letter response to Lee, R. E., Jr. The size of suspended particulate matter in air. Science 178:567–575, 1972)

803. Whitby, K. T., W. E. Clark, V. A. Marple, G. M. Sverdrup, G. J. Sem, K. Willeke, B. Y. H. Liu, and D. Y. H. Pui. Characterization of California aerosols—I. Size distributions of freeway aerosol. Atmos. Environ. 9:463–482, 1975.

804. Whitby, K. T., W. E. Clark, V. A. Marple, G. M. Sverdrup, K. Willeke, B. Y. H. Liu, and D. Y. H. Pui. Evolution of the Freeway Aerosol. Presented at the American Chemical Society Annual Meeting, Chicago, Ill., August 29, 1973.

805. Whitby, K. T., R. B. Husar, and B. Y. H. Liu. The aerosol size distribution of Los Angeles smog. J. Colloid Interface Sci. 39:177–204, 1972.

806. Whitby, K. T., D. B. Kittelson, B. K. Cantrell, N. J. Barsic, D. F. Dolan, L. D. Tarvestad, D. J. Nieken, J. L. Wolf, and J. R. Wood. Aerosol size distributions and concentrations measured during the General Motors Proving Grounds sulfate study, pp. 29–80. In R. K. Stevens, P. J. Lamothe, W. E. Wilson, J. L. Durham, and T. G. Dzubay, Eds. The General Motors/Environmental Protection Agency Sulfate Dispersion Experiment. Selected EPA Research Papers. EPA-600/3-76-035. Research Triangle Park, N.C.: U.S. Environmental Protection Agency, 1976. (Available from the National Technical Information Service, Springfield, Va.)

807. Whitby, K. T., and B. Y. H. Liu. Advances in Instrumentation and Techniques for Aerosol Generation and Measurement. Particle Technology Laboratory, P. L. Publ. No. 216. Minneapolis: University of Minnesota, 1973. 34 pp.

808. White, W. H., J. A. Anderson, D. L. Blumenthal, R. B. Husar, N. V. Gillani, J. D. Husar, and W. E. Wilson, Jr. Formation and transport of secondary air pollutants: Ozone and aerosols in the St. Louis urban plume. Science 194:187–189, 1976.

809. Whittaker, R. H., F. H. Bormann, G. E. Likens, and T. G. Siccama. The Hubbard Brook ecosystem study: Forest biomass and production. Ecol. Monogr. 44:233–254, 1974.

810. Widdicombe, J. G., D. C. Kent, and J. A. Nadel. Mechanism of bronchoconstriction during inhalation of dust. J. Appl. Physiol. 17:613–616, 1962.

811. Wilkins, E. T. Air pollution and the London fog of December, 1952. J. Roy. Sanit. Inst. 74:1–21, 1954.

812. Wilkins, E. T. Air pollution aspects of the London fog of December 1952. Roy. Meteorol. Soc. Q. J. 80:267–278, 1954.
813. Willeke, K., and B. Y. H. Liu. Single particle optical counter: Principle and application, pp. 697–729. In B. Y. H. Liu, Ed. Fine Particles. Aerosol Generation, Measurement, Sampling, and Analysis. Proceedings of a Symposium by the U.S. Environmental Protection Agency held in Minneapolis, Minnesota, May 28–30, 1975. New York: Academic Press, Inc., 1976.
814. Willeke, K., and K. T. Whitby. Atmospheric aerosols: size distribution interpretation. J. Air Pollut. Control Assoc. 25:529–534, 1975.
815. Willeke, K., K. T. Whitby, W. E. Clark, and V. A. Marple. Size distributions of Denver aerosols—a comparison of two sites. Atmos. Environ. 8:609–633, 1974.
816. Wilson, J. G. Exposed Concrete Finishes, Vol. 2., pp. 115–124. New York: John Wiley & Sons, Inc., 1965.
817. Wilson, W. E., Jr., and A. Levy. A study of sulfur dioxide in photochemical smog. I. Effect of SO_2 dioxide and water vapor concentration in the 1-butene/NO_x/SO_2 system. J. Air Pollut. Control Assoc. 20:385–390, 1970.
818. Wilson, W. E., Jr., E. L. Merryman, and A. Levy. A Literature Survey of Aerosol Formation and Visibility Reduction in Photochemical Smog. A research report for the American Petroleum Institute. Columbus, OH: Battelle Memorial Institute, 1969. 41 pp.
819. Wilson, W. E., Jr., E. L. Merryman, and A. Levy. Aerosol formation in photochemical smog. 1. Effect of stirring. J. Air Pollut. Control Assoc. 21:128–132, 1971.
820. Winchester, J. W. Trace metal associations in urban airborne particulates. Bull. Amer. Meteorol. Soc. 54:94–97. 1973.
821. Winchester, J. W., T. B. Johanson, and R. E. Grieken. Marine influences on aerosol composition in the coastal zone. J. Rech. Atmos. 8(3-4): 761–776, 1974.
822. Winkelstein, W., Jr., and S. Kantor. Respiratory symptoms and air pollution in an urban population of northeastern United States. Arch. Environ. Health 18:760–767, 1969.
823. Winkelstein, W., Jr., S. Kantor, E. W. Davis, C. S. Maneri, and W. E. Mosher. The relationship of air pollution and economic status to total mortality and selected respiratory system mortality in men. I. Suspended particulates. Arch. Environ. Health 14:162–171, 1967.
824. Winkelstein, W., Jr., S. Kantor, E. W. Davis, C. S. Maneri, and W. E. Mosher. The relationship of air pollution and economic status to total mortality and selected respiratory system mortality in men. II. Oxides of sulfur. Arch. Environ. Health 16:401–405, 1968.
825. Winkler, P. The growth of atmospheric aerosol particles as a function of the relative humidity. II. An improved concept of mixed nuclei. J. Aerosol Sci. 4:373–387, 1973.
826. Winkler, P., and C. Junge. The growth of atmospheric aerosol particles as a function of the relative humidity. Part I. Method and measurement at different locations. J. Recherch. Atmos. 6:617–638, 1972.
827. Wolff, R. K., M. Dolovich, G. Obminski, and M. T. Newhouse. Effects of sulfur dioxide on tracheobronchial clearance at rest and during exercise. In W. H. Walton, Ed. Inhaled Particles IV. London: Pergamon Press Ltd, 1977.

828. Wolff, R. K., M. Dolovich, C. Rossman, and M. T. Newhouse. Sulfur dioxide and tracheobronchial clearance in man. Arch. Environ. Health 30:521–527, 1975.
829. Wood, T., and F. H. Bormann. The effects of an artificial acid mist upon the growth of *Betula alleghaniensis* Brit. Environ. Pollut. 7:259–268, 1974.
830. Wood, T., and F. H. Bormann. Increases in foliar leaching caused by acidification of an artificial mist. Ambio 4:169–171, 1975.
831. Woodcock, A. H., and D. C. Blanchard. Tests of the salt-nuclei hypothesis of rain formation. Tellus 7:437–448, 1955.
832. Yamamoto, G., and M. Tanaka. Increase of global albedo due to air pollution. J. Atmos. Sci. 29:1405–1412, 1972.
833. Yamasaki, S., and Y. Yokoi. Accelerated atmospheric corrosion tests in polluted air. (2nd report) Influence of dewing on mild steel in sulfur dioxide environment. Boshoku Gijutsu (Corrosion Engineering) 20(11/12): 23–29, 1971.
834. Yeager, H., Jr. Tracheobronchial secretions. Amer. J. Med. 50:493–509, 1971.
835. Yeates, D. B. The Clearance of Soluble and Particulate Aerosols Deposited in the Human Lung. Ph.D. Thesis, University of Toronto, Ontario, Canada, 1974.
836. Yeates, D. B., N. Aspin, A. C. Bryan, and H. Levison. Regional clearance of ions from the airways of the lung. Amer. Rev. Resp. Dis. 107:602–608, 1973.
837. Yeates, D. B., N. Aspin, H. Levison, M. T. Jones, and A. C. Bryan. Mucociliary tracheal transport rates in man. J. Appl. Physiol. 39:487–495, 1975.
838. Yeates, D. B., J. M. Sturgess, S. R. Khan, H. Levison, and N. Aspin. Mucociliary transport in trachea of patients with cystic fibrosis. Arch. Dis. Child. 51:28–33, 1976.
839. Yocom, J. E., and R. O. McCaldin. Effects of air pollution on materials and the economy, pp. 617–654. In A. C. Stern, Ed. Air Pollution (2nd ed.), Vol. 1. Air Pollution and Its Effects. New York: Academic Press, 1968.
840. Yudine, M. I. Physical considerations on heavy particle diffusion. Adv. Geophys. 6:185–191, 1959.
841. Zeidberg, L. D., R. A. Prindle, and E. Landau. The Nashville air pollution study. I. Sulfur dioxide and bronchial asthma. A preliminary report. Amer. Rev. Resp. Dis. 84:489–503, 1961.
842. Zeidberg, L. D., R. A. Prindle, and E. Landau. The Nashville air pollution study. III. Morbidity in relation to air pollution. Amer. J. Public Health 54:85–97, 1964.
843. Zeidberg, L. D., R. J. M. Horton, E. Landau, R. M. Hagstrom, and H. A. Sprague. The Nashville air pollution study. V. Mortality from diseases of the respiratory system in relation to air pollution. Arch. Environ. Health 15:214–224, 1967.
844. Zeronian, S. H., K. W. Alger, and S. T. Omaye. Reaction of fabrics made from synthetic fibers to air contaminated with nitrogen dioxide, ozone, or sulfur dioxide, pp. 468–476. In H. M. Englund and W. T. Beery, Eds. Proceedings of the Second International Clean Air Congress, Washington, D.C., December 1970. New York: Academic Press, 1971.

Index